Sustaining Large Marine Ecosystems

The Human Dimension

Large Marine Ecosystems – Volume 13

Series Editor: Kenneth Sherman
                   Director, Narragansett Laboratory and Office of
                   Marine Ecosystem Studies
                   NOAA-NMFS, Narragansett, Rhode Island, USA
                   And Adjunct Professor of Oceanography
                   Graduate School of Oceanography, University of Rhode Island
                   Narragansett, Rhode Island, USA

On the cover

The global map of average primary productivity and the boundaries of the 64 Large Marine Ecosystems (LMEs) of the world, available at www.edc.uri.edu/lme and at www.lme.noaa.gov. The annual productivity estimates are based on SeaWiFS satellite data collected between September 1998 and August 1999, and the model developed by M. Behrenfeld and P.G. Falkowski (Limnol.Oceangr. 42(1):1997, 1-20). The color-enhanced image (provided by Rutgers University) depicts a shaded gradient of primary productivity from a high of 450 gCm^2yr^{-1} in red, to less than 45 gCm^2yr^{-1} in purple.

Sustaining Large Marine Ecosystems

The Human Dimension

Edited by

Timothy M. Hennessey
Professor
Marine Affairs and Political Science
Tucker House
University of Rhode Island
Kingston, RI 02881
USA

Jon G. Sutinen
Professor
Department of Environmental and Natural Resource Economics
216 Kingston Coastal Institute
University of Rhode Island
Kingston, RI 02881
USA

2005

ELSEVIER

Amsterdam - Boston - Heidelberg - London - New York - Oxford - Paris
San Diego - San Francisco - Singapore - Sydney - Tokyo

ELSEVIER B.V.
Radarweg 29
P.O. Box 211, 1000 AE Amsterdam
The Netherlands

ELSEVIER Inc.
525 B Street, Suite 1900
San Diego, CA 92101-4495
USA

ELSEVIER Ltd
The Boulevard, Langford Lane
Kidlington, Oxford OX5 1GB
UK

ELSEVIER Ltd
84 Theobalds Road
London WC1X 8RR
UK

First edition 2005

Library of Congress Cataloging in Publication Data
A catalog record is available from the Library of Congress.

British Library Cataloguing in Publication Data
A catalogue record is available from the British Library.

ISBN: 0 444 51026 5
ISSN: 1570-0461

⊗ The paper used in this publication meets the requirements of ANSI/NISO Z39.48-1992 (Permanence of Paper).
Printed in The Netherlands.

Series Editor's Introduction

The world's coastal ocean waters continue to be degraded from unsustainable fishing practices, habitat degradation, eutrophication, toxic pollution, aerosol contamination, and emerging diseases. Against this background is a growing recognition among world leaders that positive actions are required on the part of governments and civil society to redress global environmental and resource degradation with actions to recover depleted fish populations, restore degraded habitats and reduce coastal pollution. No single international organization has been empowered to monitor and assess the changing states of coastal ecosystems on a global scale, and to reconcile the needs of individual nations to those of the community of nations for taking appropriate mitigation and management actions. However, the World Summit on Sustainable Development convened in Johannesburg in 2002 in recognition of the importance for coastal nations to move more expeditiously toward sustainable development and use of ocean resources, declared that countries should move to introduce ecosystem-based assessment and management practices by 2010, and by 2015, restore the world's depleted fish stocks to maximum levels of sustainable yields. At present, 121 developing countries are moving toward these targets in joint international projects supported, in part, by financial grants by the Global Environment Facility in partnership with scientific and technical assistance from UN partner agencies (e.g. UNIDO, UNEP, UNDP, IOC, FAO), donor countries and institutions and non-governmental organizations including the IUCN (World Conservation Union). Many of these projects are linked to ecosystem-based efforts underway in Europe and North America in a concerted effort to overcome the North-South digital divide.

The volumes in the new Elsevier series on Large Marine Ecosystems are bringing forward the results of ecosystem-based studies for marine scientists, educators, students and resource managers. The volumes are focused on LMEs and their productivity, fish and fisheries, pollution and ecosystem health, socioeconomics and governance. This volume in the new series, "Sustaining Large Marine Ecosystems: The Human Dimension," recognizes the importance of economic activity as a driver of change in the world's LMEs.

Human Dimensions, Market Economies and the Sustainability of LMEs

This volume is the thirteenth in the Large Marine Ecosystem Series. It represents a state-of-art-focus on the new awareness by scientists, economists, resource managers and the marine academic community on the importance of strengthening links between science-based assessments of the changing states of marine ecosystems and the human dimension of actions to be taken for reversing the downward spiral of fisheries overexploitation, habitat loss, and coastal pollution. Using the case study method, innovative insights and methods are presented from the socioeconomic, governance, and policy perspectives of how to move forward in halting the downward spiral and advancing toward the recovery and sustainability of depleted fish stocks, restoration of degraded habitats, and reduction and control of pollution within the framework of an ecosystems-based approach.

The authors propose actions consistent with market driven national economies and the need to underscore the enormous economic value of marine resources in relation to public and private ownership scenarios, long term and short term values of resources at risk, and serious deliberations on issues of equity with regard to allocation of resource benefits. Important policy issues are considered that acknowledge human needs for food, shelter and livelihoods at risk from management and governance decisions on the use of marine resources that are made daily and that influence the nearly 50% of the world's population who, since 2001, live within 200 kilometers of the coasts of the world's 64 Large Marine Ecosystems (see www.oceanatlas.org.) The volume addresses how ecosystem-based management in both economically developed and developing countries can and will contribute to building the institutional capabilities to reconcile economic activities derived from the oceans with the recovery and sustainability, of structure and function of the natural state of Large Marine Ecosystems. The contributions have been peer reviewed. They have been selected for inclusion in the series as a contribution toward a continuing assessment and evaluation of the changing conditions within the World's LMEs as scientists and ocean stewardship agencies move ahead toward ecosystem improvement targets endorsed by world leaders at the Johannesburg Summit. Production of this volume was supported in part by financial assistance of the United States Sea Grant Program of NOAA.

Kenneth Sherman, Series Editor
Narragansett, Rhode Island

Acknowledgements

The editors are indebted to the willingness of the contributors to take time from their normal schedules to prepare the expert syntheses and reviews that collectively serve to move us forward toward ecosystem-based assessment and management of the world's Large Marine Ecosystems. We are pleased to acknowledge the interest and financial support of the Office of Sea Grant, the International Union for the Conservation of Nature (World Conservation Union), the U.S. National Oceanic and Atmospheric Administration, and the U.S. National Marine Fisheries Service.

This volume would not have been possible without the capable cooperation of many people who gave unselfishly of their time and effort. We are indebted to Dr. Sally Adams, North Scituate, Rhode Island for her extraordinary dedication, care and expertise in technically editing and preparing the volume in camera-ready format for publication. Also, we extend our thanks to Dr. Marie-Christine Aquarone for her meticulous review of each of the chapters to ensure that the volume was free of any inadvertent omissions or misrepresentations. Further, we extend thanks to our Elsevier editors, Ms. Mara Vos-Sarmiento, Administrative Editor of Agricultural and Biological Sciences and Dr. Christiane Barranguet, Publishing Editor for Aquatic Sciences, for their care in the final production of this volume.

The Editors

Contributors

Ronald Baird
NOAA, Sea Grant (OAR)
Silver Spring, Maryland, USA

Jung-Hee Cho
Korea Maritime Institute
Seoul, KOREA

Patricia Clay
NOAA, NMFS, EASC
Silver Spring, Maryland USA

Jerry Diamantides
David Miller and Associate
Rochambeau St.
Providence, Rhode Island USA

Christopher L. Dyer
The School for Field Studies
10 Federal Street
Salem, Massachusetts USA

Alfred Duda
GEF Secretariat
Washington, District of Columbia USA

Steven F. Edwards
USDOC, NOAA, NMFS, NEFSC
Narragansett Laboratory
Narragansett, Rhode Island USA

Frank J. Gable
NRC fellow at NEFSC
Narragansett Laboratory
Narragansett, Rhode Island USA

John M. Gates
Kingston Coastal Institute
University of Rhode Island
Kingston, Rhode Island USA

Thomas A. Grigalunas
Environmental and Natural Resource
Economics
Kingston Coastal Institute
University of Rhode Island
Kingston, Rhode Island USA

Timothy M. Hennessey
Marine Affairs and Political Science
University of Rhode Island
Kingston, Rhode Island USA

Porter Hoagland
Marine Policy Center
Woods Hole Oceanographic Institution
Woods Hole, Massachusetts USA

Di Jin
Marine Policy Center
Woods Hole Oceanographic Institution
Woods Hole, Massachusetts USA

Lawrence Juda
Marine Affairs and Political Science
University of Rhode Island
Kingston, Rhode Island USA

Andrew Kitts
NOAA, NMFS, NEFSC
Woods Hole, Massachusetts USA

Jason S. Link
NOAA, NMFS, NEFSC
Woods Hole, Massachusetts USA

Philip Logan
NOAA, NMFS, NEFSC
Woods Hole, Massachusetts USA

Meifeng Luo
Department of Environmental and Natural
Resource Economics
Kingston Coastal Institute
University of Rhode Island
Kingston, Rhode Island USA

James J. Opaluch
Department of Environmental and Natural
Resource Economics
Kingston Coastal Institute
University of Rhode Island
Kingston, Rhode Island USA

John J. Poggie, Jr.
Department of Sociology and Anthropology
University of Rhode Island
Kingston, Rhode Island USA

Barbara P. Rountree
USDOC, NOAA, NMFS
Northeast Fisheries Science Center
Woods Hole, Massachusetts USA

Kenneth Sherman
USDOC, NOAA, NMFS, NEFSC
Narragansett Laboratory
Narragansett, Rhode Island USA

Mark Soboil
Department of Environmental and Natural
Resource Economics
University of Rhode Island
Kingston, Rhode Island USA

Scott Steinback
Social Sciences Branch
NMFS, NEFSC
Woods Hole Laboratory
Woods Hole, Massachusetts USA

Jon G. Sutinen
Department of Environmental and Natural
Resource Economics
Kingston Coastal Institute
University of Rhode Island
Kingston, Rhode Island USA

Eric M. Thunberg
NOAA, NMFS, NEFSC
Social Sciences Branch
Woods Hole Laboratory
Woods Hole, Massachusetts USA

Harold Upton
Oregon Department of Fish and Wildlife
Salem, Oregon USA

John B. Walden
NOAA, NMFS, NEFSC
Woods Hole, Massachusetts 02543 USA

Hanling Wang
Center for International Law
Chinese Academy of Social Sciences
Hai Dian District
Beijing, P.R. China

Dong-Sik Woo
Department of Environmental and Natural
Resource Economics
Kingston Coastal Institute
University of Rhode Island
Kingston, Rhode Island USA

Contents

Series Editor's Introduction   v
Acknowledgements   vii
Contributors   ix
Contents   xi

**Part I. Large Marine Ecosystems, Social Theory and LME Management
Methodology**   **1**

1. The Large Marine Ecosystem Approach for Assessment and Management of
Ocean Coastal Waters   3
Kenneth Sherman

2. The Human Dimension in Ecosystem Management: Institutional Performance
and The Sea Grant Paradigm   17
Ronald Baird

3. A Framework for Monitoring and Assessing Socioeconomics and Governance of
Large Marine Ecosystems   27
*Jon G. Sutinen (lead author) with Patricia Clay, Christopher L. Dyer,
Steven F. Edwards, John Gates, Thomas A. Grigalunas, Timothy M. Hennessey,
Lawrence Juda, Andrew W. Kitts, Philip N. Logan, John J. Poggie, Jr.,
Barbara Pollard Rountree, Scott R. Steinback, Eric M. Thunberg, Harold F. Upton,
and John B. Walden*

4. Governance Profiles and the Management of the Uses of Large Marine Ecosystems   83
Lawrence Juda and Timothy Hennessey

5. A Total Capital Approach to the Management of Large Marine Ecosystems:
Case Studies of Two Natural Resource Disasters   111
Christopher L. Dyer and John J. Poggie

6. Ownership of Multi-Attribute Fishery Resources in Large Marine Ecosystems   137
Steven F. Edwards

Part II. Economic Activity and the Cost of Ownership   **155**

7. Economic Activity Associated with the Northeast Shelf Large Marine
Ecosystem: Application of an Input-Output Approach   157
Porter Hoagland, Di Jin, Eric Thunberg, and Scott Steinback

8. Portfolio Management of Fish Communities in Large Marine Ecosystems   181
Steven F. Edwards, Jason S. Link and Barbara P. Rountree

9. Fish Habitat: A Valuable Ecosystem Asset   201
    Harold F. Upton and Jon G. Sutinen

10. The Economic Values of Atlantic Herring in the Northeast Shelf Large Marine
    Ecosystem   215
    Jung Hee Cho, John M. Gates, Phil Logan, Andrew Kitts, and Mark Soboil

11. Eutrophication in the Northeast Shelf Large Marine Ecosystem: Linking
    Hydrodynamic and Economic Models for Benefit Estimation   229
    Thomas A. Grigalunas, James J. Opaluch, Jerry Diamantides and Dong-Sik Woo

12. Valuing Large Marine Ecosystem Fishery Losses Because of Disposal of
    Sediments: A Case Study   249
    Thomas A. Grigalunas, James J. Opaluch and Meifeng Luo

Part III. The Role of Governance and Institutions   **271**
13. Emergence of a Science Policy-Based Approach to Ecosystem-Oriented
    Management of Large Marine Ecosystems   273
    F. J. Gable

14. Applications of the Large Marine Ecosystem Approach Toward World
    Summit Targets   297
    Alfred Duda and Kenneth Sherman

15. The Evolution of LME Management Regimes: The Role of Adaptive Governance   319
    Timothy M. Hennessey

16. An Evaluation of the Modular Approach to the Assessment and Management
    of Large Marine Ecosystems   335
    Hanling Wang

Editors' Conclusion   **357**
    Timothy M. Hennessey and Jon G. Sutinen

INDEX   **359**

Part I:
Large Marine Ecosystems, Social Science Theory, and LME Management Methodology

Part II:

Large Marine Ecosystems, Vol. 13
T.M. Hennessey and J.G. Sutinen (Editors)
© 2005 Published by Elsevier B.V.

1

The Large Marine Ecosystem Approach for Assessment and Management of Ocean Coastal Waters

Kenneth Sherman

MOVEMENT TOWARDS ECOSYSTEM-BASED ASSESSMENT AND MANAGEMENT

During the 10-year period between UNCED in 1992 and WSSD in 2002, advances were made in introducing ecosystem-based assessment and management of natural resources and their environments. A significant milestone in the marine ecosystem assessment and management movement was achieved in the mid 1990s by the Ecological Society of America's Committee on the Scientific Basis for Ecosystem Management. The Committee concluded that the overarching principle for guiding ecosystem management is to ensure the intergenerational sustainability of ecosystem goods (*e.g.* fish, trees, petroleum) and ecosystem services or processes including productivity cycles and hydrological cycles (Christensen *et al.* 1996). From a fisheries perspective, the National Research Council (NRC 1999, 2000) concluded that sustaining fishery yields will require maintaining the ecosystems that produce the fish. These reports are supportive of a paradigm shift from the highly focused, single-species or short-term sectoral thematic approach in general practice today to a broader, more encompassing multi-thematic ecosystem-based approach that moves spatially from smaller to larger scales, and from short-term to longer-term management practices. Included in this approach is a movement away from the management of commodities toward maintaining the sustainability of marine resources to ensure benefits from ecosystem goods and services for the future (Table 1.1).

Table 1.1. Movement toward ecosystem-based management (from Lubchenco 1994)

From	→	*To*
Individual species		Ecosystems
Small spatial scale		Multiple scales
Short-term perspective		Long-term perspective
Humans: Independent of Ecosystems		Humans: Integral Parts of Ecosystems
Management Divorced from Research		Adaptive Management
Managing Commodities		Sustaining Production Potential for Goods and Services

LARGE MARINE ECOSYSTEMS (LMES)

The paradigm shift depicted in Table 1.1 is presently emerging in the applications of ecosystem-based assessment and management policies within the geographic boundaries of large marine ecosystems (LMEs). On a global scale, 64 LMEs produce 90% of the world's annual marine fishery biomass yield (Sherman 1994; Garibaldi and Limongelli 2003). Most of the ocean pollution and coastal habitat alteration also occurs within the boundaries of LMEs. LMEs are regions of ocean space encompassing coastal areas from river basins and estuaries to the seaward boundaries of continental shelves, enclosed and semi-enclosed seas, and the outer margins of the major current systems as shown in Figure 1-1. They are relatively large regions on the order of 200,000 km^2 or greater, characterized by distinct bathymetry, hydrography, productivity, and trophically dependent populations (Sherman 1994).

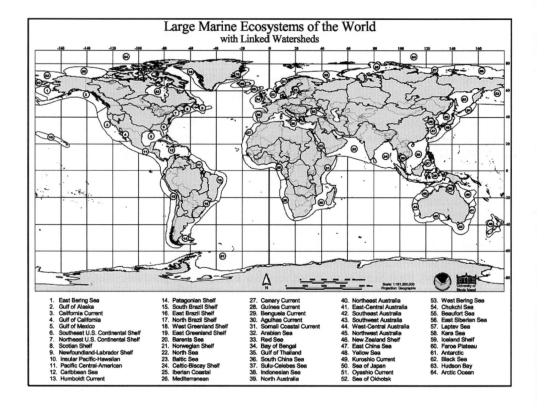

Large Marine Ecosystems of the World
with Linked Watersheds

1. East Bering Sea	14. Patagonian Shelf	27. Canary Current	40. Northeast Australia	53. West Bering Sea
2. Gulf of Alaska	15. South Brazil Shelf	28. Guinea Current	41. East-Central Australia	54. Chukchi Sea
3. California Current	16. East Brazil Shelf	29. Benguela Current	42. Southeast Australia	55. Beaufort Sea
4. Gulf of California	17. North Brazil Shelf	30. Agulhas Current	43. Southwest Australia	56. East Siberian Sea
5. Gulf of Mexico	18. West Greenland Shelf	31. Somali Coastal Current	44. West-Central Australia	57. Laptev Sea
6. Southeast U.S. Continental Shelf	19. East Greenland Shelf	32. Arabian Sea	45. Northwest Australia	58. Kara Sea
7. Northeast U.S. Continental Shelf	20. Barents Sea	33. Red Sea	46. New Zealand Shelf	59. Iceland Shelf
8. Scotian Shelf	21. Norwegian Shelf	34. Bay of Bengal	47. East China Sea	60. Faroe Plateau
9. Newfoundland-Labrador Shelf	22. North Sea	35. Gulf of Thailand	48. Yellow Sea	61. Antarctic
10. Insular Pacific-Hawaiian	23. Baltic Sea	36. South China Sea	49. Kuroshio Current	62. Black Sea
11. Pacific Central-American	24. Celtic-Biscay Shelf	37. Sulu-Celebes Sea	50. Sea of Japan	63. Hudson Bay
12. Caribbean Sea	25. Iberian Coastal	38. Indonesian Sea	51. Oyashio Current	64. Arctic Ocean
13. Humboldt Current	26. Mediterranean	39. North Australia	52. Sea of Okhotsk	

Figure 1-1. Large Marine Ecosystems are areas of the ocean characterized by distinct bathymetry, hydrography, productivity, and trophic interactions. They annually produce 90 percent of the world's fish catch. They are national and regional focal areas of a global effort to reduce the degradation of linked watersheds, marine resources, and coastal environments from pollution, habitat loss, and over-fishing.

For 40 of the LMEs, studies have been conducted of the principal driving forces affecting changes in biomass yields. Changes in biodiversity among the dominant species within fish communities of LMEs have resulted from: excessive exploitation, naturally-occurring environmental shifts in climate regime, or coastal pollution (Jackson *et al.* 2001). For example, in the Humboldt Current, Benguela Current, and California Current LMEs, the primary driving force influencing variability in fisheries yield is the influence of climate-forced changes in upwelling strength; fishing and pollution effects are secondary and tertiary effects on fisheries yields. In several continental shelf LMEs, including the Yellow Sea and Northeast United States Shelf, excessive fisheries effort has caused large-scale declines in catch and changes in the biodiversity and dominance in the fish community. In these ecosystems, pollution and environmental perturbation are of secondary and tertiary influence. In contrast, significant coastal pollution and eutrophication have been important factors driving changes in fisheries yields of the Northwest Adriatic, Black Sea, and the Baltic Sea. Following peer review, the results of these LME case studies were published in twelve volumes, listed in Table 1-2.

ROLE OF THE GLOBAL ENVIRONMENT FACILITY (GEF)

Following a three-year pilot phase (1991-1994), the Global Environment Facility (GEF) was formally launched to forge cooperation and finance actions in the context of sustainable development that address critical threats to the global environment including: (1) biodiversity loss, (2) climate change, (3) degradation of international waters, (4) ozone depletion, and (5) persistent organic pollutants. Activities concerning (6) land degradation, primarily desertification and deforestation as they relate to these threats, are also addressed. GEF projects are implemented by UNDP, UNEP, and the World Bank and expanded opportunities exist for participation by other agencies. The only new funding source to emerge from the 1992 Earth Summit, GEF today counts 171 countries as members. During its first decade, GEF allocated $US 3.2 billion in grant financing, supplemented by more than $US 8 billion in additional financing, for 800 projects in 156 developing countries and those in economic transition. All six thematic areas of GEF, including the land degradation cross-cutting theme, have implications for coastal and marine ecosystems. Priorities were established by the GEF Council in its Operational Strategy (GEF 1995) adopted in 1995. The international waters focal area was designed to be consistent with both Chapters 17 and 18 of Agenda 21. In 1995, the GEF Council included the concept of LMEs in its GEF Operational Strategy as a vehicle for promoting ecosystem-based management of coastal and marine resources in the international waters focal area within a framework of sustainable development. The Report of the Second Meeting of the UN Informal, Open-ended Consultative Process on Ocean Affairs (UN General Assembly 2001) related to UNCLOS recognized the contribution of the GEF in addressing LMEs through its science-based and ecosystem-based approach. Since the mid-1990s, developing countries have approached the GEF in increasing numbers for assistance in improving the management of Large Marine Ecosystems (LMEs) shared with neighboring nations. Processes being undertaken as part of GEF projects are focusing on Large Marine Ecosystems (LMEs) to foster country-driven commitments to policy, legal, and institutional reforms for changing the way human activities are conducted in the economic sectors that place stress on coastal ecosystems. LMEs serve as place-based, ecologically-defined areas

Table 1-2. Peer reviewed and published large marine ecosystem studies

LME	Vol.	Author(s)	LME	Vol.	Author(s)
Somali Coastal Current	7	Okemwa	East China Sea	8	Chen & Shen
Bay of Bengal	5	Dwividi	Yellow Sea	2,5,12	Tang
	7	Hazizi et al.	Kuroshio Current	2	Terazaki
East Bering Sea	1	Incze & Schumacher	Sea of Japan	8	Terazaki
	8	Livingston et al.	Oyashio Current	2	Minoda
West Greenland Shelf	3	Hovgärd & Buch	Okhotsk Sea	5	Kusnetsov et al.
	5	Blindheim & Skjoldal	Gulf of Mexico	2	Richards & McGowan
	10	Rice		4	Brown et al.
Barents Sea	2	Skjoldal & Rey		9	Shipp
	4	Borisov		9	Gracia & Vasquez Baden
	5	Skjoldal	Southeast US Shelf	4	Yoder
	10	Dalpadado et al.	Northeast US Shelf	1	Sissenwine
	12	Matishov		4	Falkowski
Norwegian Shelf	3	Ellertsen et al.		6	Anthony
	5	Blindheim & Skjoldal		10,12	Sherman
North Sea	1	Daan	Scotian Shelf	8	Zwanenburg et al.
	9	Reid	Caribbean Sea	3	Richards & Bohnsack
	10	McGlade	Patagonian Shelf	5	Bakun
	12	Hempel	South Brazil Shelf	12	Ekau & Knoppers
Iceland Shelf	10	Astthorsson & Vilhjálmsson	East Brazil Shelf	12	Ekau & Knoppers
Faroe Plateau	10	Gaard et al.	North Brazil Shelf	12	Ekau & Knoppers
Antarctic	1	Scully et al.	Baltic Sea	1	Kullenberg
	3	Hempel		12	Jansson
	5	Scully et al.	Celtic-Biscay Shelf	10	Lavin
California Current	1	McCall	Iberian Coastal	2	Wyatt &Perez-Gandaras
	4	Mullin		10	Wyatt & Porteiro
	5	Bottom	Mediterranean Sea	5	Caddy
	12	Lluch-Belda et al.	Canary Current	5	Bas
Pacific American Coastal	8	Bakun et al.		12	Roy & Cury
Humboldt Current	5	Bernal	Guinea Current	5	Binet & Marchal
	12	Wolff et al.		11	Koranteng & McGlade
Gulf of Thailand	5	Piyakarnchana		11	Mensah & Quaatey
	11	Pauly & Chuenpagdee		11	Lovell & McGlade
South China Sea	5	Christensen		11	Cury & Roy
Indonesian Sea	3	Zijlstra & Baars		11	Koranteng
Northeast Australian Shelf	2	Bradbury & Mundy	Benguela Current	2	Crawford et al.
	5	Kelleher		12	Shannon & O'Toole
	8, 12	Brodie	Black Sea	5	Caddy
				12	Daskalov

Vol. 1 1986. *Variability and Management of Large Marine Ecosystems*. Sherman & Alexander, eds. AAAS Symposium 99. Westview Press, Boulder, CO. 319p

Vol. 2 1989. *Biomass Yields and Geography of Large Marine Ecosystems*. Sherman & Alexander, eds. AAAS Symposium 111. Westview Press, Boulder, CO. 493p

Vol. 3 1990. *Large Marine Ecosystems: Patterns, Processes, and Yields*. Sherman, Alexander and Gold, eds. AAAS Symposium. AAAS Press, Washington, DC. 242p

Vol. 4 1991. *Food Chains, Yields, Models, and Management of Large Marine Ecosystems*. Sherman, Alexander and Gold, eds. AAAS Symposium. Westview Press. Boulder, CO. 320p

Vol. 5 1992. *Large Marine Ecosystems: Stress, Mitigation and Sustainability*. Sherman, Alexander and Gold, eds. AAAS Press, Washington, DC. 376p.

Vol. 6 1996. *The Northeast Shelf Ecosystem: Assessment, Sustainability and Management*. Sherman, Jaworski and Smayda, eds. Blackwell Science, Cambridge, MA. 564p

Vol. 7 1998. *Large Marine Ecosystems of the Indian Ocean: Assessment, Sustainability and Management*. Sherman, Okemwa and Ntiba, eds. Blackwell Science, Malden, MA. 394p

Vol. 8 1999. *Large Marie Ecosystems of the Pacific Rim: Assessment, Sustainability and Management*. Sherman and Tang, eds. Blackwell Science, Malden, MA. 455p

Vol. 9 1999. *The Gulf of Mexico Large Marine Ecosystem: Assessment, Sustainability and Management*. Kumpf, Steidinger and Sherman, eds. Blackwell Science, Malden, MA. 736p

Vol. 10 2002. *Large Marine Ecosystems of the North Atlantic: Changing States and Sustainability*. Skjoldal and Sherman, eds. Elsevier Science, N.Y. and Amsterdam. 449p

Vol. 11 2002. *Gulf of Guinea Large Marine Ecosystem: Environmental Forcing and Sustainable Development of Marine Resources*. McGlade, Cury, Koranteng, Hardman-Mountford, eds. Elsevier Science, Amsterdam and NY. 392p

Vol. 12 2003. *Large Marine Ecosystems of the World: Trends in Exploitation, Protection and Research*. Hempel and Sherman, eds. Elsevier Science, N.Y. and Amsterdam. 423p for which stakeholder support for integrating essential national and multi-country reforms and international agency programs can be mobilized into a cost-effective, collective response to an array of conventions and programs. Site-specific ocean concerns, those of adjacent coastal areas, and linked freshwater basins are being addressed in LMEs through GEF assistance. Operation of joint management institutions is being supported and tested in order to restore biomass and diversity to sustainable levels to meet the increased needs of coastal populations, and to reverse the precipitous declines in ecosystem integrity currently being caused by over-fishing, habitat loss, and nitrogen over-enrichment. At risk are renewable goods and services valued at $10.6 trillion per year (Costanza *et al.* 1997).

The geographic area of the LME, including its coastal area and contributing basins, constitutes the place-based area for assisting countries to understand linkages among root causes of degradation and for integrating needed changes in sectoral economic activities. The LME areas serve to initiate capacity building and to bring science to pragmatic use in improving the management of coastal and marine ecosystems. The GEF Operational Strategy recommends that nations sharing an LME begin to address coastal and marine issues by jointly undertaking strategic processes for analyzing factual and scientific information on transboundary concerns, finding their root causes, and setting priorities for action on transboundary concerns. This process has been referred to as a Transboundary Diagnostic Analysis (TDA) and it provides a useful mechanism to foster participation at all levels. Countries then determine the national and regional policy, and the legal and institutional reforms and investments needed to address the priorities in a country-driven Strategic Action Program (SAP). This allows sound science to become the basis for policy-making and fosters a geographic location upon which an ecosystem-based approach to management can be developed. This engages stakeholders in the geographic area so that they contribute to the dialogue and in the end support the ecosystem-based approach that can be pragmatically implemented by the communities and governments involved. Without such participative processes, marine science has often remained confined to the marine science community or has not been embraced in policy-making. Furthermore, the science-based approach encourages transparency through joint monitoring and assessment processes (joint cruises for countries sharing an LME) that build trust among nations over time and can overcome the barrier of false information being reported.

MODULES FOR LME ASSESSMENT AND MANAGEMENT

A five-module approach to the assessment and management of LMEs has proven useful in ecosystem-based projects in the United States and elsewhere because the approach relies on scientific information from the ecosystem under discussion The transboundary diagnostic analysis (TDA) process and the strategic action plan (SAP) development process are customized to, and then agreed upon by, all levels of the affected society. These processes integrate science into management in a practical way and establish governance regimes appropriate for the particular situation.   The Large Marine Ecosystem approach engages stakeholders, fosters the participation of the science community, and leads to the development of adaptive management institutions.

The five modules consist of three that are science-based activities focused on: Productivity, Fish/fisheries, and Pollution/ecosystem health. The other two modules, Socioeconomics and

Governance, are focused on socioeconomic benefits to be derived from a more sustainable resource base and from implementing governance mechanisms for providing stakeholders and stewardship interests with legal and administrative support for ecosystem-based management practices. The first four modules support the TDA process while the governance module is associated with a periodic updating of the Strategic Action Program or SAP. Adaptive management regimes are encouraged through periodic assessment processes (TDA updates) and through the updating of SAPs as gaps are filled (Duda and Sherman 2002). These processes are critical for integrating science-based information into management in a practical way and for establishing governance regimes appropriate for the particular situation.

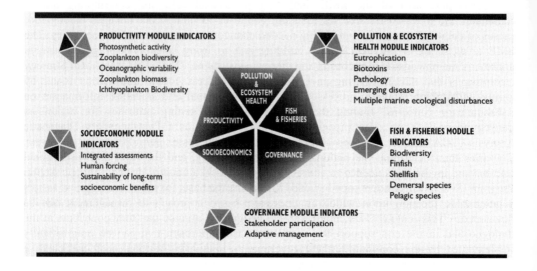

Figure 1-2. Five LME modules and lists of indicators of changing ecosystem conditions

Productivity Module
Productivity can be related to the capacity of an ecosystem for supporting fish resources (Pauly and Christensen 1995). Recently, scientists have reported that the maximum global level of primary productivity for supporting the average annual world catch of fisheries has been reached, and further large-scale "unmanaged" increases in fisheries yields from marine ecosystems are likely to be at trophic levels below fish in the marine food chain (Beddington 1995). Measuring ecosystem productivity also can serve as a useful indication of the growing problem of coastal eutrophication. In several LMEs, excessive nutrient loadings of coastal waters have been related to algal blooms implicated in mass mortalities of living resources, the emergence of pathogens (*e.g.*, cholera, vibrios, red tides, paralytic shellfish toxins), and the explosive growth of non-indigenous species (Epstein 1993).

The ecosystem parameters measured in the productivity module are zooplankton biodiversity and information on species composition, zooplankton biomass, water column structure, photosynthetically active radiation (PAR), transparency, chlorophyll-a, NO_2, NO_3, and primary production. Plankton in LMEs have been measured by Continuous Plankton Recorder (CPR) systems deployed monthly across ecosystems from commercial vessels of opportunity over decadal time scales. Advanced plankton recorders can be fitted with sensors for temperature, salinity, chlorophyll, nitrate/nitrite, petroleum, hydrocarbons, light, bioluminescence, and primary productivity. They provide the means for in situ monitoring and the calibration of satellite-derived oceanographic conditions relating to changes in phytoplankton, zooplankton, primary productivity, species composition and dominance, and long-term changes in the physical and nutrient characteristics of the LME and in the biofeedback of plankton to the stress of environmental change (Berman and Sherman 2001; Aiken *et al.* 1999).

Fish and fisheries module
Changes in biodiversity among the dominant species within fish communities of LMEs have resulted from: excessive exploitation, naturally-occurring environmental shifts in climate regime, or coastal pollution. Changes in the biodiversity of a fish community can generate cascading effects up the food chain to apex predators and down the food chain to plankton components of the ecosystem (Pauly *et al.* 1998). The Fish and Fisheries module is based on fisheries-independent information provided by bottom-trawl surveys and acoustic surveys of pelagic species. The time-series sheds light on changes in fish biodiversity and abundance levels. Standardized sampling procedures, when deployed from small calibrated trawlers, can provide important information on diverse changes in fish species (Sherman *et al.* 1998). Fish catch provides biological samples for stock assessments, stomach analyses, age, growth, fecundity, and size comparisons; data for clarifying and quantifying multispecies trophic relationships; and samples for monitoring coastal pollution. Samples of trawl-caught fish can be used to monitor pathological conditions that may be associated with coastal pollution. They can also be used as platforms for obtaining water, sediment, and benthic samples for monitoring harmful algal blooms, diseases, anoxia, and changes in benthic communities.

Pollution and ecosystem health module
In several LMEs, pollution has been a principal driving force in changes of biomass yields. Assessing the changing status of pollution and health in the entire LME is scientifically challenging. Ecosystem "health" is a concept of wide interest for which a single precise scientific definition is problematical. The health paradigm is based on multiple-state comparisons of ecosystem resilience and stability and is an evolving concept that has been the subject of a number of meetings (NOAA 1993). To be healthy and sustainable, an ecosystem must maintain its metabolic activity level and its internal structure and organization, and must resist external stress over time and space scales relevant to the ecosystem (Costanza 1992). The ecosystem sampling strategies are focused on parameters related to overexploitation, species protected by legislative authority (marine mammals), and other key biological and physical components at the lower end of the food chain (plankton, nutrients, hydrography) as noted by Sherman 1994.

Fish, benthic invertebrates, and other biological indicator species are used in the Pollution and Ecosystem Health module to measure pollution effects on the ecosystem. Pollution is measured through the bivalve monitoring strategy of "Mussel-Watch," (NOAA's National Status and Trends Program project to monitor the status of and temporal changes in metal and organic contaminants in Great Lakes, estuarine and coastal waters using bivalve mollusks as sentinel organisms), the pathobiological examination of fish, and the estuarine and near-shore monitoring of contaminants and contaminant effects in the water column, substrate, and in selected groups of organisms. The routes of bioaccumulation and trophic transfer of contaminants are assessed, and critical life history stages and selected food chain organisms are examined for parameters that indicate exposure to, and the effects of, contaminants. Effects of impaired reproductive capacity, organ disease, and impaired growth from contaminants are measured. Assessments are made of contaminant impacts at the individual species and population levels.

The US-EPA, in collaboration with NOAA, has successfully used a suite of 7 coastal condition indicators to depict the status of coastal waters and LMEs of the United States (National Coastal Condition Report 2003). The US Coastal Condition Report scheduled for release in August 2004 (National Coastal Condition Report II) includes the Northeast Shelf, Southeast Shelf and Gulf of Mexico LMEs. Where possible, bioaccumulation and trophic transfer of contaminants are assessed, and critical life history stages and selected food web organisms are examined for parameters that indicate exposure to, and the effects of, contaminants. Effects of impaired reproductive capacity, organ disease, and impaired growth from contaminants are measured. Assessments are made of contaminant impacts at the individual species and population levels. Implementation of protocols to assess the frequency and effect of harmful algal blooms, emergent diseases and multiple marine ecological disturbances (MMEDS) (Sherman 2000) are included in the pollution module.

Socioeconomic module
This module emphasizes the practical applications of its scientific findings to LME management and the integration of economic analysis with science-based assessments to assure that prospective management measures are cost-effective. Economists and policy analysts work closely with ecologists and other scientists to identify and evaluate management options that are both scientifically credible and economically practical with regard to the use of ecosystem goods and services.

Designed to respond adaptively to enhanced scientific information, socioeconomic considerations must be closely integrated with science. This component of the LME approach to marine resources management has recently been described as the human dimensions of LMEs. A framework has been developed by the Department of Natural Resource Economics at the University of Rhode Island for monitoring and assessing the human dimension of LMEs and the socioeconomic considerations important to the implementation of an adaptive management approach (Sutinen 2000). A methodology for considering economic valuation of LME goods and services has been developed around the use of interaction matrices for describing the relationships between ecological state and the economic consequences of change.

Governance module
The Governance module is evolving, based on demonstrations now underway to manage ecosystems from a more holistic perspective than generally practiced in the past. In GEF-supported projects—for the Yellow Sea ecosystem, the Guinea Current LME, and the Benguela LME—agreements have been reached, by the environmental ministers of the countries bordering these LMEs, to enter into joint resource assessment and management activities as part of building institutions. Among others the Great Barrier Reef LME is being managed from an ecosystem-based perspective. Similarly, the Antarctic marine ecosystem is

being managed under the Commission for the Conservation of Antarctic Marine Living Resources (CCAMLR). Governance profiles of LMEs are being explored to determine their utility in promoting long-term sustainability of ecosystem resources (Juda and Hennessey 2001).

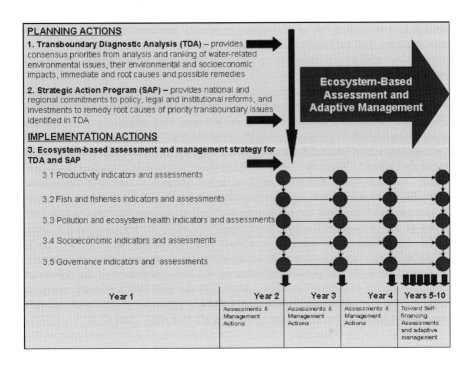

Figure 1-3. LME approach to assessment, management planning and implementation based on TDA and SAP priorities and the 5-module strategic approach

The systematic application of the 5 modules through the TDA-SAP processes fosters an adaptive management approach to joint governance. Adapted management is based on stated project goals and milestones, commitments to action, periodic progress reviews and iterative assessments of GEF monitoring and evaluation indicators (Figure 1-3). These processes, over the 5 to 10 year project period, help to integrate science into management and establish governance regimes for a collective response to site-specific priorities under various environmental conventions and action programs. The single most important consideration in the introduction of the ecosystem-based approach to the GEF-LME projects is to integrate data from the 5 modular lists of indicators, analyze it and translate the results into management actions focused on resource sustainability.

SAP goals and milestones ensure vertical interpretation across the 5 modules. Ecosystem indicators annually provide critical information for adaptive management actions (see Figure 1-3). The GEF-LME projects are generally funded for a 3- to 5- year initial phase, to be followed if successful by a second 3- to5- year grant. Thus, a 6- to 10- year time-window is provided for the participating countries bordering on the LME to have established a comprehensive ecosystem-based assessment and management system that will become a self-financed project during the second phase of project implementation. Joint monitoring surveys in GEF-LME projects provide transparency in collection of data. This builds confidence and trust among participating nations.   Achieved in GEF-LME projects are collective identification and prioritization of major transboundary environmental and living resources management issues and problems, and the adoption of common ecosystem-based strategies and policies for addressing these problems by participating countries at all levels of administration (Figure 1-3).

REVERSING BIOMASS DEPLETION IS POSSIBLE IN LMES

Recent ecosystem-based management actions are serving to reverse multidecadal declines in biomass yields. Since 1994, reductions in fishing effort increased the spawning stock biomass of haddock and yellowtail flounder, and of other species in the U.S. Northeast Shelf ecosystem. From the mid-1960s through the early 1990s, the biomass of principal groundfish and flounder species inhabiting the US Northeast Shelf ecosystem declined significantly from overfishing of the spawning stock biomass (NEFSC 2000). In response to the decline, the biomass of skates and spiny dogfish increased from the 1970s through the early 1990s (NEFSC 2000). The impact of the increase in small elasmobranches, particularly spiny dogfish, shifted the principal predator species on the fish component of the ecosystem from silver hake during the mid-1970s to spiny dogfish in the mid-1980s (Sissenwine and Cohen 1991). By the mid-1990s a newly developing fishery for small elasmobranches initiated a declining trend in biomass for skates and spiny dogfish (NEFSC 2000).

Following the secession of foreign fishing on the Georges Bank-Gulf of Maine herring complex and the Atlantic mackerel stock in the late 1970s, and after a decade of very low fishing mortality, both species began to recover to high stock sizes in the 1990s. Bottom trawl survey indices for both species showed a more than ten-fold increase in abundance (average of 1977-1981 vs. 1995-1999) by the late 1990s (NEFSC 2000). Stock biomass of herring increased to over 2.5 million metric tons by 1997 and spawning stock biomass was projected to increase to well over 3.0 million metric tons in 2000 (NEFSC 2000). The offshore component of herring, which represents the largest proportion of the whole complex, appears to have fully recovered from the total collapse it experienced in the early 1970s (NEFSC 2000). For mackerel, the situation is similar; total stock biomass has continued to increase since the collapse of the fishery in the late 1970s. Although absolute estimates of biomass for the late 1990s are not available, recent analyses concluded that the stock is at or near a historic high in total biomass and spawning stock biomass (NEFSC 2000). Recent evidence following mandated substantial reductions in fishing effort indicate that both haddock and yellowtail flounder stocks are responding favorably to the catch reductions with substantial growth reported in spawning stock biomass since 1994 for haddock and flounder. In addition, in 1997 a very strong year-class of yellowtail flounder was produced and, in 1998 a strong year-class of haddock was produced (Figure 1-4) (cf. Sherman *et al.* 2002).

At the base of the food web, primary productivity provides the initial level of carbon production to support the important marine commercial fisheries (Nixon *et al.*1986).

Zooplankton production and biomass in turn provide the prey-resource for larval stages of fish, and the principal food source for herring and mackerel in waters of the NE Shelf ecosystem. Over the past two decades, the long-term median value for the zooplankton biomass of the NE Shelf ecosystem has been about 29cc of zooplankton per 100m^3 of water strained produced from a stable mean-annual primary productivity of 350gCm^{2}yr. During the last two decades, the zooplanktivorous herring and mackerel stocks underwent unprecedented levels of growth, approaching an historic high combined biomass. This growth is taking place during the same period that the fishery management councils for the New England and Mid-Atlantic areas of the NE Shelf ecosystem have sharply curtailed fishing effort on haddock and yellowtail flounder stocks. Given the observed robust levels of primary productivity and zooplankton biomass, it appears that the "carrying capacity" of zooplankton supporting herring and mackerel stocks and larval zooplanktivorous haddock and yellowtail flounder is sufficient to sustain the strong year-classes reported for 1997 (yellowtail flounder) and 1998 (haddock) (Figure 1-4).

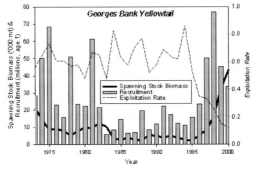

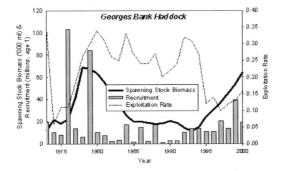

Figure 1-4. Trends in spawning stock biomass (ssb) and recruitment in relation to reductions in exploitation rate (fishing effort) for yellowtail flounder (a) and haddock (b), two commercially important species inhabiting the Georges Bank sub-area of the Northeast Shelf ecosystem.

The robust condition of the plankton components at the base of the food web of the Northeast Shelf ecosystem was important to the relatively rapid rebuilding of zooplanktivorous herring and mackerel biomass from the depleted condition in the early 1980s to a combined biomass in 1999 of an unprecedented level of approximately 5.5 million metric tons. This followed the exclusion of foreign fishing effort and the removal of any significant U.S. fishery on the stocks. The milestone action leading to the rebuilding of lost herring and mackerel biomass was the decision by the United States in 1975 to extend jurisdiction over marine fish and fisheries to 200 miles of the coastline. Recently the Fishery Management Councils of New England and the mid-Atlantic coastal states agreed to reduce fishing effort significantly on demersal fish stocks. With the reduction of exploitation rate, the spawning biomass of haddock and yellowtail flounder increased over a 4-year period and led to the production of large year-classes of haddock in 1998 and yellowtail flounder in 1997.

The Northeast Shelf ecosystem is presently showing a significant trend toward biomass recovery of pelagic and demersal fish species important to the economy of the adjacent northeast states from Maine to North Carolina. Although the recovery has not as yet been fully achieved, the corner has been turned from declining over-harvested fish stocks toward a condition wherein the stocks can be managed to sustain their long-term potential yield levels. The management decisions taken to reduce fishing effort to recover lost biomass were supported by science-based monitoring and assessment information following indicators from the productivity, fish and fisheries, pollution and ecosystem health, socioeconomics, and governance modules. These have been operational by NOAA's Northeast Fisheries Science Center for several decades in collaboration with state, federal, and private stakeholders from the region. This case study can serve to underscore the utility of the modular approach to ecosystem-based management of marine fish species. In an effort to stem the loss of fisheries biomass in other parts of the world, applications of this modular approach to LME management are presently underway by countries bordering the Yellow Sea, Benguela Current, Baltic Sea, and Guinea Current LMEs, with the financial assistance of the Global Environment Facility, collaborating UN agencies, and with the technical and scientific assistance of other governmental and non-governmental agencies and institutions (see online at <http://www.lme.noaa.gov>).

REFERENCES

Aiken J., R. Pollard, R. Williams, G. Griffiths, I. Bellan. 1999. Measurements of the upper ocean structure using towed profiling systems. In: Sherman K. and Q. Tang, eds. Large Marine Ecosystems of the Pacific Rim: Assessment, Sustainability, and Management. Blackwell Science, Inc. Malden, MA.

Beddington, J.R. 1995. The primary requirements. *Nature* 374:213–4.

Berman MS, Sherman K. 2001. A towed body sampler for monitoring marine ecosystems. *Sea Technology* 42(9):48–52.

Christensen N.L., A.M. Bartuska, J.H. Brown S. Carpenter C. D'Antonio R. Francis, J.F.Franklin, J.A. MacMahon, R.F. Noss, D.J. Parsons C.H. Peterson, M.G. Turner R.G. Woodmansee. 1996. Report of the Ecological Society of America committee on the scientific basis for ecosystem management. *Ecological Applications* 6(3):665–91.

Costanza R, R. d'Arge, R. de Groots, S. Farber, M. Grasso, B. Hannon, K. Limburg, S. Naeem, R.V. O'Neill, J. Paruelo, R.G. Raskin, P. Sutton, M. van den Belt. 1997. The value of the world's ecosystem services and natural capital. *Nature* 387:253–60.

Costanza R. 1992. Toward an operational definition of ecosystem health. In: Costanza R, B.G. Norton , B.D. Haskell, editors. *Ecosystem Health: New Goals for Environmental Management*. Island Press, Washington DC. 239–56.

Duda A.M. and K. Sherman. 2002. A new imperative for improving management of large marine ecosystems. *Ocean & Coastal Management* 45:797-833.

Epstein PR. Algal blooms and public health. World Resource Review, 1993;5(2):190–206.Food and Agriculture Organization. The State of the World Fisheries and Aquaculture. FAO, Rome, 2000. 142p.

Garibaldi, L. and L. Limongelli. 2003. Trends in Oceanic Captures and Clustering of Large Marine Ecosystems: Two studies based on the FAO capture database. FAO Fisheries Technical Paper. 435. Rome, Food and Agriculture Organization of the United Nations. 71p.

GEF (Global Environment Facility). 1995. GEF Operational Strategy. Washington, DC: Global Environment Facility.

Jackson J., M. Kirby, W. Berger, K. Bjorndal, L. Botsford, B. Bourque, R. Bradbury, R. Cooke, J. Erlandson, J. Estes, T. Hughes, S. Kidwell, C.B. Lande, H. Lenihan, J. Pandolfi, C. Peterson,, R. Steneck, M. Tegner, R. Warner. 2001. Historical over fishing and the recent collapse of coastal ecosystems. *Science* 293:629–638.

Juda L., T. Hennessey. 2001. Governance profiles and the management of the uses of large marine ecosystems. *Ocean Development and International Law* 32:41–67.

Lubchenco J. 1994. The scientific basis of ecosystem management: framing the context, language, and goals. In: Zinn J. and M.L. Corn, editors. Ecosystem Management: Status and Potential. 103rd Congress, 2d Session, Committee Print. U.S. Government Printing Office, Superintendent of Documents. 33–9.

NEFSC. 2000. Status of Fishery Resources off the Northeastern United States for 1999, Steve Clark, Editor. NOAA Tech. Memo. NMFS-NE-115.

Nixon S.W., C.A. Oviatt, J. Frithsen, B. Sullivan. 1986. Nutrients and the productivity of estuarine and coastal marine ecosystems. *Journal of the Limnology Society of South Africa* 12:43–71.

NRC (National Research Council). 2000. Clean coastal waters: understanding and reducing the effects of nutrient pollution. National Academy Press, Washington, DC.

NRC (National Research Council). 1999. Sustaining marine fisheries. National Academy Press, Washington, DC.

NOAA (National Oceanic and Atmospheric Administration). 1993. Emerging Theoretical Basis for Monitoring the Changing States (Health) of Large Marine Ecosystems. Summary Report of Two Workshops: 23 April 1992, National Marine Fisheries Service, Narragansett, Rhode Island, and 11–12 July 1992, Cornell University, Ithaca, New York. NOAA Technical Memorandum NMFS-F/NEC-100.

Pauly D, V. Christensen. 1995. Primary production required to sustain global fisheries. *Nature* 374:255–7.

Pauly D, V. Christensen, J. Dalsgaard, R. Froese, F. Torres Jr. 1998. Fishing down marine food webs. *Science* 279:860–3.

Sherman B. 2000. Marine ecosystem health as an expression of morbidity, mortality, and disease events. *Marine Pollution Bulletin* 41(1–6):232–54.

Sherman K. 1994. Sustainability, biomass yields, and health of coastal ecosystems: an ecological perspective. *Marine Ecology Progress Series* 112:277–301.

Sherman, K., J. Kane, S. Murawski , W. Overholtz , A. Solow. 2002. The US Northeast Shelf Large Marine Ecosystem: zooplankton trends in fish biomass recovery. In: *Large Marine Ecosystems of the North Atlantic: Changing States and Sustainability*. Amsterdam: Elsevier Science. 195-215. 449p.

Sherman K., E.N. Okemwa, M.J. Ntiba, eds. 1998. *Large Marine Ecosystems of the Indian Ocean: Assessment, Sustainability, and Management.* Cambridge: Blackwell Science, Inc. 394p.

Sutinen J. editor. 2000. A framework for monitoring and assessing socioeconomics and governance of large marine ecosystems. NOAA Technical Memorandum NMFS-NE-158. 2000. 32p.

United Nations General Assembly. 2001. Report on the work of the United Nations Open-ended Informal Consultative Process established by the General Assembly in its resolution 54/33 in order to facilitate the annual review by the Assembly of developments in ocean affairs at its second meeting. Report A/ 56/121, 22 June, New York, 2001. 62p

Large Marine Ecosystems, Vol. 13
T.M. Hennessey and J.G. Sutinen (Editors)
© 2005 Elsevier B.V. All rights reserved.

2

The Human Dimension in Ecosystem Management: Institutional Performance and The Sea Grant Paradigm

R. Baird

ABSTRACT

Sustainability has become the focus and organizing principle for the reconciliation of economics and the environment. The concept has led to the emergence of holistic or ecosystem-based management approaches where maintenance of the environment and associated ecosystems in acceptable condition indefinitely is the goal. Application in management contexts, however, remains an immense challenge. Globally, environments continue to degrade, especially in third world countries. While the scientific basis for such management approaches remains an issue, many of the impediments to acceptable performance in governance lie in the human dimensions. Performance is a function of institutional capacity, the complex of people, infrastructure, education, resources and legal framework that determines societal response to environmental decisions.

The Sea Grant Paradigm, whereby universities partner with federal governments to create centers that conduct programs of research, education and outreach on management critical problems, holds great promise as an institution for building long-term global capacity to manage marine environments. The paradigm is particularly powerful in developing the human dimensions so critical to good marine governance. Sea Grant is culturally and administratively adaptable to both developed and developing countries. Nine specific human dimension elements inherent in the Sea Grant model are described in the context of holistic management of large marine ecosystems.

INTRODUCTION

The last several decades have seen the emergence of a concept known as sustainability, which has become the central focus and organizing principle for reconciling fundamental human needs with the preservation of the natural environment. To sustain is to support, nourish and prolong, according to Webster's Dictionary. Today, the concept has been broadly applied to natural resource management, and provides the conceptual framework driving contemporary resource management policy and its associated scientific framework (Kates *et al.* 2001; Pauly 2002; Costanza *et al.* 2001).

The application of sustainability to management contexts, however, remains an immense challenge at virtually all relevant scales from global to local. Sustainability involves a holistic consideration of the environment, at many geographic scales, and the maintenance of that environment and associated ecosystems within acceptable conditions for an indefinite period of time. Societal values, however, are integral to any practical, management-related definition of acceptable ecosystem states (Baird 2002). To satisfy the sustainability equation, then, requires management approaches based on ecosystem concepts, together with a complex of interrelated socioeconomic/governance methodologies and infrastructure necessary to support coherent and effective long-term natural resource management (Kates *et al.* 2001).

Much has been written about the economic importance of coastal marine environments, their economic utility (*e.g.*, transportation, recreation) and ecological value (*e.g.*, fisheries) derived from the high productivity of associated ecosystems (Welsh 1988; Nixon 1992). Likewise, it is clear that adverse environmental impacts to such environments continue at significant rates (Pew Commission Report 2003). While an adequate science basis for marine ecosystem management remains a lengthy and complex challenge, the general precepts for such an approach are now better understood as are ecosystem processes in general (Levin 1999; Verity *et al.* 2002; Sherman and Skjoldal 2002).

The predicted global population growth, especially in coastal areas, and the rate and magnitude of observed human-induced impacts on marine environments, highlight the urgency for rapid transition to sustainable development. To do this well will require profound changes at all levels of society and in our institutions of governance. That means our institutions must be capable of high levels of performance—and performance largely involves human dimensions. A management context for sustainability is defined below, as is the importance of rapidly developing the human resource base and infrastructural capacity to implement ecosystem-based management worldwide. Developing and maintaining the essential ties between science, society and institutions of governance must also be a high priority for any global sustainability agenda.

THE MANAGEMENT CONTEXT

Coastal and contiguous marine environments provide the geographic focus for the management of natural resources and the consideration of the human dimension in the transition to ecosystem management. Several examples used here are drawn from fisheries management since this area is the most advanced in global terms and in regulatory jurisdictions. Since fish stocks are integral components of marine ecosystems, they also represent convenient indicators of broader ecosystem conditions and human impacts (Rosenberg 2003).

Our domain of concern then includes environments that are dynamic, highly complex and geographically heterogeneous. These environments have been subjected to prolonged human disturbance. Moreover, such areas are generally fragmented in terms of institutions of governance and political jurisdictions. In concert with population growth, human-induced impacts are escalating at alarming rates in many dimensions, such as habitat degradation, pollution loadings and over-fishing (Duda and Sherman 2002).

Existing regulatory mechanisms concerning the management of human activities in these areas have been inadequate. Adler (2002) notes nearly three decades of federal regulation have failed to provide for the sustainable utilization of the USA's marine resources. Duda and Sherman (2002) speaking specifically about the management of Large Marine Ecosystems (LMEs) globally, list a host of institutional shortcomings ranging from jurisdictional fragmentation to inadequate legislation and enforcement.

From an ecosystem management perspective, however, the issues are conceptually succinct. Determine whether the current environmental state is within acceptable limits, whether it can be maintained and at what societal costs. If not acceptable, determine whether mitigation policies can restore the system to acceptability. If the latter is not politically feasible, determine whether the degraded system can be contained or reduced geographically. That is the management context for sustainability and ecosystem management going forward. Implementing effective solution strategies is a very different matter (Link 2002).

Finally, definitions of acceptable states are politically mediated and therefore rest on a foundation of human beliefs, values and preferences that are reflected in institutions of governance (Orbach 2002). Kennedy (2002) points out that if we ask the question in a poor, developing country, we are likely to get an answer quite different from the one we would get in the U.S. And even if the answers were the same, the capacity of local institutions of governance to perform effectively is generally inadequate — much more so in developing countries even though such countries are the most vulnerable to cumulative impacts on coastal environments (Kates *et al.* 2001). There is also the phenomenon of the "ratchet effect" (Baird 1996) where human adaptability can increase the tolerance (including political acceptance) of increasing degradation of ecosystem states with each generation. Societies do not stand still. The sustainability equation must deal with these changes in a rational and consistent manner. Improved equity and capacity for economic development are critical ingredients to a successful transition to global sustainability and effective ecosystem management.

INSTITUTIONAL PERFORMANCE AND CAPACITY

The transition to ecosystem based management now in progress is well short of realization. Significant impediments must be overcome in a timely manner to make that transition a reality. There is a voluminous literature emerging on integrated management approaches (Cicin-Sain and Knecht 2000; Sherman and Anderson 2002). Likewise, there is a rich literature on the complex interplay of science, institutions of governance and non-government agencies in setting public policy (Lee 1993; Hahn 1994). Recent essays speak to the need for major changes in human governance institutions (Orbach 2002) and our associated legal framework (Adler 2002).

The bottom line, however, is that appropriate management decisions must be made, then translated through a complex institutional infrastructure into public policy that elicits effective responses from society at large. It is this diverse array of institutional entities and individuals that collectively determines the tempo and mode of societal responses to environmental issues and management decisions (Baird 1996). The effectiveness of that response dynamic in terms of sustainability is defined here as institutional performance.

Recent reports by the National Council for Science and the Environment (2002, 2003) list a number of specific recommendations on how to achieve sustainable communities. Embedded in those recommendations are the following concepts: communities, stakeholders, science and technology, conflict resolution, timeliness, cost, integration of economy, society, environment, education, information dissemination, long term focus, place-based activities and collaborative governance. Kates *et al.* (2001), in addressing requirements for progress in sustainability science, explicitly name institutions and infrastructure as essential components in the development of capacities adequate to support a sound science basis. Included here are observational and reporting systems and long term, integrated research on natural systems and human interactions at various geographic scales.

What is evident from this literature is that better institutional performance is critical to sustainability and that performance is tied directly to the adequacy of institutional capacity to support a high level of performance. Capacity as defined here as the complex infrastructure of public education, research, training, experience, capital assets, mechanisms of communication and coordination, and governance framework that collectively determine long-term ecosystem management performance.

Capacity building is an often under-appreciated and under-capitalized element in the global ecosystem management equation. The diffuse nature of the process, accountable to multiple agencies, jurisdictions, universities, resource bases and local cultural differences, makes holistic approaches difficult. The situation is particularly acute in developing countries where the infrastructure and knowledge base is often poorly developed while environmental degradation is acute. Nonetheless, success rests on a substantial, ongoing investment in capacity building.

For effective ecosystem-based management, then, an enormous demand for environmental knowledge, mitigation technologies, environmental literacy, public awareness and informed decision making must be satisfied. Currently, there must be a large premium on mechanisms that effectively build capacity to enhance management performance.

THE SEA GRANT PROGRAM AND PARADIGM

The National Oceanic and Atmospheric Administration (NOAA)'s National Sea Grant College Program (NSGCP) is described elsewhere (NAS 1994) and its legislative basis is encompassed in the Sea Grant Act of 2002 (P.L. 107-299). The NSGCP is a national organization headquartered at NOAA and consisting of 30 university-based "centers" that conduct locally based programs.

The program's purpose is to engage the capabilities of universities in addressing marine resource management issues consistent with the Agency's mission. The NSGCP is a partnership among the federal government, universities and state/local governments. Direct federal appropriations contribute about 60 percent of program costs, and local programs are encouraged to seek other funding sources. The basic organizational paradigm is based on the Land Grant concept of mission-directed research, extension of science based information to user constituencies and capacity building educational programs.

The outcome is a powerful, enabling infrastructure that is highly networked nationally and locally. Planning is national in scope but locally implemented in accordance with local and regional needs. The infrastructure allows for a high degree of responsiveness and rapid identification of problems at multiple scales, plus timely transfer of relevant information to users/policy makers in a useful form. Well developed relationships among federal agencies, universities and stakeholders have emerged.

Local programs have considerable autonomy and administrative flexibility yet are held to high standards of performance. The decentralized management structure is quite effective in directing resources to solve problems and in conducting place-based, management-critical science. The program has had 30 years of evolution and is a proven, effective and efficient organizational model with an impressive array of assets and core competencies. Over the years the program has made seminal contributions to marine and coastal resource management (NAS 1994).

LME MANAGEMENT AND THE SEA GRANT PARADIGM

An outcome of the 1992 Conference on Environment and Development has been the establishment of an international concept that has shown great promise in promoting the transition to adaptive and holistic management. The concept is based on the worldwide identification of geographically specific areas labeled Large Marine Ecosystems or LMEs. The approach has the great advantage of delineating specific areas in which to apply ecosystem-based management approaches, especially in fisheries, for which a considerable regulatory infrastructure already exists (McGlade *et al.* 2002). The approach has had significant global success, engaging stakeholders, encouraging participation by scientists and promoting multi-jurisdictional adaptive management (Duda and Sherman 2002).

The LME program has helped demonstrate that environmental issues are global in scale and require multi-national responses. Yet we are far from an effective, politically acceptable system of international environmental governance. We do not yet have the capacity to respond if such a system existed. In the absence of both adequate governance and capacity, a necessary initial step to effective transition to ecosystem-based management of LMEs is to strengthen the marine research and resource management capabilities of individual countries and regions.

The Sea Grant Paradigm (SGP) has been advanced as a possible international role model in building management capacity (Baird and Eppi 2002). Here, these earlier arguments are extended specifically to the human dimensions of LME management where the SGP is particularly powerful in building capacity to put adaptive ecosystem-based techniques into effective practice. Described below are several human dimensions important to ecosystem-based management; they show how the SGP can make contributions to their positive enhancement.

Mission congruence: Sustainability and adaptive management require novel environmental management systems, innovative technology, dedicated research, education and strong focus on a broad array of complex, interrelated issues. Fragmentation of mission among various, sometimes competing institutions, agencies and programs often reduces critical focus and the integration needed for holistic practical management. Sea Grant's specific mandate is to address practical marine resource management issues.

Continuity: Progress in sustainability is a highly complex endeavor of indefinite length and is comparable in many respects to human health management. The institutional capacity is far less well developed in sustainability science than in medicine, but the supporting infrastructure is no less important. That infrastructure must be geared toward the long term, should have appropriate financial support and be reasonably free of political encumbrances. Effective infrastructures have corporate memories and cultures that are results-oriented and invested for the long term. The SGP involves the establishment of quasi-permanent, university-based centers in partnership with government agencies. Such centers help create the institutional permanence and stability in changing political climates so necessary to sustained progress.

Adaptability: Virtually every serious treatise on the subject acknowledges that a successful transition to sustainability must involve recognition of indigenous cultures, of local to national institutions of governance and of regional socio-economic conditions. The SGP is inherently flexible both culturally and administratively. It can be adapted and customized to a diverse set of university, government, cultural milieus (*i.e.,* customized) yet retain critical programmatic strengths.

Effectiveness: Management of LMEs, like any human endeavor, requires institutions that have the right programmatic strengths and the ability to deliver products and services that enhance management performance. The SGP is a proven, effective paradigm that combines research, education and extension of science-based knowledge—all designed to enhance informed decision making. Embedded in effectiveness is the notion of high standards and a system of performance accountability.

Engagement: Effective resource management decisions rest on a firm foundation of environmental literacy and the constructive engagement of many disparate societal elements critical to the decision making process. The SGP is designed to specifically engage those diverse constituencies in resource management issues. The foremost is the university system and the associated scientific and educational assets found nowhere else. By extension, regional and local communities, decision makers and industries are engaged through advisory boards and information transfer. Partnerships among Federal, State and local agencies are incorporated in financial and information networks. Sea Grant managers are leaders, experts and participants in their own right. Education programs target the educational system promoting environmental literacy and provide the trained human resource base required of each generation.

Objectivity: SGP is not a regulatory institution yet it provides information on sensitive, regulatory issues and engages in conflict resolution. Objectivity and a reputation for unbiased transfer of information are essential to institutional effectiveness. As a university-based institution, Sea Grant has considerable latitude in prioritizing programmatic objectives on sensitive issues while maintaining a reputation for objectivity.

Efficiency: The SGP creates a supportive institutional environment through its unique financial structure. Direct federal appropriations must be matched (at least 1/3 in U.S.) by non-federal sources. This provides an economical enabling infrastructure while enhancing local and university partnerships. Once in place, resources can be solicited by the infrastructure from a diverse array of funding sources for specific activities. This creates substantial financial leverage together with the administrative competence that instills confidence in investors. Because the SGP has both permanent and project elements (about 60% projects in U.S.) it is highly flexible and can re-program investment resources to highest priority and emerging problems. Because of its regional based centers, efficient transfer of resources to problems from national to local scales is possible.

Regionality: Engagement at regional and local scales is critical simply because people and problems vary with geography and scale. Effective environmental decision making must take into account the decision makers, the jurisdictions at various regional scales, and place-based knowledge. To do this well requires a regional infrastructure which the SGP can provide.

Networked Organization: Rapid advances in communication technology have greatly advanced our potential for effective global management of LMEs. Given the disparity in capabilities and institutional capacity among countries, mechanisms that increase communication and the flow of management-critical information are urgently needed. In the SGP, one has an emerging international network of immense collective capability that is geographically diverse and is on the "cutting edge" of virtually every facet of marine and coastal resource management. That is an invaluable resource that is just beginning to emerge on the international scene but is an increasingly effective organizational strength.

CONCLUDING REMARKS

As Duda and Sherman (2002) so aptly point out, there is indeed a new imperative for improving management of LME's. That improvement rests in large measure with human dimensions. The temporal urgency of the issues means that any promising mechanism to improve management performance is critically important and worthy of our close attention. The SGP is such an institution, one that has the potential for developing and implementing ties between scientific assessment of LME, state and socio-economic and governance systems. The SGP holds great promise for rapidly introducing ecosystem-based management in both developed and developing countries while enhancing regional coordination and cooperation so necessary for effective LME management. There is much to learn in adapting the SGP to LME management globally but the movement has begun. Several countries (*e.g.*, Korea, Indonesia, Honduras, Nicaragua) have initiated programs based on the SGP. There is already in place a foundation, the NSGCP in the U.S., providing a wealth of institutional capability and expertise upon which to build a global network of independent but connected programs.

REFERENCES

Adler, J.H. 2002. Legal obstacles to private ordering in marine fisheries. *Roger Williams University Law Review* 8(1):9-42.

Baird, R.C. 1996. Toward new paradigms in coastal resource management: linkages and institutional effectiveness. *Estuaries* 19(2A):320-335.

Baird, R.C. 2002. Sustainable coastal margins. In: *A Sustainable Gulf of Mexico: Research, Technology and Observations 1950-2050.* Proceedings of the 125th Anniversary Celebration Symposium, February 19-21, 2002. Texas A&M University, College Station, Texas. 46-49.

Baird, R.C., and R.E. Eppi. 2002. Building global capacity in coastal resource management: the emerging Sea Grant model. *Fisheries* 27(6):32-35.

Cicin-Sain, B. and R.W. Knecht. 2000. *The Future of U.S. Ocean Policy: Choices for the New Century.* Island Press, Washington, DC.

Costanza, R., B.S. Low, E. Ostium, and J. Wilson, eds. 2001. *Institutions, Ecosystems and Sustainability.* Lewis Publishers. Boca Raton, Fl.

Duda, A.M. and K. Sherman. 2002. A new imperative for improving management of large marine ecosystems. *Ocean and Coastal Mgt.* 45:797-833.

Hahn, R.W. 1994 United States environmental policy: past, present and future. *Natural Resources Journal* 34:305-348.

Kates, R.W. and 22 co-authors. 2001. Sustainability science. *Science* 292:641-642.

Kennedy, D. 2002. Sustainability: problems, science and solutions. *Report of the Second National Conference on Science, Policy, and the Environment. National Council of Science and the Environment*, Washington, D.C. 35-39.

Lee, K.N. 1993. *Compass and Gyroscope: Integrating Science and Politics for the Environment.* Island Press, Washington, D.C.

Levin, S. 1999. *Fragile Dominion: Complexity and the Commons*. Perseus, Cambridge, MA.

Link, J.S. 2002. What does ecosystem based management mean? *Fisheries* 27 (4):18-21.

McGlade, M., P. Cury, K.A. Koranteng and N.J. Hardman-Mountford, eds. 2002 . *The Gulf of Guinea Large Marine Ecosystem: Environmental Forcing and Sustainable Development of Marine Resources.* Elsevier, Amsterdam.

NAS. 1994. *A Review of the NOAA National Sea Grant College Program.* National Academy Press, Washington, D.C.

NCSE. 2002. Recommendations for achieving sustainable communities, science and solutions. *Report of the Second National Conference on Science, Policy and the Environment.* National Council of Science and the Environment, Washington, D.C.

NCSE. 2003. Recommendations for education for a sustainable and secure future. *Report of the Third National Conference on Science, Policy and the Environment.* National Council for Science and the Environment. Washington, D.C.

Nixon, S.W. 1992. Quantifying the relationship between nitrogen input and productivity of marine ecosystems. *Advances in Marine Technology* 5: 57-83.

Orbach, M.K. 2002. *Beyond Freedom of the Seas: Ocean Policy for the Third Millennium.* National Academy of Sciences, Washington, D.C.

Pauly, D. 2002 . Sustainable ecosystems. In: *A Sustainable Gulf of Mexico: Research, Technology and Observations 1950-2050.* Proceedings of the 125th Anniversary Celebration Symposium, February 19-21, 2002:16-17. Texas A&M University, College Station, TX.

Pew Oceans Commission. 2003. America's Living Oceans: Charting a Course for Sea Change. *A Report to the Nation.* Pew Oceans Commission, Arlington, VA.

Rosenberg, A.A. 2003. Multiple uses of marine ecosystems. In: Sinclair, M. and G. Valdemarrson. eds. *Responsible Fisheries in the Marine Ecosystem.* 189-200. CABI Publishing, Cambridge, MA.

Sherman, K. and E.D. Anderson. 2002. A modular approach to monitoring, assessing and managing large marine ecosystems. In: McGlade, J.M., P. Cury, K.A. Koranteng and N.I. Hardman-Mountford, eds. *The Gulf of Guinea Large Marine Ecosystem.* 9-26. Elsevier, Amsterdam.

Sherman, K. and H.R. Skjoldal, eds. 2002. *Large Marine Ecosystems of the North Atlantic: Changing States and Sustainability.* Elsevier, Amsterdam.

Verity, P.G., V. Smetacek and T.J. Smayda. 2002. Status, trends and the future of the marine pelagic system. *Environmental Conservation* 29(2):207-237.

Walsh, J.J. 1988. *On the Nature of Continental Shelves.* Academic Press, New York

Large Marine Ecosystems, Vol. 13
T.M. Hennessey and J.G. Sutinen (Editors)

3

A Framework for Monitoring and Assessing Socioeconomics and Governance of Large Marine Ecosystems[1]

Jon G. Sutinen (lead author) with Patricia Clay, Christopher L. Dyer, Steven F. Edwards, John Gates, Tom A. Grigalunas, Timothy Hennessey, Lawrence Juda, Andrew W. Kitts, Philip N. Logan, John J. Poggie, Jr., Barbara Pollard Rountree, Scott R. Steinback, Eric M. Thunberg, Harold F. Upton, and John B. Walden

INTRODUCTION

The ecosystem paradigm is emerging as the dominant approach to managing natural resources in the U.S. as well as internationally. The shift away from the management of individual resources to the broader perspective of ecosystems has not been confined to academia and think tanks where it first began; it also is beginning to take root in government policy and programs. Since the late 1980s, many federal agency officials, scientists and policy analysts have advocated a new, broader approach to managing the nation's natural resources. The approach recognizes that plant and animal communities are interdependent and interact with their physical environment to form distinct ecological units called ecosystems. These systems contribute to the production of fish, marine birds, and marine mammals that cross existing jurisdictional boundaries. The approach also recognizes that many human actions and their consequences, including marine pollution, extend across jurisdictional boundaries.

Emergence of this paradigm is a response to the failure of the single sector/single species approach to achieve sustainable development of interdependent natural resources and effective protection of the natural environment. There is now a pronounced trend towards more integrated ecosystem management. US administration and legislation are increasingly requiring an ecosystem approach to natural resource research and management. The September 1993, Report of The National Review: Creating a Government That Works Better and Costs Less recommended that the President issue an executive order establishing ecosystem management policies across the federal government.[2]

---

[1] First published by Sutinen *et al.* in 2000: NOAA Technical Memorandum, NMFS-NE-158 (under contract # ENN F7 00378)

[2] The policies are based on the following principles: 1) managing along ecological boundaries, 2) ensuring coordination among federal agencies and increased collaboration with state local and tribal governments, the public and congress; 3) using monitoring, assessment and the best science available; and 4) considering all natural and human components and their interactions.

To implement an ecosystem approach for environmental management, the White House Office of Environmental Policy established an Interagency Ecosystem Task Force to implement an ecosystem approach to environmental management. To date, the movement toward ecosystem management is reflected in, for example, the Magnuson-Stevens Fishery Conservation and Management Act (as amended through October 11, 1996), NOAA's Marine Sanctuaries Program, the National Estuary Program, the National Estuarine Research Reserves System, the 1990 Amendments to the Coastal Zone Management Act and also in the actions of federal agencies with resource management responsibilities. [3] Further, NOAA's Strategic Plan (1997) is based, in large part, on the ecosystem approach to living marine resource management.

Ecosystem management is defined as a system 'driven by explicit goals, executed by policies, protocols, and practices, and made adaptable by monitoring and research based on our best understanding of the ecological interactions and processes necessary to sustain ecosystem structure and function' (Ecological Society of America 1995: 1). Ecosystem management necessitates intergovernmental and intersectoral management. This is why federal agencies will have to identify barriers to interagency coordination and why they must develop alliances and partnerships with non-federal agencies and private sector stakeholders (Hennessey 1997). Ecosystems management must be able to cope with the uncertainty associated with the complexity of ecosystems as natural systems, and the organizational and institutional complexity of the implementation environment (Hennessey 1997, Acheson 1994).

The fit between the spatial and temporal scales of government jurisdictions on the one hand and ecosystems on the other requires investigation of ways to connect 'nested' ecosystems through 'networked institutions' at federal, state, local and NGO levels (Hennessey 1997). How these institutions must adapt to deal with the complexity of the ecosystem and the complexity of the governance system in order to achieve an optimal mix of benefits and costs is a fundamental issue (Creed and McCay 1996).

The need for improved management of living marine resources is critical. The livelihood of coastal populations and national economies has depended, for many decades, on coastal and marine resources. As indicated in NOAA's Strategic Plan, 'over half of the US population now lives on the coast. Between one-third and one-half of US jobs are located in coastal areas. About one-third of the nation's GNP is produced there through fishing, transportation, recreation and other industries dependent on healthy coastal ecosystems for growth and development. Rapid population growth and increasing demand for recreation and economic development in many coastal areas have degraded natural resources and led to declines in both environmental integrity and general productivity.   Coastal areas provide essential habitats for the majority of commercially valuable marine species. But habitat loss, pollution and overfishing have reduced populations of coastal fish and other species to historically low levels of abundance and diversity. Maintaining coastal ecosystems health and biodiversity is essential to the sustainable development of coastal resources and economies, and to the future welfare of the Nation' (NOAA 1997).

---

[3] Since 1992 all four of the primary land management agencies (the National Park Service, the Bureau of Land Management, the Fish and Wildlife Service and the Forest Service) have independently announced that they are implementing or will implement an ecosystem approach to managing their natural resources, and each has been working to develop it own strategy (GAO 1994). Several other agencies, including the Soil Conservation Service, the Department of Defense, Department of Energy, Bureau of Indian Affairs, Bureau of Mines, Bureau of Reclamation, Minerals Management Service, USGS, EPS, and NASA, have engaged in significant ecosystem management activities (CRS 1994).

The complex interplay of socioeconomic, ecological, political and legislative processes underscores the need for an integrated approach to the management of drainage basins, coastal areas and linked continental shelves and dominant current systems. In this report, we develop an integrated approach to these problems based on the large marine ecosystems concept.

The concept of Large Marine Ecosystems (LMEs) is a science-based method for dividing the world's oceans, developed 15 years ago by Kenneth Sherman and Lewis Alexander. LMEs are geographic areas of oceans that have distinct bathymetry, hydrography, productivity, and trophically dependent populations. The geographic limits of most LMEs are defined by the extent of continental margins and the seaward extent of coastal currents. Among these are the Northeast US Continental Shelf, Southeast US Continental Shelf, Gulf of Alaska, Gulf of Mexico, Eastern Bering Sea and the California Current. Some LMEs are semi-enclosed seas, such as the Caribbean, Mediterranean and Black Seas. LMEs can be further divided into subsystems such as the Gulf of Maine, Georges Bank, Southern New England, and the Mid-Atlantic Bight in the case of the Northeast US Continental Shelf (Sherman et al. 1988). Approximately 95 percent of all fish and other living marine resources produced are taken from the world's 64 LMEs. Unfortunately, many LMEs are currently stressed from overexploitation of marine resources, habitat degradation, and pollution.

The LME management approach links the management of drainage basins and coastal areas with continental shelves and dominant coastal currents. The approach (i) addresses the many-faceted problem of sustainable development of marine resources; (ii) provides a framework for research monitoring, assessment and modeling to allow prediction and better management decisions; and (iii) aids in focusing marine assessments and management on sustaining productivity and conserving the integrity of ecosystems.

The World Bank and the Global Environment Facility (GEF) have adopted the LME approach to marine ecosystem research and management, viewing it as 'an effective way to manage and organize scientific research on natural processes occurring within marine ecosystems [and] to study how pollutants travel within these marine systems ...' (World Bank 1995: Annex A). The World Bank's Operational Guidelines for LME research requires social science as well as natural science investigations, since many of the problems of the marine environment are human induced. The GEF LME initiative has 5 modules:
1. Productivity
2. Fish resources and fisheries
3. Pollution and ecosystem health
4. Socioeconomics
5. Governance

The first three modules are natural resource science-based and well developed. During the past 15 years, extensive scientific work has resulted in methods for monitoring and assessing the productivity, fish resources and fisheries, pollution and ecosystem health of LMEs. Sustained, accurate and efficient assessments of changing ecosystem states are now feasible because of the advent of advanced technologies applied to coastal ocean observation and prediction systems. Such systems can now measure ocean productivity, changes in fish stocks, and changes in water and sediment quality and general health of the coastal ocean.

Consideration of the socioeconomic and governance modules has been more limited,[4] despite the fact that work on these modules is essential to achieving effective ecosystem management. Management of LMEs requires not only knowledge of changing states of the system but also the effects of change on socioeconomic benefits to be derived from using the LME resources. To provide sustainable, optimal use of marine resources, the services they provide must be identified and valued, the sources of market failure must be understood, and policy instruments to correct market failures and move toward sustainability must be adopted.

This report presents a methodology for determining what is known of the socioeconomic and governance aspects—the human dimensions—of LME management. The following sections describe a basic framework for identifying the salient socioeconomic and governance elements and processes of an LME. Methods for monitoring and assessing the various elements and processes are also discussed.

HUMAN DIMENSIONS OF LMES

Monitoring and assessment are prerequisites to effective management of LMEs threatened by pollution, over exploitation and other misuses of these important resources. Furthermore, management involves altering human behavior, especially behavior that threatens, directly or indirectly, the sustainability of LME resources. Therefore, we need to understand the human system and its relationship to the sustainability of LME resources and their services.

Human and ecological systems are both composed of complex webs of interrelated components and processes. Interactions occur both within each respective system and between systems. We view the natural environment and related human dimensions as a set of interrelated components and processes rather than isolated elements that act independently.

Ecological components of an LME can be viewed as, among other things, biophysical capital (*i.e.*, stocks of valuable natural resources). The various forms of the biophysical capital generate flows of goods and services, many of which are directly or indirectly used by humans (*e.g.*, in fishing and shipping activities). Some ecological goods and services are transformed into commodities that are cycled through the economy. These flows also include outputs of processes that are returned to the environment, sometimes as wastes.

Traditionally, property rights are poorly defined in the coastal zone and marine areas.[5] Externalities impact fishermen, recreation, and other activities relying on the natural system for flows of commodities and opportunities from these capital assets.[6] Human activities that use or impact the biophysical capital of a typical LME may occur on land, in the coastal zone or in off shore areas. High human population densities in the coastal regions and associated

---

[4] The only work on this is a brief sketch by Broadus (Sherman 1997, and Sherman *et al.* 1993) and an unpublished white paper, 'LME Socioeconomic Module (Sociocultural Submodule),' prepared for the NMFS Office of Science and Technology.

[5] As explained below, we define property as a benefit (or income) stream, and a property right is a claim to the benefit stream that government agrees to protect from those who may covet, or somehow interfere with, the benefit stream (Bromley 1992: 2).

[6] Externalities are unintended harmful or beneficial effects incurred by a party that is not directly involved with exchange, production or consumption of the commodity in question.

manufacturing, transportation and extractive activities often result in environmental degradation and overexploitation. Municipal sewage and industrial waste disposal in coastal waters often overwhelm the assimilative capacity of marine areas. Nutrient pollution may result in large increases in phytoplankton production and microbial activity and reduced oxygen—eutrophication. Fish and shellfish populations that are dependent on estuaries as essential habitat may be harmed, displaced or rendered unfit for human consumption. In virtually all of these examples the five LME modules are interdependent—a change in one module will have impacts on other modules.

MONITORING AND ASSESSMENT

We anticipate several steps in the process of monitoring and assessing the human dimensions of an LME and the use of its resources. These steps are summarized in Table 1 and they provide information to management authorities, especially with regards to the efficacy of management policies. Most of these steps should be repeated periodically to update the information on the status of the LME. This information is an essential ingredient of the adaptive management approach, which requires frequent evaluation and feedback to take full advantage of experience and learning (Hennessey 1994; Lee 1993; Walters 1986).

Table 3-1. Steps for Monitoring and Assessment

1. Identify LME resource users and their activities
2. Identify governance mechanisms influencing LME resource use
3. Assess the level of LME-related activities
4. Assess interactions between LME-related activities and LME resources
5. Assess impacts of LME-related activities on other users
6. Assess the interactions between governance mechanisms and resource use
7. Assess the socioeconomic importance of LME-related activities and economic and sociocultural value of key uses and LME resources
8. Identify the public's priorities and willingness to make trade-offs to protect and restore key natural resources
9. Assess the cost of options to protect or restore key resources
10. Compare the benefits with the costs of protection and restoration
11. options
12. Identify financing alternatives for the preferred options for protecting and restoring key LME resources

Step 1: Identify principal uses of LME resources

The first step in the monitoring process involves identifying principal uses of LME resources. Management of large marine ecosystems requires comprehension of a variety of relationships within the natural and human environment and also the effect of human uses on the environment. In other words, policy makers need to be aware of and sensitive to the pattern of interaction resulting from their policy decisions if the sustainability of the environment, which supports human needs, is to be maintained.

Use is an important concept and requires careful definition. We define several types of use as follows:

- **Direct use** refers to the physical use of a resource on site or *in situ*. Common examples of direct use include commercial and recreational fishing, beach use, boating, and wildlife viewing. Most direct use is targeted by participants who visit a beach, fish at a particular location, and so forth. Direct use also may be *incidental*, for example, when a person traveling by boat unexpectedly sees whales or marine birds while en route to a destination (Freeman 1993).

- **Indirect Use** occurs when, for example, wetlands or other LME habitats contribute to the abundance of wildlife or fish observed or caught elsewhere in the LME. In effect the ecological services of the wetland or habitat help 'produce' the wildlife or fish concerned, although the link between the direct use and the ecological services provided by the wetland or habitat may not be apparent to the recreational participant.

- **Non-use (or 'Passive Use')** refers to the enjoyment individuals may receive from knowing that the resources exist ('existence value') or from knowing that the resources will be available for use by one's children or grandchildren ('bequest value') or others even though the individuals themselves may not actually use the resources concerned.

Another useful distinction is between *consumptive* use and *non-consumptive* use:

- **Consumptive use** occurs when one person's use of a resource prevents others from using it. For example, the shellfish, finfish, or waterfowl I take in the LME are unavailable for others to catch. Hence, consumptive use of natural resources in this sense is like consumptive use of common private goods exchanged on markets, such as a pizza or a pair of shoes.

- **Non-consumptive use** refers to cases where one person's enjoyment does not prevent others from enjoying the same resource. For example, my viewing of marine mammals, other wildlife, or attractive views in the LME does not prevent you from enjoying the same resources.[7]

In this report, the uses include *direct consumptive* and *non-consumptive use*, such as shipping, commercial and recreational fishing, mining, boating, beach use, and wildlife viewing. We emphasize that many activities, such as fishing and viewing of wildlife, rely upon the ecological productivity of LMEs; hence, these activities also involve the *indirect* use of these ecosystems.

---

[7] Some recreational uses, for example beaches, are intermediate cases. An individual's use of a beach may not interfere with others, at least *up to a point*, after which beach (or parking) congestion would make use of a beach like a consumptive use in that my use may prevent you from gaining access to the resource.

Step 2. Identify LME resource users and their activities

LME-related activities play a major role in the livelihood of numbers of coastal state residents who own, operate or are employed by thousands of businesses in many sectors. These sectors engage in or support such activities as fisheries, marine transportation, and particularly tourism and recreation.

Determining use sectors that are LME-related is not always straight-forward, and judgment necessarily plays an important role in making such decisions (*e.g.*, Rorholm, *et al.* 1967; King and Story 1974; Grigalunas and Ascari 1982; Crawford 1984). Certain sectors are clearly LME-related, such as commercial fishing, marinas, ferries, and specialized retail stores, such as, bait and tackle shops. These are primary activities, which by their nature operate on or around the water; or they supply goods and services clearly related to consumptive and non-consumptive uses of LME resources.

In a broad sense, however, much if not most activity along a coast is LME-related, at least in part. For example, restaurants, hotels and motels, retail shops, real estate, and gasoline stations serve seasonal visitors to the coastal resources of the LME, as well as year-round residents and businesses.[8] Thus, these sectors are also LME-related to a large extent (although some activity in these sectors may also be dependent, in part, on the inland, terrestrial resources). Moreover, many residents may view the quality of the LME environment as an important factor attracting them to the area. In short, the dependence of human activity on LME resources and their quality is much broader (and more subtle) than might be suggested by first impressions.[9]

We recommend a pragmatic approach by defining two broad use sectors that are LME-related: directly-related and indirectly-related use sectors. Both sectors are involved with the consumptive or non-consumptive uses of LME resources.

- *Directly-related use* sectors are relatively distinct and include primary activities or those that operate on or in the LME. These marine-related sectors are considered to be 100 percent LME-related. Examples include commercial fishing ports, marinas, and ferries that are physically located along, or that operate within, the LME.[10]

- *Indirectly-related use* sectors include tourism and recreation activities such as hotels, motels, restaurants, and sport facilities (e.g., public golf courses and membership sports clubs) and retail sectors that service tourists and coastal residents such as, gas stations, bakeries, grocery stores, general merchandise stores, etc. Other indirectly-related sectors may include land-based agriculture, manufacture and forestry, which may indirectly affect the health of the LME via pesticide runoff, waste water discharge or soil erosion up stream. These use sectors are considered not fully LME-related since the link between the LME and the level of these activities is weak or less clear.

---

[8] If they were available for the LME, input-output studies could be used to identify linkages among sectors within an area (see, *e.g.*, Rorholm *et al.* 1967; Grigalunas and Ascari 1982; King and Story 1974; Tyrrell *et al,* 1982 for examples of such studies for other marine areas).

[9] Property value studies also give general support for the relatively high residential demand for, and the correspondingly relatively high value of, shoreline proximity (*e.g.*, Edwards and Anderson 1984).

[10] Other LME-related sectors, such as retail or wholesale seafood, may also be somewhat dependent on activities in the LME.

Step 3: Identify governance mechanisms influencing LME Resource use

As conflict of use and negative environmental consequences of human use become more obvious, collective responses at a variety of levels begin to emerge—in short, governance efforts evolve. We recommend developing a governance profile for each LME (Juda and Hennessey 1998). It should be noted that in the case of most of the identified LMEs, governance involves governments and people of more than one State since political and LME boundaries typically do not coincide. This reality has significant implications and could provide either a rationale for interstate cooperation or, alternatively, an abandonment of national efforts, since if they are undercut by the actions of others they will be rendered ineffective.

Just as natural ecosystems vary from one another so too do governance systems. Governance arrangements already exist in areas encompassed by LMEs; they are not, however, presently organized around the concept of LMEs. Institutional, sociocultural, and economic factors are of substantial significance in the use and management of the natural environment; like aspects of the natural environment, they are also 'site specific.' In seeking to move toward a governance system that is more appropriate for ecosystem based management, it is necessary to understand how existing institutional and cultural systems operate, their implications for the natural environment and its resources, and how needed change may emerge, given societal structures and norms.

Why is governance important? The answer to this question lies in the fact that attempts to manage resources and the environment are really about managing human behavior and encouraging patterns of conduct that are in accord with the operation of the natural world. Governance affects human uses of LME resources and may be conceived of as:

> the formal and informal arrangements, institutions, and mores which determine how resources or an environment are utilized; how problems and opportunities are evaluated and analyzed, what behavior is deemed acceptable or forbidden, and what rules and sanctions are applied to affect the pattern of resource and environmental use. (Juda 1999).

As suggested by this definition, the concept of governance is not equivalent to government but rather incorporates other mechanisms and institutions (both formal and informal) that serve to alter and influence human behavior in particular directions.

Reflecting the notion that governance is not the same as government, there are three key, general mechanisms of governance: markets, government, and non-governmental institutions and arrangements. These mechanisms interact with one another in a pattern of dynamic interrelationships. Through the forces they generate, they individually and collectively impact use behaviors. (Figure 3-1)

Markets generate prices, which structure the incentives faced by firms and households, affecting how environmental resources are utilized. Resources for which no markets exist in effect have zero prices (*e.g.*, fish in the sea), artificially deflating the cost of using such resources. In other

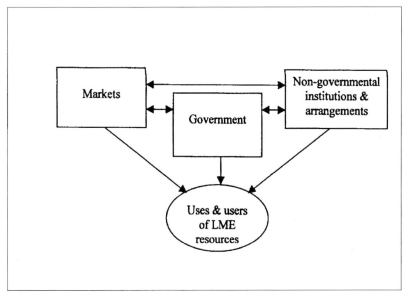

Figure 3-1. Governance mechanisms

words, users do not face the full social and environmental cost of fishing, habitat destruction, waste disposal, etc., when these resources are not priced. Lower cost of use, in turn, tends to encourage excessive use and results in depleted LME fish stocks, too little essential habitat, and too much pollution.

Government regulations and requirements, whether at a local, regional, national, or international level, affect resource use. In general, government sets a wide array of rules and enforces them, recognizes and protects property rights, and produces goods and services. The rules regulate the use of environmental resources and affect the way goods and services are produced. The protective function of government is to maintain security and order by enforcing a set of rules within which people can interact peacefully with one another. These include rules against theft, fraud, and physical harm to person and property. Without protection, property rights are not secure and externalities arise. The government also produces goods and services that cannot be efficiently organized by the market. These activities and outputs include a system of jurisprudence (an example of a pure public good), fisheries and oceanographic research (quasi-public goods), fishing license and boat registries (regulatory services), guaranteed loans, and vessel buy-back programs (transfer payments to users of the marine environment). These and other government activities tend to have a profound influence on how LME resources are used.

Social forces that are generated by non-governmental institutions and arrangements also influence use patterns. These forces are shaped by norms, values and beliefs that rationalize cognition of self and other members of society (ICGP 1993). They are dependent on the importance people attach to their community and neighborhoods, traditions, and long standing social networks. Failure to heed the pressures from these factors may lead to sanctions that range from economic loss, to incarceration or monetary penalties, or to expulsion from the community.

The principal task of this step is to identify and describe the salient forces (markets, governmental, and non-governmental institutions and arrangements) that influence users and their uses of LME resources. Practical, applied field methods will have to be developed to insure a complete inventory of such forces is compiled.

Step 4: Assess the level of LME-related activities

This step involves assessing the nature and extent of all LME-related activities identified in step 2 above. The tasks include measuring the quantity and value[11] of the goods and services produced, the employment and income generated, use rates of LME resources and other significant inputs used by these sectors.[12] These levels of LME-related activities should be calculated for the LME as a whole and disaggregated by appropriate sub-regions and user/producer groups. Recent trends and patterns in these activities should be described in as much detail as the data allow. Historical uses should also be incorporated in order to provide a context with regard to present activities and arrangements.

Step 5: Assess the Interactions Between LME-related activities and LME Resources

The notion that human use alters the natural environment is not new; what is relatively new is the degree to which that environment and its natural process may be affected by human actions. If future sustainability is a matter of concern to decision makers, then it is necessary for them to consider the nature and character of the interactions between human activities and natural systems. Therefore, our monitoring and assessment framework must fully integrate the human and ecological systems related to an LME.

The need for human-ecological system integration is readily understandable due to the similarities and interaction between the two systems. This is reflected in many government and development agency policies, which advocate the use of ecological management and principles. However, the complexity with respect to the number of components and relationships make this a difficult task. Most ecological studies have not fully integrated human activities, and most approaches have considered only one or a few sectors at a time. There is also a broad body of integrated environmental and ecological economics studies, but relatively few of these attempts have been successful (van den Bergh 1996).

Early attempts to integrate the two systems utilized input-output models to construct matrices of economic and ecological components and processes (Cumberland 1966, Daly 1968, Isard 1971, Victor 1972). Although the format of each attempt varied, the framework generally followed work by Isard (1971), depicting both ecological and economic processes. A variety of matrices also have been used by others in conjunction with generic coastal or ocean use (Couper 1983,Vallega 1992).

---

[11] These values should include market and non-market values, gross and net values, and net benefits to consumers and producers.

[12] Excellent guidelines on fisheries data needs are Brainerd *et al.* (1996); Kitts and Steinback and FAO 1999 Guidelines for the Routine Collection of Capture Fishery Data. Fisheries Technical Paper. (See sections 4.3.3 Economic indicators and 4.3.4 Socio-cultural indicators.) (FAO Fisheries Technical Paper No. 382. Rome, FAO 113p.)

Input-output models have disadvantages that limit their usefulness. Input-output models are composed of a system of linear equations that are dependent on technical coefficients that symbolize the amount of an input required for each dollar of output. This assumes constant proportions with no substitution or economies or diseconomies of scale. Unfortunately most socioeconomic and ecological systems involve component relationships that are neither static nor linear. Additional problems are related to the complexity of ecological systems. Most models of ecosystems consider the transfer of energy through the food chain. However, ecosystems resemble a web with multiple connections rather than a linear chain. Many species are generalists that change diet according to season, prey availability or life history stage while each prey item has a different energy transfer efficiency. Decomposers that utilize dead organisms and other unused organic matter also add another layer of complexity to the structure of food webs. In addition, there are other interactions within the system such as competitive and mutualistic relationships that are especially difficult to quantify. Given more realistic, albeit complex modeling alternatives, input-output models may be too restrictive and simplistic for quantitative assessments of these systems. However, the input-output framework is a good foundation for our purposes, especially with respect to the inventory, organization and exploration of the myriad relationships related to an LME.[13]

We believe it is important to organize data and information about the interplay of human activities with natural processes in such a manner as to illuminate interrelationships, with the hope that consideration of highlighted interactions will foster behavior appropriate to the goal of sustainability. The framework that follows seeks to promote understanding of relationships and to encourage the utilization of adaptive management approaches (Hennessey 1994; Lee 1993; Walters 1986) that take full advantage of experience and learning. For these purposes, this study proposes the use of interaction matrices that can serve as diagnostic tools and provide a framework for analysis and consideration of management problems and possibilities. These matrices have the capacity to inventory human uses of the LME and ecological processes that are related to the LME, organize human activities, commodities and processes within a framework, and to explore linkages and relationships among sectors. The interaction matrices can be modified to depict different geographic zones and different industrial or species groupings. The geographic designation of land and marine can be further broken down into economic and ecological subgroupings. Therefore regions or ecological groups can be subdivided depending on the desired method of classification, the desired scale and the functional relationships that are being investigated. Given the complexity and need for a comprehensive approach, interaction matrices can provide a description of the current situation and provide a basis for predicting the consequences of changes to the system. Readily showing relevant sectors and linkages, the matrices also are useful education and communication tools for policy makers.

The study of these systems relies on common descriptive characteristics such as scale, spatial and temporal distribution, linkages, thresholds, resiliency and diversity. Ecosystems can be assessed on different levels that include the individual organism, population, community and the ecosystem. Human systems also include different levels such as the individual, household and family, community, business enterprise, use-sector, region and society. Scale is essential to

---

[13] It appears certain that further work in this area will require incorporation of a great deal of complexity, nonlinear relationships and dynamics. Previous work using general equilibrium models (Ayres and Kneese 1969) and systems approaches may be productive.

understanding both human and ecological systems. An ecosystem may range in scale from a cubic foot of soil to thousands of square miles. Human systems also range from the household to the national and global economy. Spatial and temporal factors are also important considerations for both systems. Many socio-economic activities and ecological distributions may be seasonal, patchy, and migratory in nature. Ecosystems with greater diversity are likely to be the most stable (Caddy and Sharp 1986). This is also likely to be true of a regional economy with diverse economic sectors compared to one dependent on only one or a few commodities.

Perhaps the most important characteristics are the linkages and interdependencies between components within both systems. In both cases a change in one element has repercussions for other elements within the system. Figure 3-2 illustrates a matrix that could be used to show the potential linkages among land-based and marine-based processes. For example, nutrients from agriculture can affect productivity in estuaries and near shore areas. Land-based processes that affect marine-based processes potentially involve similar modules that may need to be taken into account. Figure 3-2 divides the processes in each region into ecological and use sectors, and provides a few examples of such processes (*e.g.*, fishing, aquaculture, marine transportation). The matrix contains cells wherein the relationships between, say, aquaculture and ecosystem health are described in terms of the degree of compatibility and nature and extent of impact.

This and the other matrices illustrated in this report are recognized as being general and simplistic. Clearly, it is necessary that broad categories of activity be subdivided appropriately. In the case of fishing, operations are conducted in many different ways around the world and even among fishermen of a particular State. Commercial, industrial, artisanal, recreational and subsistence fishing, and the use of different gear and techniques, while all coming under the general rubric of fishing, may have varying impacts on the biomass and the physical environment and, accordingly, must be differentially assessed.

Contained in the elements of the matrices is information on the interactions between human use activities and the LME environment and its resources. It may well be that the effects of human use are not well understood or fully documented and a degree of precaution may be called for to avoid irreversible damage or long term costs as decisions are made. Indeed, it would be useful for decision makers if some explicit assessment could be made as to data availability and the degree of understanding of natural processes both of which could be factored into decisions about the application of precaution. Consideration of interplay based on experience may be suggestive of priorities for future study where data or understanding are deemed insufficient.

To a considerable extent, human use of and effects on the ocean/coastal natural environment have been generally described. For instance, water quality has been monitored and evaluated, wetland loss has been studied, the introduction of alien species has been noted, and coastal demographic changes have been documented. But in addition to studying changing conditions of the environment, greater consideration must be given to the consequences of those changes. The scientific community needs to highlight, in terms understandable to the lay person, the consequences of those changes for human well being, a step that goes beyond observing the relationships of the type noted in the interaction matrices.

| | | | Land | | | | | | | | Marine | | | | | | |
| | | | Use Sectors | | | | Ecological | | | | Use Sectors | | | | Ecological | | |
			Agriculture	Manufacturing	Services	Forestry	Productivity	Wildlife	Plants	Eco. Health	Aquaculture	Fishing	Transportation	Recreation	Productivity	Fish	Eco. Health
Land	Use Sectors	Agriculture															
		Manufacturing															
		Services															
		Forestry															
	Ecological	Productivity															
		Wildlife															
		Plants															
		Ecosystem Health															
Marine	Use Sectors	Aquaculture															
		Fishing															
		Transportation															
		Recreation															
	Ecological	Productivity															
		Fish															
		Ecosystem Health															

Figure 3-2. Interactions among LME-related activities and LME resources

The finding of depleted oxygen in coastal waters, for example, needs to be attached to the more practical consequences of fewer opportunities for commercial and recreational fishermen since it is such considerations that may serve to motivate public concern and appropriate actions. Accordingly, a subsidiary matrix that reflects the impact of ecosystem effects on outcomes of interest to stakeholders and the wider public is warranted. A more sophisticated version would provide for contemplation of impacts of human uses on the environment and its ecosystem resources.

A 'vulnerability assessment' of specific environmental conditions is needed for coastal management (Laurens et al. 1997), since variance in a number of natural conditions may alter the significance of possible threats. Indeed, the process of International Maritime Organization special area designation (MARPOL 1973/1978), the establishment of marine sanctuaries (Marine Protection, Research, and Sanctuaries Act), the inclusion of the concept of Essential Fish Habitat in the 1996 Sustainable Fisheries Act, and other special area zones (Coastal Zone Management Act 1972) indicate recognition of vulnerabilities of particular areas.

Step 6: Assess the Impacts of LME-related Activities on Other Users

The lack of strong and complete property rights to the biotic and abiotic elements of an LME is the fundamental cause of externalities and threats to the sustainable use of an LME. (For more on this see the following section, 'Property Rights Entitlements and Regimes for LME Management.') The users of LME resources usually have free and open access to those resources (*e.g.*, transportation of goods, harvesting of food and industrial raw material and for leisure activities). The resulting negative externalities tend to affect the marine ecosystem in injurious ways: overharvesting of wild species, destruction of habitat, pollution, etc. These effects are costs (harms) imposed on the environment and on users other than (external to) those causing the damage. The methodologies developed for assessing the damages to natural resources can be employed to estimate the external costs of pollution, habitat destruction, etc. We explain below (Example 1) how the monetary damages from oil spills and other transboundary marine pollution can be assessed.[14]

This step aims to assess the impacts that activities directly or indirectly have on others, especially the extent and nature of those impacts. The task is to identify and measure the benefit or harm imposed and the compatibility or incompatibility of particular uses in relation to other uses. Contained in the elements of an interaction matrix is information on the nature and extent of the interactions among the users of the LME environment and its resources. The information should characterize the interactions in terms of (1) degree of impact, (2) compatibility, and (3) desirability.

Compatibility implies either that the uses do not interfere with one another or, possibly, that they may serve to enhance one or both of the uses through positive externalities. Incompatibility indicates detrimental effects of one use on another or both on each other through negative externalities. The compatibility of particular uses in relation to other uses may be measured or described in terms such as compatible, conditionally compatible, incompatible. The amount or frequency of activity (*e.g.*, high, moderate, or low) needs to be considered as it relates to compatibility. At low levels of use, uses may be compatible, while this might not be the case with high levels of an activity.

The concept of compatibility and conflict of use is basic to the fields of coastal zone management (Clark 1996) and land planning; as ocean uses increase and intensify, it has been recognized in sea or ocean use management (Vallega 1992). Some activities are mutually exclusive while others are compatible to varying degrees. Often incompatibility is demonstrated in practice as implemented, sectorally based, decisions from which negative externalities are generated. In the face of such experience, planners and coastal managers, have resorted to devices such as zoning to keep apart activities that have significant incompatibilities.

Social conflict may take several forms both within specific sectors and between sectors. Allocation decisions that favor a specific sector may be economically efficient, but detrimental to a specific user group.[15] This may result in high social costs at the household or community level

---

[14] Also see Dyer, Gill and Picou (1992) for an examination of sociocultural considerations in assessing oil spills.

[15] Public policies are also shaped by political expediency that may favor the minority that is composed of a specific user group over society. Special interest effects occur when an issue is important to a specific group while most members of society are unaware of, or disinterested in, its outcome. In these cases it becomes

if alternatives are unavailable. Social costs may take the form of higher crime rates, poor diet, drug use or the breakup of households and families. These effects include less easily quantified social considerations such as community stability, maintenance of social networks and traditions, and the distribution of benefits.

The normative characterization (*e.g.*, desirable or undesirable) of the interplay among users and uses is essential to management decisions. We note that normative characterizations are determined largely in a cultural context, a factor which once more underscores the need for a site specific analysis of human interaction with the environment (Juda and Hennessey 1998).

How should these elements be measured; and what scale should be used? The characterizations suggested above require operationalization; that is, terms such as 'compatible,' 'high,' 'substantial,' and 'desirable' need to be given definition. As suggested by McGlade 'fuzzy logic' may be of assistance in this regard (McGlade 1995). But beyond the matter of assessing each of the four elements, the question remains as to how the data will be aggregated (Underdal 1980). Whatever device or procedure is used to organize and evaluate data, there can be no escape from a significant element of subjectivity. Moreover, values and preferences aside, the fact is that decisions will be made under conditions of imperfect knowledge and uncertainty.

Step 7: Assess the Interactions Between Governance Mechanisms and Resource Use

Traditionally, governance arrangements have developed along sectoral lines on an *ad hoc*, piecemeal basis. As noted in the classic 1969 report of the Stratton Commission, in governmental contexts, a problem is brought to light one way or another and some department or agency is given the responsibility to address that particular problem (Stratton Commission 1969). Over time responsibilities are spread among levels of government and among departments over a host of activities and areas. Eventually, interactional problems become evident since decisions are being taken without due regard to externalities: the lack of coordination leads to mutual interference, inefficiencies, and uncoordinated management.

While substantial attention has been given to mapping ecosystems, the mapping of governance systems, too, deserves attention. The 'mapping of LME governance' can be facilitated by filling in the cells of a matrix such as the Gulf of Maine example in Figure 3-3. There is no question but that the governance system affects the pattern of use of coastal/ ocean areas. It is important to know who is responsible for what and how the elements of governance, like those of ecosystems, interrelate and interact. They, too, are part of the 'working environment' and must be taken into account as efforts are made to provide for effective use and protection of ecosystems.

As noted above, the concept of governance involves more than government and its dimensions include (a) levels of governance (*e.g.*, international, national, regional, local), (b) sectoral areas *(e.g.* fisheries, offshore mining, waste disposal, recreation), and (c) stakeholders (*e.g.*, fishermen, corporations, real estate interests, port authorities). As is the case with particular LMEs,

---

politically expedient for politicians and policy makers to agree with the minority -- even when the decision is detrimental to society or economically inefficient. Often these conflicts take the form of a tug-of-war between stake-holders -- each of which claim to represent what is best for society -- and between stake-holders and the actual interests of society.

Governance Units:	General Purpose Regional Intergovernmental		Single Purpose Regional Intergovernmental		Central Government		State/Provincial Governments		Local Authorities		NGOs
	International	National	International	National	US	Canada	US	Canada	US	Canada	
	GOM Council	Canada Atlantic Action Plan	NAFO				Maine, Mass., N.H.	Nova Scotia, New Brunswick		Halifax Port Authority	Save Casco Bay
Use Sectors											
Shipping/ports											
Fishing											
Aquaculture											
Industrial siting											
Military uses											
Recreation											
Waste disposal											
Offshore oil/gas											
Mining											
Housing											
Agriculture											
Forestry											

Figure 3-3. Governance-use matrix: Gulf of Maine example.

governance arrangements have site specific characteristics that need to be recognized and understood.

The level at which a problem should be addressed needs consideration. The principle of subsidiarity suggests that authority belongs at the lowest level capable of effective action (von Moltke 1997). In fact, the European Union in its Integrated Coastal Zone Management Programme has adopted this principle and calls for problems to be addressed in the order of local, regional, national, and EU levels (EU 1998). And in its consideration of a needed framework for managing activities in the ocean and coastal areas of the United States, a recent report of the National Research Council entitled, Striking a Balance: Improving Stewardship of Marine Areas, emphasizes the need for a federalist approach in which power is placed at the level appropriate to achieving desired objectives (National Research Council 1997). In this context, different levels of governance share responsibility and coordination is provided at higher levels. The subsidiarity principle is suggestive, then, of another matrix, one which relates level of governance to issues and ponders what is the appropriate level of governance to treat identified problems.

Governance Interactions

The manner of organization of governance arrangements can certainly affect resource use and ecosystem health (Costanza 1992). As long noted by political scientists and office holders, bureaucratic arrangements can be instruments of delay and introduce the element of 'turf' into all decisions (Downs 1967). But the interplay of different elements of government and governance can also play a positive role by widening perspectives and forcing consideration of externalities.

In looking to the future and considering how ecosystem-based management efforts may be improved, it is necessary to take the current governance system as a given and the point of departure. Changes will be needed in terms of institutions, mores, and values if there is to be a shift away from sectoral approaches to management of natural systems and their resources. Identification of incremental modifications would be desirable since such changes are easier to adopt and implement than more radical changes and, cumulatively, may still have substantial effects.

Government Programs and Use Interactions

As the problems associated with sectoral approaches become increasingly manifest, efforts are made to overcome them. One approach is through the adoption of legislation and the development of governmental programs that reach across sectoral divides, with crosscutting effects, and force consideration of externalities. The National Environmental Policy Act (NEPA 1969) provides one such legislative example (Juda 1993). The requirement for the use of an environmental impact statement, mandates attention to the subject of externalities.

In the United States major federal, state, and local programs have the potential to impact LMEs. Such programs now encompass all of the coastal watershed leading to areas of fisheries and marine habitat. Watershed management emerged through the passage of section 6217 of the Coastal Zone Act Reauthorization Amendments of 1990, which mandates efforts to control non-point source pollution in coastal waters. Coastal States are required to use a watershed planning and control approach to deal with sources of pollution from agriculture, forestry, urban development, marinas, recreational boating, and hydromodifications. Plans must address the preservation and restoration of wetlands and riparian areas. States are to develop enforceable management measures to treat these sources of pollution (Imperial and Hennessey 1993, 1996).

The National Estuary Program, established in 1987, complements the above efforts by providing funding to States to develop a comprehensive planning process to improve water quality and enhance living resources. There are currently 30 estuary programs in the United States, including four in the Gulf of Maine watershed (Imperial and Hennessey 1996).

Coastal habitat issues have recently come to the fore and have been addressed through the Sustainable Fisheries Act of 1996, that re-authorized and modified the Magnuson Fishery Conservation and Management Act (now the Magnuson-Stevens Fishery and Conservation and Management Act) and required the National Marine Fisheries Service to specify 'essential fish habitat' for all managed species and fisheries. Each fishery management council must amend its fishery management plans to: identify and describe the essential fish habitat for each managed species, identify the fishing and non-fishing related threats to the habitat, and develop management and conservation alternatives for that habitat.

Exploring the legislative or programmatic mandates relevant to LME management is worthwhile for several reasons. First, it is important to understand the program interactions with different LME uses and interaction with existing governance structures. Second, management decision-makers need to understand how they may alter traditional agency activities and how they may serve to contribute to more holistic management approaches. Figure 4 indicates selected programmatic initiatives in the United States that merit specific attention in this context. Many other programmatic examples are available that also merit evaluation.

		Program		
		CZMA Section 6217	National Estuary Program	FCMA Habitat Provisions
Use Sector	Fisheries			
	Aquaculture			
	Shipping			
	Military Uses			
	Recreation			
	Industrial Siting			
	Housing			
	Agriculture			
	Forestry			
Governance	GeneralPurpose Regional			
	Single Purpose Regional			
	Central Government			
	United States			
	Canada			
	State/Provincial Government			
	Local Authorities			
	NGOs			

Figure 3-4. Program-use and governance matrix

Step 8: Assess the Socioeconomic Importance of LME-related Activities and Economic and Sociocultural Value of Key Uses and LME Resources

The coastal and marine natural resources of an LME are capital assets—in effect representing wealth embodied in its marine natural resources. Capital assets, natural or otherwise, can provide valuable services ('interest') over time if maintained, much like savings in a bank provide a flow of interest income.

Underlying much of environmental economics is the notion of resource valuation (*i.e.*, valuing nature's services). Resource valuation involves the use of concepts and methods to estimate the economic value the public holds for natural resource services.[16] These services may be direct or indirect; and they may or may not be bought and sold in the marketplace.

Direct services include on site use of marine parks, beaches, commercial fishing, exploitation of marine minerals, or harvesting of fish, shellfish or wood from mangroves. Indirect services occur off site, for example, when fish 'produced' by a mangrove are harvested many miles away. Some natural resources services are exchanged in organized markets, such as commercial fisheries, oil and other minerals, some coastal property, or tourism. However, a

---

[16] A succinct explanation of these methods is provided by the National Research Council (1997: 21-24).

central feature of many, if not most, marine resource issues is that the services provided are not traded on markets. The services provided, as for example, by mangroves, corals, and sea grasses, water quality, recreation, scenic amenities and biodiversity are not bought and sold on markets and, as a result, often are given inadequate attention in public policy.

Four types of value are associated with resource services. First, *use value* is the benefit received from on site or physical use, such as harvesting of fish, exploitation of oil or beach use. Second, *passive use value* is the enjoyment one gets from a resource above and beyond any direct use.[17] Passive use losses may arise if individuals feel worse off when they learn of the loss of an endangered species, closure of beaches or other adverse impact on other natural resources, even if they do not use these resources themselves. People might be willing to pay to prevent such losses, much as they might pay to preserve, say, an historically or culturally significant building or site, even if they never actually visit it. Third, *total value* is the sum of use and passive use value. Fourth, individuals also may think of a resource as having an *option value* when either supply (*e.g.*, threat of extinction; the outcome of a policy) or demand is uncertain. Option value may be thought of as what would be paid to keep the opportunity open to later use a site or resource.

Resource valuation usually is not an end in itself, except in the case of commodities such as oil or other minerals, or of fish, where the government might lease public resources to private businesses.[18] Instead, estimates of the value of particular resource services normally are more useful as contributions to policy for improving resource management. Most policy decisions involve specific proposals affecting resources and their services 'at the margin'; hence, resource valuation most often will involve assessments of the marginal value of resource services rather than the aggregate value.

Social and cultural factors correspond to and reinforce the need for economic valuation, but their focus and the use of sociocultural analysis is also quite different. Indicators such as income, employment, and economic sector performance are elements of both types of investigation. However, sociocultural analysis takes a step away from strict enumeration of these elements and focuses on people's knowledge and views (norms and values) about their work, and how this affects their perceptions and actions towards LME resources (Brainerd *et al.* 1996). Although this is not easily measured on a monetary scale, these factors are considered significant by those involved in resource use. Sociocultural analysis has the capacity to contribute to management by considering the values of cultural and social elements of the community, and the potential costs associated with social and economic disruption and dislocation.

Social and cultural factors are closely linked to governance, users and uses of LME resources. One way to account for these linkages is to view human action within the context of Natural

---

[17] Many improvements in methods have been made, but reliable quantification and acceptability of passive use value as a measure of damages is still a subject of lively debate (*e.g.* Portney, 1994; Diamond and Hausmann, 1994; Hanemann, 1994).

[18] Common examples of resource valuation involve determining the value of oil and gas leases, tradable fishing licenses, and government concessions. These cases are relatively straightforward in that they involve the use of market information on anticipated revenues and costs over time. An important issue, discussed further below, concerns estimating the change in the value of a marine asset due to changes in pollution or fishery regulations, for example.

Resource Communities (NRCs) (Dyer *et al.* 1992). The interface between a regional system of extractive NRCs, their service flows and the associated LME is here defined as a Natural Resource Region (NRR).[19]   Dyer *et al.* (1992) define NRCs as populations whose sustainability depends upon the utilization of renewable natural resources. By broadening the definition to include those dependent on non-renewable aspects of the marine environment as well, they and their aggregations as NRRs represent the LME-dependent communities within a coastal region.

The Natural Resource Region (NRR) includes social, cultural, human, economic and biophysical capital and their interactions within networks of LME-dependent communities (Dyer and Poggie 1998). These forms of capital are defined as follows:

- *Social capital* refers to the interactive networks of humans that occurs within and between natural resource communities. Social capital is key to the flow of other forms of capital, as well as central to the dynamics of governance and resource utilization.

- *Cultural capital* refers to the behaviors, values, knowledge and culturally transmitted behavior and ideas of a population, applied to the transformation and utilization of natural resources.

- *Human capital* refers to the human population and the knowledge and skills they acquire from formal and informal education associated with the occupational roles of natural resource extraction.

- *Manufactured capital* refers to long-lasting manufactured goods (*e.g.*, buildings, machines, tools, fishing vessels and gear) that enhance the ability to produce other goods and services.

- *Biophysical capital*, as explained above, is used to denote those natural resources of an LME that directly or indirectly generate flows of goods and services used by humans. The value of these natural resources is derived from the dynamic between human action and the natural environment. These include potential resources, identified but not actively utilized in extractive processes, or those having primary value in passive recreational activities (*e.g.* the whale as resource to the whale watching industry).

Fishing is a good example of the interactions of some of these forms of capital. A fishing boat out at sea is a production-extraction unit of the NRR, relying directly on the productivity of the fish resources of the LME (the NRR biophysical asset). The fishing boat is thus an extension of the NRC from which it came, carrying with it social, cultural, human, and manufactured capital in its hunt for fish resources.

The conceptualization of capital interactions within an NRR network lends understanding to the occupational valuation placed on 'way of life.' For example, Doeringer, Moss and Terkla (1986) show how kinship support systems—a form of social capital in our formulation—allow fishermen to maintain labor linkages to the fishing industry in defiance of seemingly

---

[19] An NRR is thus formally defined as a network of Natural Resource Communities whose existence is defined by the interactions among the social, cultural, human, economic and biophysical capital that are part of or closely linked to the resources of a Large Marine Ecosystem. (Dyer and Poggie 1998)

debilitating economic conditions, usually associated with declines in volume and value of fish catch, as well as severe management restrictions on fishing.

In the interface with LMEs, primary units of human-environment interaction—individuals, families, households or communities—are to be viewed as interconnected within regional networks held together by shared values and forms of capital. The NRC is a nodal form of human organizational structures, regional and capital interactions, and provides for points of spatial reference by which to study the LME-NRR dynamic.

Networks of Natural Resource Communities within NRRs act as conduits through which total capital is exchanged, shared, and transformed by human action. For example, we can consider the NRC[20] as a regional contributor to whatever commerce is stimulated by LME-related activities, and as a means of providing sustainable support to LME-related households and families as they contribute products and services to the region and nation in which they are embedded. While only a subset of the NRC interact directly with the marine environment and its resources (*e.g.* fishermen, shipping vessel operators), these individuals are nevertheless connected to more differentiated communities and towns, contributing to the economic and food security of those communities and towns and buffering coastal development in a way that contributes to social and economic diversity.

Social impact assessment variables point to measurable change in the human population, communities, and social relationships resulting from policy change (ICGP 1993). The Interorganizational Committee on Guidelines and Principles (1993) identified a list of social variables under the general headings of (1) population characteristics, (2) community and institutional structures, (3) political and social resources, (4) individual, household and family changes, and (5) community resources. Definitions of each heading considered by the Committee are given below.

- *Population characteristics* mean present population and expected change; ethnic and racial diversity, influx and outflows of temporary residents as well as the arrival of seasonal or leisure residents.

- *Community and institutional structures* mean the size, structure and level of organization of local government to include linkages to the larger political systems. The historical and present patterns of employment and industrial diversification. The size and level of activity of voluntary organizations and interest groups and finally, how these institutions relate to each other.

- *Political and social resources* refer to the distribution of power and authority, the identification of interested and affected parties as well as the leadership capability and capacity within the community or region.

- *Individual, household and family changes* refer to factors which influence the daily life of the individuals, households and families, including attitudes, perceptions, family and household characteristics and social networks. These changes range from attitudes toward the policy to an alteration in family and household relations and social networks to perceptions of risk, health and safety.

---

[20] The NRC may encompass more than one port.

- *Community resources* include patterns of natural resource and land use; the availability of housing and community services to include health, police and fire protection and sanitation facilities. Key to the continuity and survival of human communities are their historical and archaeological cultural resources. Under this collection of variables we also consider possible changes for indigenous, ethnic and religious sub-cultures.

Sociocultural elements may also be assessed by performance indicators related to equity issues such as the distribution of benefits among stakeholders, the nature of access to LME resources, and the reliance of communities on LME resources (Clay, per. com., 1998). The distribution of income is a measure of equity within natural resource communities and between communities and wider society. Benefits distribution can take other forms such as the pattern of fish consumption and distribution, and allocation of and/or access to resources. The nature of access to LME resources considers property rights as well as the local involvement in resource management. Community reliance on LME resources may take several forms including employment and other economic factors, food security and cultural factors. The relative importance of different social variables will vary depending on the specific community and its relationship to the resource in question.

Dyer and Griffith (1996) isolated five variables that help identify fishing community dependence on an LME. It will become obvious that the five variables overlap somewhat; thus, they must be considered together. These are:

- *Relative isolation or integration of LME resource users into alternative economic sectors.* To what extent have users (*e.g.*, fishermen, processors) segmented themselves from other parts of the local political economy or other fisheries?

- *User types and strategies of users within a port of access to LME resources.* What impact does the mix of types (*e.g.*, fixed fishing gear—weirs, fish corrals—versus mobile fishing gear) across ports and States have on the long-term sustainability of LME resource stocks?

- *Degree of regional specialization.* To what extent have users from related areas and use-sectors moved into the region? Clearly, those users who would have difficulty moving into alternative use-sectors are more dependent on LME resources than those who have histories of moving among several sectors in an opportunistic fashion.

- *Percentage of population involved in LME resource-related industries.* Those communities where between five and ten percent of the population are directly employed in LME resource-related industries are more dependent on the LME than those where fewer than five percent are so employed.
- *Competition and conflict within the port, between different components of use sectors.* Competition between smaller scale and industrial scale users can create conflict between users within the same port—as well as between different actors in a use-sector (such as boat owners, captains, and processors). Dependence may have a strong perceptual dimension, with users perceiving the resources they are extracting to be scarce and that one user group's gain (*e.g.* industrial trawling, purse seining) is another user group's loss (*e.g.* gill netting).

These five variables can be adapted and broadened to cover the full range of LME-related activities. A fundamental assumption of the NRR model is that there is some degree of reliance on the natural resources (*i.e.*, biophysical capital) of an LME. In an LME-linked NRC, biophysical capital reliance manifests itself as learned social behaviors of LME-related activities. The combined social, cultural and economic interactions arise from the conditions that increase or decrease access to the LME and its biophysical capital. Furthermore, dependence on natural resources limits the occupational roles of community members, and can intensify cultural assimilation for those immigrating into an NRC.

Disruption of LMEs is occurring more frequently as NRRs are stressed by human factors that push resources beyond their ability to renew themselves and permanently degrade physical structures such as bottom topography. Such resource degradation patterns in an NRR can be found in conditions of severe poverty, overpopulation, the practice of destructive extraction techniques (*e.g.* blast fishing in Philippine reef systems), or the development of overcapacity in a fishery (*e.g.* the groundfish fisheries of New England, Dyer and Griffith 1996). In an idealized condition, an effective state of environmental awareness is generated among NRC residents and NRR networks that allows for sustainable utilization of biophysical capital in an LME. Less idealized conditions—most real world ecosystems and their human actors—require some form of management appropriate to the political ecology and cultural and environmental history of the region in question. Thus, although a generic LME/NRR management framework for the Bay of Bengal and the Gulf of Maine may be conceptually similar, operationalizing the model cannot proceed without considering site-specific human-environmental dynamics.

The interdependence of economic, social, cultural and governance elements is readily apparent. They overlap, complement, and conflict with one another in different situations. Their relative importance and tradeoffs between different sociocultural and economic values will depend on the interplay of the community, LME resources, and larger society.

Step 9: Identify the Public's Priorities and Willingness to Make Tradeoffs to Protect and Restore Key Natural Resources

An implicit assumption underlying social science research in this document is that what people want (their preferences) matters in public policy decisions concerning LMEs. In economies where markets work reasonably well, market prices are a good indicator of the marginal value individuals attach to incremental units of a good or service. However, widespread market failure in LMEs makes the connection between market prices and preferences tenuous or nonexistent for many major problems. An important issue in the absence of reliable market data, then, is how to obtain useful information on public priorities and preferences that can be used in decision making in LMEs.

One possibility is greater use of opinion polls and general attitude surveys on LME resource issues. However, most members of the public, when asked, will identify 'the environment' as an important concern and will indicate that, at an abstract level, we should 'do more' for the environment. Such general attitudes, however, are not very informative of actual values people hold for resources and their services. This is because value is indicated by what one is willing to give up to keep or get more of something and general opinion polls do not confront respondents with the costs of their decisions. It is not surprising, for example, that when asked to assign priorities to management actions to improve coastal environments, survey respondents will recommend actions that impose little or no direct costs on themselves, but are less favorably disposed to measures that would require them to bear costs (Opaluch *et al.*, 1998). Choices by definition imply tradeoffs and values. Real policy actions are not free; opinion polls and general surveys that do not require respondents to recognize costs of actions are unlikely to provide useful information to LME decision makers about public preferences. For this purpose, more structured surveys are needed that specifically ask respondents to make trade-offs.

Stated preference methods, such as contingent choice and contingent valuation, are potentially valuable frameworks for assessing public priorities, their willingness to make tradeoffs, and the public's economic values. These methods involve the use of carefully developed surveys that are then administered to a random sample of the population of interest. Stated preference methods are one way to assess resource priorities for public goods and to potentially estimate passive use values for LME resources.

Ethnographic fieldwork can provide in-depth assessment of values and the degree to which they are strongly or weakly held. This type of research is more labor intensive, but can be especially important when dealing with site-specific decisions or where a decision must be made that may go against particular local values and thus require public education or remediation.

Hence, the development of socioeconomic and governance elements for LMEs may well draw heavily upon advances in the use of survey and other methods (*e.g.*, ethnographic interviews, focus groups, panels) for obtaining information on public preferences for resource management decisions, information that otherwise may be unavailable for decision makers to consider.

Step 10: Assess the Cost of Options to Protect or Restore Key Resources

Typically many alternatives will be available for addressing any problem within an LME, and each can be accomplished at different scales. Consider nutrients, for example, a serious coastal water quality issue in many areas. This issue can be addressed in many ways, including: expanding or upgrading public wastewater treatment facilities, encouraging measures to reduce application of fertilizers in agriculture, using buffers for agricultural lands along water bodies to reduce runoff, introducing measures to control runoff of animal wastes from farms and roads into coastal waters, and investing in sewage lines to avoid use of septic systems for household residences. Pollution trading between sources (*e.g.*, wastewater facilities and farmers) also is possible.

Each of the above alternatives is technically feasible and will be effective in varying degrees. However, the investment and recurring costs of the alternatives will vary substantially. Selecting among them is not straightforward and requires information not only on costs over time but also on their relative efficiency in reducing nutrient discharges, *i.e.* cost effectiveness. Cost effectiveness involves selecting the alternative(s) with lowest cost per unit treated. At one level this can be viewed as a technical, engineering-economic problem.

However, effective policy requires implementation, and thus it is critical that management mechanisms and institutional structures be in place that will allow alternatives to be considered with their cost-effectiveness used as an important criterion.

Step 11: Compare the Benefits with the Costs of Protection and Restoration Options

As noted often, many technically feasible alternatives are available to address resource management problems. Cost effectiveness, outlined above, ensures activities will be done at least cost. However, cost effectiveness presumes that an activity is a worthwhile investment of society's scarce resources. In fact, there are many good potential societal investments that compete implicitly or explicitly for limited public resources, and an important issue concerns whether a particular proposal is a good investment in the sense that the resulting benefits justify the costs (*i.e.*, what society must give up in other goods and services to realize the benefits).

Increasingly, international agencies and others require benefit-cost analysis be conducted to help decide whether, and to what extent, to undertake projects. In carrying out such analyses, agencies are concerned about considering not only narrow, commercial transactions, but environmental benefits and costs as well.

Benefit-cost analysis can be a valuable decision tool, for several reasons. First, it puts public investments on the same footing as private investments in that they must meet the same standard: the costs of a policy, program, or activity should be justified by the resulting benefits. Further, a well done benefit-cost analysis makes all calculations and assumptions explicit and by that, transparent for all stakeholders. This may help add legitimacy to a process, an important consideration in many situations.

Of course, benefit-cost analysis raises several issues, as well. One is whether all important benefits and costs can be quantified. Many advances have been made in natural resource valuation, and the opportunities and limitations of resource valuation are becoming increasingly well understood. But it is also true that many difficulties remain, and data problems are always an issue, especially in developing countries. Furthermore, equity—the distribution of the benefits and costs of a proposed policy action—is an important issue influencing whether actions will be taken and what form they will take (*e.g.*, Zeckhauser 1985). However, distributional effects can always be included in an analysis. For example, different groups and/or regions can be assigned different weights, provided one knows the relative importance (weight) assigned to them; indeed evaluation of such issues is a strength of analytical methods commonly used in economics. Beyond this, even the best benefit-cost analysis is not a substitute for good policy making; decision-makers as a matter of course take into consideration the distribution of benefits and costs when deciding whether and how to implement a program or action. Looked at this way, to the extent good social science data are available on distributions and types of impacts (*e.g.* through social and economic impact analyses based on the framework established here), the equity problem in practice is not as serious an issue for benefit-cost analysis as some may believe.

Step 12: Identify Financing Alternatives for the Preferred Options for Protecting and Restoring Key LME Resources

The results of cost-effectiveness analysis, benefit-cost analysis and social impact analysis can help select the preferred option from among several technically feasible alternatives.   To implement the preferred option, however, sustainable financing must be available. Financing is often viewed as simply a distributional issue but, in fact, sustainable financing has become an increasingly important issue to ensure that revenues cover costs and to affect incentives that encourage favorable behavior and discourage unfavorable actions.

Many alternative financing approaches are available, depending upon the issue and area. Broad principles to be employed may include the user or beneficiary pays and polluter pays principles. The user-pays or beneficiary-pays principle has strong appeal on fairness grounds in many if not most cases, but is less useful and may need modification for cases where a program is provided specifically to achieve an equity objective. The polluter-pays principle also has a strong basis in fairness, but additionally, when effective, provides incentives for operators to avoid pollution by internalizing costs. The polluter-pays principle also works to place at least some of the costs of such actions on the consumer of the polluting good. In sum, the polluter-pays principle ensures that operators and consumers face the full social costs of producing and using the good involved.

The user- or beneficiary-pays principle is especially challenging to invoke in practice for resources that have widely dispersed and significant non-use benefits.   For example, preservation of unique marine parks (*e.g.*, the Great Barrier Reef) or marine mammals (*e.g.*, sea manatees or whales) will likely provide major benefits far beyond those to people who actually use these resources and may extend to the public nationally and internationally. A user-pays or beneficiary-pays principle obviously is difficult to invoke on non-users in such cases. This suggests that for such unique, widely appreciated resources with strong non-use value, international donations must play a critical role rather than reliance on access fees.[21]

Criteria for selecting the type of financing might include adequacy of revenues, transactions costs, distributional effects, political feasibility, effects on behavior, and conflicts with other objectives.   Examples of the last criterion include actions by some countries (1) to provide subsidies to fisheries while at the same time trying to limit catch, (2) to impose taxes on imports of construction materials, while trying to protect corals (which are mined in some countries as a source of construction materials), or (3) to encourage agriculture while at the same time attempting to protect or restore water quality.

APPLICATIONS OF THE MONITORING AND ASSESSMENT FRAMEWORK[22]

Example 1: Assessing Monetary Damages from Oil Spills and Other Transboundary Marine Pollution[23]

---

[21] Of course this argument, in part, provides the justification for international donors and programs including the Global Environment Facility.

[22] The following examples are excerpted from Grigalunas (1998).

Oil spills and other transboundary pollution in an LME are important concerns due both to (1) the risk of accidents and (2) the many important resources, activities, and ecosystems that are vulnerable to injury from pollution. Managing the risk of spills raises two inter-related issues. One is the appropriate scale of measures to prevent and control spills. A second issue—the focus of this section—has to do with the institutional framework, methods, and standards that might be used to assess the monetary value of natural resource-related damages when spills occur.

When oil spills or other pollution incidents occur, it is necessary to decide whether to assess damages, which losses can be compensated for, the best method(s) to be used to assess damages, and the institutional framework within which such assessments take place. This is where natural resource damage assessment becomes important.

Natural resource damage assessment (NRDA) is a methodology that applies legal, scientific, and economic principles to assess monetary damages caused to natural resources by pollution and other human actions. NRDA provides measures for sustainable financing in the form of compensating for injuries and lost natural resource services due, for example, to transboundary pollution. NRDA, as applied in the U.S., consists of a formalized process and an institutional regime within which allowable losses from covered incidents can be quantified and collection of claims can be undertaken and enforced. NRDA is a relatively new area of research, and the concepts and approaches being used have been evolving relatively quickly.

The intended outcome of an NRDA (Natural Resource Damage Assessment) is a claim against a responsible party, and the scope of items included by governments as damages has grown, as has the size of settlements. As a result, NRDAs necessarily involve tensions and adversarial debate between government, which is responsible for implementing and enforcing NRDAs, and industry, which must respond to and pay legitimate claims. Critics of NRDA question the reliability and, in some cases, the appropriateness of NRDA assessments. Supporters of NRDA make comparisons with the many empirical challenges and imprecisions addressed as a matter of course when assessing damages such as the value of intellectual property rights, anti-trust issues, and losses from personal injury in work-related accidents.

In spite of controversies surrounding NRDA throughout its evolution, establishing liability for damages due to oil and hazardous substance marine pollution is of increasing interest as a practical method based on economic incentives (*i.e.* the polluter- pays principle) in environmental policy. An improved understanding of the scientific, economic and legal concepts used in NRDA. NRDA is of interest to many parties, because:

- Littoral States must decide the adequacy of NRDA measures for compensation for losses due to spills. Particularly important are losses to publicly controlled or managed resources, such as open sea fisheries, wildlife, and ecosystems.

- Owners and operators of mariculture, fishing, tourism and other coastal businesses at risk from spills are concerned about recovering lost earnings.

---

[23] This section draws heavily from a recent report prepared for the Regional Programme for Preventing Pollution in East Asia Seas (Grigalunas and Opalluch 1998)

- Industry is concerned about the legitimacy of claims against them for losses, about transactions costs for legal and expert reports and proceedings, and about avoiding double counting of losses (paying twice or more, for the same loss). They are especially troubled about the potential for damage claims based on speculative losses or losses based on unreliable or 'theoretical' methods. Of particular worry is the potential for major claims, if damages are expanded to include non-market and other, hard-to-quantify losses, especially passive use value,[24] as they have in the United States, (*e.g.*, NOAA 1992; Hanemann 1994).

- Insurance companies are concerned about the nature and size of claims they will face for response, cleanup, assessment, and damages. In many respects, their concerns are similar to those of industry.

Interest in NRDA by public bodies stems from its promise in helping to achieve two important environment policy goals. First, it provides an organized framework for pursuing compensation for the many costs that can result when natural resources, coastal activities, and property are adversely affected by oil and other marine pollution. Many types of pollution damages currently are not compensated for and, as a result, these costs are borne by coastal States. Second, polluter liability under NRDA requires the responsible party to bear the costs of marine pollution (*i.e.* polluter pays principle). Liability provides built-in incentives for polluters to avoid incidents and by that, plays to their self interest as a matter of course (*e.g.*, Opaluch and Grigalunas 1984; Grigalunas and Opaluch 1988).   This is consistent with worldwide trends toward the use of market mechanisms to address environmental issues as recommended, for example, in Agenda 21 of the 1992 Rio Convention. At the same time damage assessment raises several issues, including:

- the nature of liability
- the scope of incidents covered
- the scope of impacts ('injuries') for which damages can be assessed
- allowable damages
- allowable methods for estimating damages
- standards to apply in weighing the results of such methods
- means for limiting transactions costs.

A recent survey paper by Grigalunas *et al.* (1998) presents concepts and issues in NRDAs and summarizes several case studies to illustrate different types of losses and efforts to estimate these losses. Any attempt to develop an LME-wide approach for NRDA in an LME would need to address these (and other) issues in great detail.

---

[24] As note earlier, Passive Use losses may arise if individuals feel worse off when they learn of the adverse effects of a spill on wildlife, beaches and other natural resources--even if they do not use these resources themselves. People might be willing to pay to prevent such losses, much as they might pay to preserve, say, an historic or culturally significant building or site, even if they never actually visit it. Many improvements in methods have been made, but reliable quantification and acceptability of passive use value as a measure of damages is still a subject of lively debate (e.g. Portney 1994; Diamond and Hausmann 1994; Hanemann 1994).

This example is based on a series of economic studies for the Peconic (New York) Estuary System as part of the National Estuary Program in the United States. The estuary and surrounding watershed are very attractive and are used intensively, particularly during the peak summer season. The estuary itself has generally good water quality. However, pollution exists and threatens some uses; for example, extensive shell fishing grounds have been lost due to pollution. Also, development has caused the loss of important habitats/ecosystems and threatens the scenic amenities of the area. Thus, many market and non-market valued resource services are at issue in this case— as is true in most other coastal and marine cases.

Working closely with program mangers, scientists, and citizen advisory groups, we have carried out, or have ongoing, a series of studies (Grigalunas and Diamantides 1996; Opaluch *et al.* 1998) to:

- Estimate the economic importance of estuary-related activities
- Identify coastal users, their activities, and concerns, using a carefully prepared survey
- Identify the public's priorities and willingness to make tradeoffs to protect and restore key natural resources using a second carefully developed survey
- Estimate the economic value (benefits) of key recreational uses and coastal amenities
- Assess wetland productivity and habitat services
- Assess the cost of options to preserve or restore key resources
- Compare the benefits with the costs of preservation and restoration options
- Help select financing alternatives for the preferred options for preserving/restoring key natural resources

Preliminary results indicate that estuarine-related activities play a major role in the livelihood of several thousand residents who own, operate or are employed by over 1,000 business in some 24 identified sectors. These sectors engage in or support such activities as fisheries, marine transportation, and particularly tourism and recreation.

It was also found that over a hundred thousand people annually engage in millions of days of marine recreational activities, and preliminary estimates of the value of key recreational activities range from $8.59 per trip for beach use to $38 for a recreational fishing trip. The total annual value across the three recreation activities studied to date is over $50 million per year, again based on preliminary results. These are economic benefits to users above the costs they incur (*i.e.* unpaid-for benefits).

Users of coastal areas are clearly affected by water quality. A link between objective water quality measures and subjective measures of quality as perceived by recreationists, has been estimated. This allows joint work with scientists who estimate the changes in various measures of water quality due to policies being considered to control pollution. Given the cost of such control measures and of preservation and protection measures, we will be able to compare the benefits with the costs of these policies.

Preliminary survey results also suggest that the public holds strong values for preserving key area natural resources. These results are supported by preliminary results from a separate, housing value study. This study suggests that residents are willing to pay more for property located near coastal waters, parks, and open space.

A wetland productivity study of the value of eelgrass, inter-tidal salt marsh and mud flats yielded preliminary results for the marginal value (asset value) ranging from $12.7 thousand per acre for eel grass to $4.4 thousand per acre for mud flats. These estimates include the estimated market value of fish and shellfish produced in these ecosystems and harvested; the value of waterfowl hunted and birds viewed. The value estimates include only food web effects and habitat values, and hence are conservative in that such services as shoreline erosion protection and storm protection services provided by salt marsh, for example, were not considered. Our estimates of economic value (benefits) of these three types of wetlands will help us do benefit-cost studies of management proposals for restoration of habitats.

As noted, ongoing work will examine the cost of options for preserving and restoring resources, compare the benefits with the costs for different options and help select financing alternatives for the preferred options. Again, an important aspect of this work is the willingness and commitment among the program managers and participants to work together to link socioeconomics, natural resource science and policy.

Environmental programs to prevent or control pollution may require major investments. Benefit-cost analyses of public projects often do not consider how projects will be financed, nor do they usually present the implications of different financing and institutional alternatives for implementation.[25] Yet, to be successfully implemented and maintained, attention must be given to financing, to important institutional measures, and to the distribution of benefits and costs in general. Financing in particular is important for several reasons:

- Inadequate funding will limit the implementation of effective pollution prevention measures.

- Mechanisms used to finance projects, e.g., user fees versus general revenues, or different formulae for cost-sharing, have important distributional effects, which often are a major factor influencing how, even whether, a policy is adopted (Zeckhauser 1985).

- Financing options can affect users' incentives, thus influencing behavior and the resulting size of benefits.

- Financing options may differ with respect to: Ease of administration (transactions costs), political feasibility, stability of revenues, or in other important respects, all of which influence whether and how measures are adopted, as well as their effectiveness.

For all of these reasons, sustainable financing of pollution management actions is a significant issue for LMEs.

Sustainable financing mechanisms include: (1) user fees and related, cooperative mechanisms, when available and appropriate (and allowed under UNCLOS), (2) natural resource damage assessment, (3) potentially attractive investments in Private Public Partnerships, including

---

[25] See Musgrave (1969) for further discussion of benefit-cost analysis and financing when capital market are not perfect, when social and private discount rates differ, and the distribution of benefits and costs are important.

potential investments under the Buy-Transfer-Operate and related public-private programs, (4) international donors and (5) international trusts.

User fees and, more generally, mechanisms employing incentive-based approaches, have considerable appeal. They are based on the user-pays and the polluter-pays principles and reflect commonly shared notions of fairness. They also can work to harness the power of the market to sustain pollution prevention and control measures, in an efficient manner, in effect using private interest to serve the broader public interest (Schultz 1975; Grigalunas and Opaluch 1988).

To be effective, however, markets must work, or appropriate institutional arrangements must exist to allow markets to function. Major problems may arise in devising mechanisms to prevent and control pollution, though, because of market failure and institutional failure. For example, many navigational aids and safety measures are public goods; other safety measures (*e.g.*, use of pilots and vessel transit systems) may create important external benefits not captured in the market; and in still other cases, institutional problems prevent effective reliance on user fees. As a result, developing methods to promote greater reliance on user fees for sustainable financing of anti-pollution measures often is not a straightforward exercise.

Many measures are available to prevent or control sea-based transboundary pollution:

- Best management practices to control agricultural wastes
- Sewerage treatment facilities in critical areas
- Compulsory pilotage
- Salvage operation
- Vessel traffic information service (VTIS)
- Navigational aids/services
- Electronic charts (marine electronic highway)
- Shore reception facilities
- Contingency planning and oil spill response

These measures are, or can be, taken by private parties (*e.g.*, vessel and cargo salvage, shore reception facilities, sewerage facilities), governments (*e.g.*, navigational aids), or a combination of the two (*e.g.*, VTIS), to prevent or control spills or promote port efficiency. The above list is not exhaustive and omits, *e.g.*, efforts for further cooperation and training among the coast guards of littoral States.

Mechanisms currently used to finance programs in most areas rely primarily on national sources, but also include user fees, international donations, and other support through international organizations, notably the International Maritime Organization, in the case of pollution from shipping. Liability used to compensate for response, control, and cleanup of spills, as well as for payment for certain economic losses and for restoration actions, is another funding source for managing pollution by restoring the environment. Individual companies also spend considerable (but unknown) amounts on pollution prevention and response training, as well as on purchase of equipment to prevent and control spills and avoid other sources of marine pollution.

Financing mechanisms to prevent transboundary pollution include:

- Penalties, fines, and taxes
- Subsidies
- User fees
- Port dues
- Revolving funds
- Public-private partnerships
- Privatization
- Natural resource damage assessment (NRDA)

Briefly, the Revolving Fund is a source of money that the littoral countries can draw upon (*i.e.*, borrow from) to finance response and cleanup activities in the event of a spill. Natural resource damage assessment ('NRDA') is a process to (1) identify categories of costs and losses due to oil spills for which compensation would be paid and (2) provide appropriate methods and standards to be used to quantify losses in monetary terms.[26] Port dues are self explanatory. Public-private partnerships involve various cooperative approaches the private and public sectors might take to jointly address pollution from shipping or other pollution.

Measures can be evaluated using several criteria or factors, such as administrative efficiency, effectiveness as a region-wide instrument, revenue generating potential, behavioral change potential, fairness and equity among users and beneficiaries, and political acceptability among the littoral States.

PROPERTY RIGHTS ENTITLEMENTS AND REGIMES FOR LME MANAGEMENT

Marine resource management is fragmented in many coastal States by policies that pay little attention to environmental, institutional, social and economic scale, or to interactions and tradeoffs. In fisheries in particular, the single-species (stock) approach to management does not adequately account for ecological interactions (Larkin 1996) or for what factors influence harvesting and investment decisions (Hanna 1998). Recent research on environmental management is attempting to integrate natural and human systems in order to sustain benefits that humans derive from fishery and other natural resources (Costanza *et al.* 1997; Larkin 1996; McGlade 1989; Sherman *et al.* 1996).

This section investigates the implications of ecology, technology, and what are known as transactions costs (see below) for the structure of property right entitlements in LMEs, and it comments on the characteristics of concordant property rights regimes that structure human behavior vis-à-vis an LME. This line of inquiry has received serious attention recently in the ecological economics literature (Costanza and Folke 1996; Hanna 1998; Hanna *et al.* 1996), but it was introduced mid-century by economist H. Scott Gordon who explained why the absence of property rights to fishing 'grounds' caused fishery resources to be overfished and

---

[26] Note, however, that NRDA can also be considered a pollution prevention measure, to the extent that it provides an incentive for vessel operators to exercise more care (Grigalunas and Opaluch, 1988; Grigalunas, 1997).

their value dissipated through investment in too much fishing capital. Although the subsequent literature developed around disaggregated fish stocks, by 'grounds' Gordon (1954: 129) actually referred to 'shallow continental shelves' where upwelling waters support 'marine-food chains' of resident demersal and migratory pelagic species. He emphasized that 'it is necessary to treat the [collective] resource of the entire geographic region as one.' In a later, also seminal work, Steven N. Cheung (1970: 50) asked: 'What resource in marine fisheries is non-exclusive [accessible with little or no effective restriction]—the ocean bed, the water, or the fish? The answer is that any productive resource is multi-dimensional, and the term 'fishing ground' is chosen to include all of them.'

Related anthropological literature on common property regimes (*e.g.* McCay and Acheson (eds.) 1987) and Territorial Use Rights Fisheries (TURFs) (*e.g.* Pollnac 1984)) describe the frequency of geographically-based folk management (*e.g.* McGoodwin 1990) and their applicability to modern management (*e.g.* Cordell 1984; Dyer and McGoodwin 1994), as well as discuss the implications and benefits of property held under group versus individual tenure (*e.g.* Hunt 1997).

Although shrouded by confusion, bias, and emotion, property rights and their institutional context are the foundation of economic activity and are therefore essential to sustainable management of the goods and services supplied by marine ecosystems. LME management will be improved by scientific research, but we risk repeating the mistakes of single-species fishery management in particular if humans continue to be regarded as exogenous agents of regulatory regimes. Ecosystem management also requires structures of property rights that reflect environmental and economic principles, and it requires governance institutions that reflect the goals and values of a society.

THE STRUCTURE OF PROPERTY RIGHTS ENTITLEMENTS IN AN LME

Property Rights

In his book on the evolution of property rights in natural resource sectors, Gary Libecap explained that '[b]y defining the parameters for the use of scarce resources and assigning the associated rewards and costs, the prevailing system of property rights establishes incentives and time horizons for investment, production and exchange' (Libecap 1989: 227). Different property rights structures lead to different rewards (or penalties) and thereby create incentives that influence how people use the natural environment. For clarity's sake, we adopt the definitions of property and property rights used by Bromley:

> *Property is a benefit (or income) stream, and a property right is a claim to the benefit stream that some higher body—usually a government—will agree to protect through the assignment of duty to others who may covet, or somehow interfere with, the benefit stream.* [Bromley 1992: 2]

It is useful to identify five dimensions of property rights that affect the size and duration of benefits owners can expect to receive from economic resources:

1. Entitlements

2. Divisibility
3. Exclusion
4. Right to transfer entitlements
5. Enforceability

Entitlements are the ways that owners—government, commons or private entities—are allowed to use and derive benefits from assets, including attributes of the environment. For example, the U.S. federal and state governments own marine resources on behalf of the public. In contrast, fishermen own vessels and fishing permits, energy companies own leases to Outer Continental Shelf Lands above pools of petroleum and natural gas, and shipping companies own access rights to shipping lanes, to name a few. Virtually all entitlements are attenuated, however. Thus, it is against the law for fishermen to use their vessels to smuggle contraband into the USA, and their fishing activities tend to be regulated by a host of gear restrictions and time and area closures.

Divisibility involves the richness of entitlements to complex resources with multiple attributes. The scope of this property right is suggested in a quote from Alchian (1977: 132-33) who wrote about partitioning land: '[A]t the same time several people may each possess some portion of the rights to use the land. A may possess the right to grow wheat on it. B may possess the right to walk across it. C may possess the right to dump ashes and smoke on it. D may possess the right to fly an airplane over it. E may have the right to subject it to vibrations consequent to the use of some neighboring equipment.' We can obviously substitute fishing ground or a marine environment such as Georges Bank or the Gulf of Maine in the Northeast Shelf Ecosystem for 'land' and illustrate with separate entitlements to harvest (or preserve) populations of Atlantic cod, Atlantic sea scallop and Atlantic lobster, to extract minerals such as sand, gravel and petroleum from the seabed, to sail a boat or to ride a jet ski, to transport cargo in shipping corridors, to patrol using military craft, to conduct scientific research on benthic communities and habitat requirements, to dump sewage, and so on. The ecology and economics of divisibility will be a major consideration in designing property rights structures for multi-attribute LMEs as discussed below.

The remaining three dimensions of property rights are mentioned here, but they are most relevant to the discussion of regimes in the next section. *Exclusion* concerns whether others are prevented from using or damaging your entitlements. In the papers by Gordon and Cheung that were quoted above, non-exclusiveness, or even extreme attenuation of this right, shortens time-horizons, giving rise to short-term profit motives. In fisheries, harvesters invest in technologies that facilitate rapid catches of target species independent of the technologies' impacts on discards or habitat. Sustainability is further undermined by the absence of investment in resource productivity, including in enhancing the survival of prey and controlling predator populations.

The *right to transfer* entitlements to other entities increases the time-horizon beyond the owner's lifetime or generation. Transfer increases property value by making it available to others who value it more highly and by implicitly including demands of future generations.

Finally, without enforcement the other rights have no practical value. In addition to being a property right, enforcement must also be affordable otherwise it won't be practiced.

Enforcement is the bane of fisheries management by governments, and it is infeasible for resource claimants when resources are non-exclusive.

Virtually all property rights are attenuated by private contracts, laws or government regulations that protect public safety and social values. For example, you are entitled to drive your car to work if you are licensed and your car is registered; however, you may not exceed speed limits or to violate other motor vehicle laws. Likewise, fishermen are entitled to use their vessels to harvest fish stocks, but their landings might be restricted in terms of overall weight or fish size, or time or area closures might be imposed to protect marine mammals and endangered species, or their gear might be restricted to configurations that reduce discarding of uneconomic species. Attenuations reduce the value of a property right to the owner, but they are justified to protect public welfare.

Bundled Entitlements

The remainder of this section attempts an objective analysis of the implications of ecology, technology and transactions costs for partitioning LME resources into bundles of entitlements. Transactions costs are a collection of costs involved with gathering information on and otherwise delineating a resource, establishing contracts (formal or informal) that define the entitlement(s), and monitoring and enforcing the entitlement(s). The three other property rights reflect society's preferences for an environmental property rights regime and are therefore normative.

From an economics perspective, an ecosystem such as a LME can be viewed as a matrix of environmental and biological attributes, some of which either yield or contribute to as 'inputs' (*e.g.*, prey, habitat) a variety of goods, such as food, and services, such as assimilation of waste, over time that benefit humankind (Costanza *et al.* 1997). The variety of resources in an LME stems from heterogeneity in biological (*e.g.*, fish populations), physical (*e.g.*, sediments, currents, space) and chemical (*e.g.*, dissolved nutrients and salts) attributes and their structure (*e.g.*, trophic relationships, current systems) and function (*e.g.*, regulate prey populations, recycle nutrients). From this perspective, an LME is a differentiated capital asset that provides humans with flows of environmental goods and services not unlike what Rosen (1974) described for human-made capital assets (*e.g.*, houses, automobiles, vessels, docks).

In theory, each LME attribute is potentially a resource when it contributes to goods and services valued by humans, sometimes indirectly. For example, the megafaunal prey of commercially important Atlantic cod stocks across the Northeast Shelf Ecosystem (*e.g.*, polychaetes and shrimps) and their biogenic (*e.g.*, bryozoan colonies, sponges) and sedimentary (*e.g.*, sand waves, glacial gravel deposits) habitats are not themselves in demand by seafood consumers, but they do contribute to Atlantic cod production. Likewise, microorganisms in sediments are essential to primary productivity because they recycle nutrients in detritus.

However, there are several reasons why not all resources are candidates for property right entitlements at a point in time. First, there is no need to conserve resources that are not scarce because there is more than enough to satisfy demand at zero price. For example, salinity and the concentration of dissolved carbon in the open ocean do not limit photosynthesis. In other cases, the cost of gathering information on a resource may be too great (*e.g.*, population dynamics of deep-sea fishes), or there may be relatively cheaper substitutes (*e.g.*, production of manganese from deep-sea nodules or of energy from temperature gradients). Finally, where resource attributes can be delineated, it might cost too much to monitor and enforce entitlements, or the institutions that govern use might preclude ownership. Many resource attributes thereby remain in the public domain until such time that technological innovations or changes in people's preferences make them economical (Barzel 1989; Cheung 1970).

We can contemplate which of an LME's many resource attributes are not viable candidates for property rights at this time. Diffuse and fluid resources, such as water temperature, concentrations of dissolved nutrients (nitrogen, phosphorus, trace metals) and currents and related oceanographic phenomena such as warm-core-rings and El Niño all limit the survival or transport of fish larvae and, therefore, eventual recruitment to harvestable stocks, but they are indivisible and non-exclusive. Such resources are sometimes classified as public goods (or public bads).[27] Plankton communities— phytoplankton, copepods, microorganisms, fish larvae—are similarly off limits at this time.

We can also nominate a class of LME resources that is suitable for property rights definitions given today's information, technology and demand. These include the measurable stocks of renewable fishery resources, mineral deposits, ocean space, and, conceivably, highly migratory species such as herring, tunas, salmon, marine mammals and sea turtles. The latter present problems due to relatively high transactions costs, but recall the 1911 Pacific Fur Seal Treaty in which Russia, Japan, and Canada contracted harvest rights to the United States in return for annual compensation.

Finally, and importantly, entitlements to resource attributes can be bundled using geographical coordinates as implied by Gordon (1954) Cheung (1970) and Pollnac (1984). Doing so would include many of the resource attributes that currently defy divisibility/partitioning but which are known to contribute to a good or service in demand. A spatial orientation is critical to LME property rights structures because it moves benefits out of the public domain where they will be dissipated by too rapid use and depletion into the calculus of a government, commons or private owner. Of special importance to fisheries is management of discards and habitat, but interactions--and tradeoffs—with other resource attributes, including marine mammals, minerals deposits and ocean space are important too.

Design Principles for Property Rights Structures

Bundling LME resource attributes within spatial boundaries raises several questions regarding geographical scale and what logically belongs in a bundle from the perspectives of ecology, economics and sociocultural theory. Co-existence and co-evolution of marine species and the chemical and physical surroundings are important considerations so as to control losses from 'externalities,' or 'spillover effects.' By externalities, economists are referring to situations in which interdependence between production practices and people's welfare ('utility') are exogenous to the decision makers.[28] In other words, entitlements intermingle owing to an

---

[27] Pure public goods are non-rival and non-excludable; and pure private goods are rival and excludable. Examples of pure public goods include national defense where food and clothing are examples of private goods.
[28] Anthropological theory can be especially helpful in teasing out the different utility functions that may be held by different stakeholder groups due to social or cultural variances in preferences.

inability to completely delineate property rights because it is too costly or because of institutional constraints (Cheung 1970, Russell 1994). Although positive externalities are equally germane, externalities that damage the property interests of others receive the most attention. For example, otter trawl and dredge fishing tear up lobster pots and other types of fixed gear. However, groundfish and scallop fishermen have no incentive to restrict their fishing practices or to invest in different technologies (hook-and-line fishing for cod or cage culture of scallop) when fishing grounds and/or lobster stocks are non-exclusive.

The Coase Theorem (Coase 1960) teaches us that externalities do not result in a misallocation of resources (aside from wealth effects) in a utopian world of perfect knowledge, zero transactions costs and complete property rights assignments to all resources. Where externalities crop up due to changes in technology or peoples' preferences, property rights are exchanged in order to maximize total net value. In reality—and this is certainly the case for the nascent assignment of property rights in an LME—the transactions costs of delineating resources (*e.g.*, costs of information) and negotiating and enforcing new contracts can preclude exchange. Thus, it is prudent to consider ways to bundle LME resources that are consonant with today's ecological and socioeconomic information but are also flexible to change.

First, the 64 vast LMEs probably can be subdivided into smaller areas in order to incorporate principal interactions among attributes and reduce externalities. The division could be based on physical features that 'enclose' enduring species assemblages of marine species (marine communities) over time and that largely entrain energy flow and nutrient recycling across trophic levels. The physical features may be geologic, such as trenches or deep water slopes that limit seasonal migrations, or oceanographic, such as areas where upwelling or eddies occur. For example, scientists divide the Northeast Shelf Ecosystem into four sub-systems: Georges Bank, the Gulf of Maine, Southern New England, and the Mid-Atlantic Bight (Sherman *et al.* 1996). Competitive or mutually exclusive uses of the same areas—potentially, minerals extraction, transportation and/or endangered species protection—could be accommodated through contracts, litigation or 'combination sales' (Demsetz 1967). For example, the National Audubon Society has managed bird sanctuaries, cattle grazing and oil production at its Rainey Wildlife Sanctuary in Louisiana where the public is excluded (Baden and Stroup 1981).

Moving in the other direction on the spatial scale, whole sub-systems might be sub-divided into parcels of same uses or zoned for different uses based on smaller landscape or seascape features, but geopolitical lines (*e.g.,* state waters in the United States) probably are arbitrary criteria, and fragmentation of communities and habitat requirements would be counterproductive if it substantially interfered with the basic ecosystem functions of energy flow and nutrient cycling (Costanza and Folke 1996).

Scale also has socioeconomic and geo-political determinants that will affect design. The cost of monitoring and enforcing property rights will be a function of scale, and monopoly power in markets for LME goods and services would be illegal. Dividing LMEs into smaller units might help resolve or minimize transboundary disputes with other countries where resource attributes are mostly fixed in their location and can be zoned or otherwise allocated among users.[29]

Coexistence is only a necessary, not a sufficient, condition for bundling LME resource attributes. Strong complementarity and separability should also be guides when deciding what resources to bundle in a geographic area. User groups' local classifications (*e.g.* commonly known fishing grounds) also need to be taken into consideration. Joint ecologic relationships that have co-evolved to a high degree of specificity over time, such as species-specific predation, commensalism, and habitat dependence, are strongly complementary and, therefore, should justify inclusion in a bundle. Special attention should be given to possible cascade effects (see Christensen *et al.* 1996).

Unions made on the basis of joint ecological relationships should be overlaid by technology to see where joint production by fishing gear or interactions with other technologies (*e.g.*, sand and gravel mining) might combine ecological bundles. For example, in their study of the New England multi-species trawl fishery, Kirkley and Strand (1988: 1291) rejected the hypothesis of non-joint production of several groundfish species, including Atlantic cod and haddock, and concluded that '[m]anagement of one species independent of other species or of an aggregate output will not prevent overfishing or economic waste.'

In contrast to jointness, strong separability implies ecologic independence among resources and, in economics, an ability to substitute environmental inputs in order to produce different outputs. Ecological separation of species populations with closely related niches—*i.e.*, competitive exclusion—is a criteria for separate bundles provided technologic interactions are minimal. In stark contrast, 'regulatory bycatch'—a political economy artifact of single-species thinking (*e.g.*, groundfish and lobster caught in sea scallop dredge gear)—defies any ecologic-economic rationale for separability.

The rationale for using observed technology to determine resource bundles needs to be qualified. The technologies we observe in fisheries reflect the mostly non-exclusive history of marine resources that created incentives for rapid capture of target species. Institutional change that includes property rights will change investment decisions, conceivably to technologies that are more selective and less destructive of habitat. For example, production of the Japanese scallop increased over 30-fold after fisheries cooperatives in Japan substituted fixed gear culture technology for dredging.

---

[29] The basic idea here involves zoning certain resource attributes the way that real estate is zoned on land. For example, does a timber stand that stretches across the US-Canadian border have to be jointly managed? Probably not if lumber production is the only commodity—but maybe so if the most valuable use of the forest is habitat for a wandering endangered species. Likewise, does the Hague Line across Georges Bank present a problem for sustainable use of sea scallop resources? Probably not, since scallop resources are relatively sedentary. The Line does present a problem for migratory groundfish. However, zoning or some other means of allocating shares even in the migratory-type cases may or may not result in larger authorities (government or private), depending on scale efficiencies and other things. Small Authority A could purchase Small Authority B (or be joined by Congress or an international management authority). Or A and B may be able to negotiate contracts (treaties) that specify levels/locations of activities—*e.g.*, unitization of common pool oil fields.

PROPERTY RIGHTS REGIMES AND MANAGEMENT OF LME RESOURCES

Institutions

North (1992: 5) succinctly defines institutions as 'the rules of the game in a society,' or 'the humanly devised constraints that shape human interaction.' We are especially interested in how property rights regimes influence use of the natural environment (Hanna *et al.* 1996).

Hanna (1998) appears to have been the first to discuss at length the role of institutions in marine ecosystem management. Economic development and sustainability depend on an institutional environment that promotes the following management functions: integrate multiple objectives, control transactions costs, promote socially appropriate time horizons, engender legitimacy among users, and be flexible to change. No specific type of property right regime is endorsed (state, common, private), but decentralized decision-making is favored over centralized economic planning on these grounds. The Conference of the Parties to the Convention on Biological Diversity has likewise recommended 'decentralized systems' among its dozen principles of ecosystem management (UNEP 1998).

Functions of an environmental property rights regime

In contrast to structures of property rights that stem primarily from the nexus of ecological and economic interactions, a discussion of institutions addresses how a society's 'rules' for exclusion, transfer and enforcement rights influence use of entitlements and affect long-run economic performance and resource sustainability. There are several considerations.

First is the notion of sustainability itself. Nobel laureate economist, Robert Solow, expressed a perspective that is likely shared by most economists when he remarked that preservation of an individual species or habitat is 'fundamentally the wrong way to go in thinking about [sustainability]' (Solow 1991: 180). Instead, he emphasizes the important fact that people substitute goods and services for one another—'If you don't eat one species of fish, you can eat another species of fish.' (ibid.: 181)—and thereby defines sustainability as 'an obligation to conduct ourselves so that we leave the future the option or capacity to be as well off as we are.' (ibid.: 181)

The Ecological Society of American (ESA) also embraced 'long-term sustainability' when it defined ecosystem management as follows:

'Ecosystem management is management driven by explicit goals, executed by policies, protocols, and practices, and made adaptable by monitoring and research based on our best understanding of the ecological interactions and processes necessary to sustain ecosystem composition, structure, and function. Ecosystem management does not focus primarily on 'deliverables' but rather on sustainability of ecosystem structures and processes necessary to deliver goods and services' (Christensen *et al.* 1996: 668-669).

Although one wonders about its view on substitution, the ESA did downplay notions of constancy when it endorsed homeorhesis (tendency of a system to return to its previous trajectory) over homeostasis (return to a predisturbed state) and cautiously suggested 'biomanipulation' as a means to enhance 'deliverables.' Biomanipulation includes predator control and the selective removal of close competitors, artificial habitats that enhance survivorship or productivity, and 'fertilizing' waters with inorganic or organic nutrients (*e.g.*, sewage) to increase primary productivity or the growth of detritovores that are prey for target species.

Economists, anthropologists, sociologists and ecologists are likely to agree about many of the ecosystem structures and functions that need to be sustained, if not about specific components. For example, instead of MSYs determined for individual species (stocks), the aggregate biomass yields from species in a community that are in demand by consumers (whether for food, ritual or other uses) might be sustained although not necessarily at the natural maximal level. Predator control and culture could increase yields beyond that observed for conventional fisheries. On the other hand, alternative uses of the same area— recreation, preservation, minerals extraction—may prove more valuable than only commercial fishing in some areas.

At a higher level of ecologic organization, biological diversity as it relates to ecosystem functions (photosynthesis, nutrient recycling, energy flow) and responses to disturbance (resistance and resilience) probably should be maintained. This is not an endorsement of community-type diversity indices that would maintain a constant number or kind of species or their abundance (*e.g.*, the Shannon-Wiener index). It concerns ecosystem 'health.' Systems require redundancy to be homeorhesically stable; therefore, entitlements to harvest close competitors might be attenuated. Likewise, predator control should not proceed to the point of trophic cascades. 'Fishing down food webs'—*i.e.*, depleting long-lived, high trophic level fish and then transitioning to low trophic level invertebrates and planktivorous fish—is not a sustainable fisheries policy because piscivore populations do not recover (Pauly *et al.* 1998).

A second consideration with direct ties to institutions is uncertainty and variability. The environmental influences of temperature, currents, and food supply on commercially and recreationally valuable fish populations is poorly understood and highly variable year-to-year and over longer periods of time (McGlade 1989). Fishing technologies and markets are also difficult to predict. Institutions that are able to adapt quickly to change and to experiment and innovate to gain new knowledge—*i.e.*, adaptive management (Walters 1986; McGlade 1989; Larkin 1996)—would be consonant with LME management.

The mention of commercial and recreational fishing in one sentence raises a third important function of institutions, namely resolving multiple use or goal conflicts cost-effectively. The myriad resources of an LME—renewable, non-renewable, space—have scores of uses and values that can be competitive or mutually exclusive. In addition to seafood and recreation, a short list includes energy, waste disposal, transportation and preservation of marine mammals. Addressing conflicts through the political process by rent-seeking is costly for a society because it uses scarce productive resources to transfer or otherwise alter the distribution of benefits, not to increase economic growth (Rowley *et al.* 1988). Market exchange or direct negotiations among affected parties (Coase 1960; Demsetz 1964; Pollnac 1992) or resort to courts to settle liability claims (compensation) will resolve problems with minimum transaction costs. Divisibility, exclusion and transfer rights are important here.

Related to multiple-use conflicts are a host of transboundary problems that most LME property rights regimes will confront. To be of any value, an LME, such as the Northeast Continental Shelf (Sherman *et al.* 1996), will be large enough to subsume the principal ecosystem and human dynamics. Yet, no LME is insulated from the rest of the world, and vice versa. Surface currents pass through LMEs and carry with them nutrients, plankton, and pollutants. Many species of marine finfish (*e.g.*, tunas), mammals, and sea turtles migrate through several LMEs in a year. Fishing effort is derived from consumer demand for fish products, including foreign demand, and therefore from people's preferences and incomes. Technological advances in power, food processing, transportation and electronics eventually find uses in marine fisheries. Development and other economic activities within coastal watersheds impact nursery grounds and generate pollutants that are carried to LMEs. LMEs straddle geopolitical boundaries.

Accountability for management decisions is widely cited as a principle of ecosystem management (Christensen *et al.* 1996; Hanna 1998; UNEP 1998). Here, the theory of agency, which is a branch of the economics of transactions costs and therefore involves property rights regimes (Eggertsson 1990), is germane. In an agency relationship, an owner delegates or transfers some rights to an agent for their mutual benefit. Such relationships exist in all types of human organizations, including those involving natural resource management. Accountability is weakened by the costs of gathering information to monitor the agent. The scope of agency problems might be a function of the scale of an LME holding if resource management is centralized.

Scale has other economic and social ramifications. Economic ramifications relate to competition in markets for goods and services and to the transaction costs of exclusion and enforcement. Monopoly power is not in society's best interest when compared to competition because too little is produced. Regarding transactions costs, it is unclear whether there are scale economies to owning large parcels (Demsetz 1967). Scale, along with the degree of social and cultural diversity within any given stratum, also has a strong influence on factors such as enforcement and mutual monitoring (Ostrom 1990).

Finally, there is the issue of temporal scale. Much of the misguided criticism about sacrificing the environment for short-term profits stems from situations where resource attributes are non-exclusive (Cheung 1970) or property rights are 'incomplete, inconsistent, or unenforced' (Hanna *et al.* 1996: 3), including the right to transfer entitlements to others. Coupled with exclusivity, transferability allows entitlements to move into the hands of people who value the resources more highly; it also increases the value of property entitlements in the present by factoring in demands by future generations.

CONCLUSIONS

This report has described a framework for assessing and monitoring the salient socioeconomic and governance elements of LMEs. The assessment and monitoring framework consists of 12 steps that, if applied, are expected to produce the essential information required for adaptive ecosystem management (Ecological Society of America, 1995: 1). The 12 steps are:

1. Identify principal uses of LME resources
2. Identify LME resource users and their activities
3. Identify governance mechanisms influencing LME resource use
4. Assess the level of LME-related activities

5. Assess interactions between LME-related activities and LME resources
6. Assess impacts of LME-related activities on other users
7. Assess the interactions between governance mechanisms and resource use
8. Assess the socioeconomic importance of LME-related activities and economic and sociocultural value of key uses and LME resources
9. Identify the public's priorities and willingness to make tradeoffs to protect and restore key natural resources
10. Assess the cost of options to protect or restore key resources
11. Compare the benefits with the costs of protection and restoration options
12. Identify financing alternatives for the preferred options for protecting/restoring key LME resources

One of the most challenging aspects of ecosystem management, especially for LMEs, is '[t]he mismatch between the spatial and temporal scales at which people make resource management decisions and the scales at which ecosystem processes operate' (Christensen *et al.* 1996: 678). Christensen and his co-authors, writing for the Ecological Society of America, went on to lament that 'we have identified few mechanisms to translate the actions occurring within individual forest ownership or local fishing communities into strategies to reconcile competing demands for resources or promote a regional vision for sustainability.'

We have argued in this report that the property rights paradigm could very well be the framework necessary to design LME resource management policies for long-term economic growth and resource sustainability. Property rights establish the incentives and time-horizons for resource use and investment (Libecap 1989). Property rights structures could be designed using ecological, economic and sociocultural principles related to jointness and separability on spatial scales that bundle resource attributes instead of leaving them exposed to overexploitation in the public domain. Property rights regimes need to translate a society's legitimate goals for use of its LMEs into concordant rules for exclusion, transfer, and enforcement.

GLOSSARY

Abiotic – reference to the nonliving portion of the environment.

Adaptive management – regulation or control of resource use that adapts in response to the results of management actions. It is also a learning process as managers observe environmental responses to actions and learn how the system reacts to a given set of measures.

Alien species (a.k.a. introduced, exotic or nonindigenous species) – a species that has been transported by human activity, intentional or accidental, into a region where it does not naturally occur.

Asset (a.k.a. capital asset) -- a physical entity with embodied wealth (such as a house or fishing vessel) that provides a flow of valuable services over time.

Assimilative capacity – capacity of the ocean to dilute, absorb or modify wastes such as sewage or heat.

Bathymetry – the measurement of the depths of oceans, seas or other large bodies of water.

Beliefs – opinions or convictions that shape social relationships or perceptions.

Benefit-cost analysis – a comparison of the economic benefits of using a productive resource with the opportunity cost of using the resource. Projects or regulations are evaluated based on how they change net economic value.

Biodiversity (biological diversity) – the diversity of life that occurs at several hierarchical levels of biological organization. Usually defined by three levels: genetic, species and ecosystem.

Biotic – the living portion of the environment.

Commensalism – having benefit for one member of a two-species association but having neither positive nor negative effect on the other.

Complementarity – LME resources that are closely linked such as organisms that have mutualistic relationships.

Consumptive use – occurs when one person's use of a resource prevents others from using it. For example, the shellfish, finfish, or waterfowl I take in the LME are unavailable for others to catch. Hence, consumptive use of natural resources in this sense is like consumptive use of common private goods exchanged on markets, such as a pizza or a pair of shoes.

Contingent choice – a direct economic valuation technique that is dependent on choices that respondents make in response to hypothetical questions or situations such as the ranking of environmental options.

Contingent valuation – a direct economic valuation technique that ascertains value by asking people their willingness to pay for a change in environmental quality. The information that is sought from respondents is conditional on some hypothetical market context such as the nature of the change, how it will be implemented, and what it will cost.

Cost effectiveness – minimization of costs in order to achieve a given end, such as the selection of the alternative(s) with the lowest cost per unit.

Direct use – refers to the physical use of a resource on site or in situ. Common examples of direct use include commercial and recreational fishing, beach use, boating, and wildlife viewing. Most direct use is targeted by participants who set out to visit a beach, to fish at a particular location, and so forth. Direct use also may be incidental, for example, when a person traveling by boat unexpectedly sees whales or marine birds while en route to a destination (Freeman, 1993).

Ecology – the branch of biology that involves the study of the relationships among organisms and the interaction between organisms and the physical environment

Economics – is the study of the choices people make to allocate scarce resources among alternative uses to satisfy human needs and desires.

Economic value – the most people are willing to pay to use a given quantity of a good or service; or, the smallest amount people are willing to accept to forego the use of a given quantity of a good or service.

Ecosystem – the biotic components of a community, and the abiotic elements of the environment that interact with these components.

Ecosystem health – the state of ecosystem metabolic activity levels, internal structure and organization, and ability to resist external stress over time and space.

Environmental resources – in the most general sense all renewable and non-renewable resources of the LME. Sometimes used to refer to water and air resources as opposed to natural resources such as fisheries or oil.

Estuaries – a semi-enclosed body of water that has a free connection with the open sea and within which seawater is diluted measurably with fresh water that is derived from land drainage.

Eutrophication – enrichment of a water body with nutrients, resulting in excessive growth of phytoplankton, seaweeds or vascular plants, and often oxygen depletion from decomposition of plant material.

Externalities – unintended harmful or beneficial effects incurred by a party that is not directly involved with production or consumption of the commodity in question.

Fishery – commonly used to include the interaction between fishermen using gear to catch fish in a certain area at a certain time. One or more stocks of fish and fishing for such stocks including the resource, and the active use of gear by fishermen to capture fish..

Fuzzy logic – mathematic technique capable of using qualitative, linguistic, and imprecise information. Relationships are based on a linguistic implication between an antecedent and its corresponding consequent.

Goods and services – any commodities or nonmaterial goods (services) such as assistance or accommodations that yield positive utility.

Governance –the formal and informal arrangements, institutions, and mores which determine how resources or an environment are utilized; how problems and opportunities are evaluated and analyzed, what behavior is deemed acceptable or forbidden, and what rules and sanctions are applied to affect the pattern of resource and environmental use.

Governance (Module) – considers the formal and informal efforts to manage human behaviors that affect the LME and encourages patterns of conduct which are in accord with the natural world.

Government – the political direction and control exercised over actions of the members, citizens, or inhabitants of communities, societies and States.

Government or institutional failure – inefficient delivery or use of scarce resources by the public sector.

Habitat – the place where an animal or plant lives.

Homeorhesically stable – tendency of a system to return to its previous trajectory or a return to normal dynamics rather than some undisturbed state.

Hydrography – measurement of water body arrangement and movements, such as currents and water masses.

Hydromodifications – modification of ocean, estuarine and riparian areas and features such as currents, depth or the configuration of the coastline and adjacent waters.

Indirect use – occurs when, for example, wetlands or other LME habitats contribute to the abundance of wildlife or fish observed or caught elsewhere in the LME. In effect the ecological services of the wetland or habitat help 'produce' the wildlife or fish concerned, although the link between the direct use and the ecological services provided by the wetland or habitat may not be apparent to the recreational participant.

Input-output models – a systematic method that both describes the financial linkages and network of input supplies and production which connect industries in a regional economy, and predicts the changes in regional output, income, and employment. Input-output analysis generally focuses on economic activity and the self-sufficiency of an economy, unlike cost-benefit analysis which focuses on changes in net national benefits from use of a productive resource.

Institutions – the humanly devised constraints that shape human interaction, or the rules that govern human

Integrity of ecosystems – when subjected to disturbance, the ecosystem sustains an organizing, self-correcting capability to recover toward an end-state that is normal and 'good' for that system.

Interaction matrices – the use of a matrices to organize LME activities and resources by listing them on each axis. Matrix cells represent potential interactions or linkages between the components listed in a particular column and row of the matrix. These matrices have the capacity to inventory, organize and explore relationships or linkages among human uses, ecological components and processes of the LME.

Jointness – interdependence among as system's components such as spatially or temporally related species of an LME.

Large Marine Ecosystem – a geographic area of an ocean that has distinct bathymetry, hydrography, productivity, and trophically dependent populations.

Legitimacy – perception of conforming to established social rules or standards.

Management – the act of influencing, directing, or controlling use of a resource.

Market failure – the inability of the market to allocate resources efficiently. The major causes of market failure include: 1) imperfect competition (monopoly), 2) imperfect information, 3) public goods, 4) inappropriate government intervention, and 5) externalities.

Markets – a collection of buyers and sellers who interact, resulting in the exchange of goods and services.

MARPOL – international convention for the prevention of pollution from ships.

Modeling – a simple representation or abstraction of a feature of the real world that reveals important relationships, processes or elements of the feature.

Monitoring – to observe and record changes with regard to physical, biological, or social conditions related to an LME.

MSY– Maximum Sustainable Yield from a fishery or fish stock. Largest long-term average yield or catch from a stock under prevailing ecological and environmental conditions.

Non-consumptive use – refers to cases where one person's enjoyment does not prevent others from enjoying the same resource. For example, my viewing of marine mammals in the LME, other wildlife, or attractive views does not prevent you from enjoying the same resources.

Non-governmental institutions and arrangements – informal norms and rules of behavior embodied in social arrangements and organizations such as non-governmental organizations (NGOs)

Non-joint production – production of a single or few elements of an LME without regard to relationships to other elements.

Nonuse (Passive use) – refers to the enjoyment individuals may receive from knowing that the resources exist ('existence value') or from knowing that the resources will be available for use by one's children or grandchildren ('bequest value') or others even though the individuals themselves may not actually use the resources concerned.

Normative – analysis leading to a recommendation or prescription that is based on value judgements or reflects society's preferences.

Norms – the 'understood' rules for appropriate behavior. This is broader than social organization and includes non-social behavior.

Nutrient pollution – pollution such as sewage, agricultural runoff or atmospheric deposition which increases nutrients available for primary production. Increased nutrient levels may lead to eutrophication.

Open access – access to the resource is free to anyone who wants to use or harvest it because there is no ownership of the resource.

Overexploitation – level of exploitation where the resource level has been drawn down below the level that on average would support the long term maximum yield of the fishery.

Passive use – refers to the enjoyment individuals may receive from knowing that the resources exist ('existence value') or from knowing that the resources will be available for use by one's children or grandchildren ('bequest value') or others even though the individuals themselves may not actually use the resources concerned

Phytoplankton – passively drifting or weakly swimming usually microscopic plant organisms. They are the most important community of primary producers in the ocean.

Productivity – usually in reference to primary productivity, the rate of assimilation of energy and nutrients by green plants (photosynthesis) and other autotrophs (chemosynthesis). Usually expressed as grams of carbon per square meter per year.

Property – is a benefit (or income) stream associated with a property right.

Property right – a claim to the benefit stream that some higher body – usually government – will agree to protect through the assignment of duty to others who may covet, or somehow interfere with the benefit stream.

Property rights structures – various types of property rights arrangements, all of which exhibit attributes of the five dimensions of property rights. For example private and common property each exhibit the following dimensions to varying degrees. The five include: 1) the goods and services that owners can derive benefits from, 2) divisibility, the richness of entitlements to complex resources with multiple attributes, 3) exclusion of others from using or damaging the owner's entitlements, 4) the right to transfer ownership of entitlements, and 5) enforcement of property rights.

Public good – a good that can be used by anyone and for which one person's use does not diminish the good's value for others.

Rent seeking – actions by individuals and interest groups designed to restructure public policy in a manner that will either directly or indirectly redistribute more benefits to themselves.

Resource valuation – calculation or estimation of the economic value of a natural resource.

Resources – 1) anything that has value. 2) living and nonliving components of nature such as fish, oil, water and air.

Separability – independence among resources and in economics, an ability to substitute environmental inputs in order to produce outputs. Therefore these resources may be managed or utilized separately due to the lack of strong linkages to other LME components.

Social and cultural factors – in addition to factors related to economics such as benefits, capital and labor, considerations such as social structure and social organization, peoples' knowledge and views (norms and values) about their work and how this relates to the resource.

Social conflict – when the existing order is perceived as oppressive or unfair, parties try to meet their needs by destroying their opposition and by replacing those who make the rules.

Social costs – costs associated with the disruption of communities, households and related social structures resulting in the loss of human potential.

Social forces – factors related to human behavior as shaped by group life, including both collective forces (group construction) and the ways in which people given meaning to their experiences (self-reflection).

Social impact assessment – an evaluation of the likely outcomes and impacts of a specific policy or regulation on a designated target group or groups, as well as likely ripple effects to other groups.

Social networks – comprises the sum total on one's group membership and relationships.

Social systems – represents an arrangement of statuses and roles that exist apart from the people occupying them.

Socioeconomic – pertaining to the combination or interaction of social and economic factors and involves topics such as distributional issues, labor market structure, social and opportunity costs, community dynamics and decision making processes.

Socioeconomic Module – application of economic and social science analysis to LME management. Six major element of analysis include: human forcing functions—ways in which human activities affect the natural marine system; assessing impacts; feedback of impacts to human forcing functions; the value of ecosystem services/biodiversity; estimation of non-market values; and integration of economic and social science and natural science assessments.

Socioeconomic benefits – benefits to humans gained through utilization of resources, including both economic and social benefits.

Spillover effects – sometimes referred to as externalities, an unintended effect (positive or negative, benefit or cost) imposed on others and not borne by the party responsible for the effect.

Stated preference methods – general category of indirect valuation methods that includes contingent choice or contingent valuation methods. Individuals are asked to makes choices regarding their willingness to pay or to rank environmental options.

Stocks – generally referred to as LME assets such natural resources that can be utilized as inputs for economic processes. In fisheries science a fish stock is used as a unit for fisheries management.

Subsidiarity principle – as related to levels of governance, suggests that authority belongs at the lowest level capable of effective action.

Sustainability – resources are managed so that the natural capital stock is non-declining through time while production opportunities are maintained for the future.

Sustainable development – development that recognizes the need to maintain capital stock and future production opportunities.

Tradeoffs – compromises among resource uses that are required because some bundles of entitlements defy divisibility/separability.

Transaction costs – a collection of costs involved with gathering information on and otherwise delineating a resource, establishing contracts (formal or informal) that define the entitlement, and monitoring and enforcing the entitlement.

Transboundary – resources or economic impacts such as pollution that straddle political boundaries, usually national borders. For example, transboundary stocks occur on both sides of a given national border.

Tropic (tropic level) – position in food chain determined by the number of energy-transfer steps to that level. Plant producers constitute the lowest level, followed by herbivores and a series of carnivores at the higher levels.

User group – a group of individuals that utilize a resource is a specific manner such as inshore lobster fishermen.

Utility – the level of satisfaction that a person gets from consuming a good or undertaking an activity.

Value – market and non-market values, gross and net values, and net benefits to consumers or goods and services.

Values – ideals, customs and beliefs of a given society.

Welfare – the prosperity or, more broadly, the well being of a person or group.

REFERENCES

Acheson, J.M., ed. 1994. *Anthropology and Institutional Economics*. Lanham, MD: University Press of America.

Alchian, A.A. 1977. Economic forces at work. Liberty Press, Indianapolis, Indiana.

Alexander, L.M. 1993. Large Marine Ecosystems: A New Focus for Marine Resources Management. *Marine Policy* 17:186-198.

Ayres, R.U. and A.V. Kneese. 1969. Production, consumption and externalities. American Economic Review, Vol. 59: 282-297.

Baden, J. and R. Stroup. 1981. Saving the wilderness: a radical proposal. *Reason* 13:29-36.

Braden, John and Charles Kolstat. 1991. *Measuring the Demand for Outdoor Recreation*. New York: North Holland publishing Co.

Bradley, Paul. 1974. 'Marine Oil Spills: A Problem in Environmental Management'. Natural Resource Journal 14(July): 337-359.

Brainerd, T., P. Clay, D. Hakserver, M. Hall-Arber, C. Kellogg, A. Kitts, and D. McCarron. 1996. Report to the ASMFC committee on economics and social sciences; Commercial sector reference document on identification and prioritization of economic and sociocultural data elements. ASMFC, Washington DC, 41p.

Brans, Edward H.P. 1994. 'Liability for Ecological Damage Under the 1992 Protocols to the Civil Liability Convention and the Fund convention, and the oil pollution Act of 1990, Part I', TMA '94-3, 61-84.

Brans, Edward H.P. 1994. 'Liability for Ecological Damage Under the 1992 Protocols to the Civil Liability Convention and the Fund convention, and the oil pollution Act of 1990, Part II', TMA '94-3, 85-91.

Brans, Edward H.P. 1995. 'The Braer and the Admissibility of Claims for pollution Damage Under the 1992 protocols to the civil Liability Convention and the Fund Convention. *Environmental Liability* (3)4: 61-69.

Caddy, J.F. and G.D. Sharp. 1986. An ecological framework for marine fishery investigations. *FAO Fish. Tech. Pap.*, (283):152 p.

Calow, P. and V. Forbes, Initial Risk Assessment Manila, The Philippines: UNDP IMO Regional Programme for the Prevention and Management of Marine Pollution in East Asian Seas.

Cheung, S.N.S. 1970. The structure of a contract and the theory of a non-exclusive resource. *Journal of Law and Economics* 13: 49-70.

Christensen, Norman et al. 1996. 'The Report of the Ecological Society of America Committee on the Scientific Basis for Ecosystem Management,' *Ecological Applications* 6: 665-691.

Clark, John R. 1996. *Coastal Zone Management Handbook*. Boca Raton: Lewis Publishers.

Coase, R. 1960. The problem of social cost. *The Journal of Law and Economics* 3:1-44.

Coastal Zone Management Act, Public Law 92-583.

Committee on the Scientific Basis for Ecosystem Management. 1996. *Ecological Applications* 6: 665-691.

Cordell, John. 1984. *Defending Customary Inshore Sea Rights*. Senri Ethnological Studies 17: 301-326.

Costanza, R. and 12 others. 1997. The value of the world's ecosystem services and natural capital. *Nature* 387:253-260.

Costanza, R. and C. Folke. 1996. The structure and function of ecological systems in relation to property rights regimes. In S. Hanna, C. Folke and K-G. Maler (eds.), *Rights to*

Nature: Ecological, Economic, Cultural, and Political Principles of Institutions for the Environment. Island Press, Washington, D.C. 13-34.

Costanza, R., *et al.*, (eds.). 1992. *Ecosystem Health: New Goals for Environmental Management.* Washington, D.C.: Island Press.

Couper, A. 1983. *The Times Atlas of the Oceans. New York.* Van Nostrand Reinhold Company.

Crawford, J.S. 1984. An Economic Input-Output Analysis of the Marine-Oriented Industries of New London County, Connecticut. Masters Thesis. Storrs: Univ. Of Connecticut.

Creed, C. F. and B.J. McCay. 1996. Property rights, conservation and institutional authority: Policy implications of the Magnuson Act reauthorization for the Mid-Atlantic Region. *Tulane Environmental Law Journal* 9(2): 245-256.

Cumberland, J.H. 1966. A regional inter-industry model for analysis of development objectives. *Regional Science Association Papers,* Vol. 17: 65-95.

Daly, H.E. 1968. On economics as a life science. *The Journal of Political Economy* 76: 392-406.

Demsetz, H. 1964. The exchange and enforcement of property rights. *The Journal of Law and Economics* 7:11-26.

Demsetz, H. 1967. Toward a theory of property rights. *American Economic Review* 57:347-359.

Doeringer, P., P. Moss, and D. Terkla. 1986. *The New England Fishing Economy: Jobs, Income and Kinship.* Amherst: University of Massachusetts Press.

Downs, A. 1967. *Inside Bureaucracy.* Boston: Little, Brown and Company.

Dyer, C., and J. Poggie. 1998. Application of the natural resource region as a unifying human ecosystem theory. unpublished draft. University of Rhode Island.

Dyer, C. L., and J. J. Poggie, Jr. 1998. Integrating socio-cultural variables into large marine ecosystems: The natural resource region model. NOAA Workshop on large marine ecosystems. University of Rhode Island. Kingston.

Dyer, C.L. and D. Griffith. 1996. An Appraisal of the Social and Cultural Aspects of the Multispecies Groundfish Fishery in New England and the Mid-Atlantic Regions. Aguirre International/ NOAA. Bethesda, MD.

Dyer, C.L., D.A. Gill and J. S. Picou. 1992. Social disruption and the Valdez oil spill: Alaskan natives in a natural resource community. *Sociological Spectrum* 12(2):105-126.

Dyer, C.L. and J.R. McGoodwin. 1994. *Folk Management in the World's Fisheries: Lessons for Modern Fisheries Management.* Boulder: University Press of Colorado.

Edwards, S.F. and G.A. Anderson. 1984. Land use conflicts in the coastal zone: An approach for the analysis of the opportunity costs of protection of coastal zone resources. *J. of the Northeast Agric. Econ. Council* 13 (1):73-82.

Eggertsson, T. 1990. *Economic Behavior and Institutions.* Cambridge University Press, New York.

European Union (EU). 1998. Overview of the Programme [Integrated Coastal Zone Management   Demonstration   Programme], http://europa.eu.int/en/comm/dg11/iczm/overview.html Accessed September 1998.

FAO. 1999. Guidelines for the Routine Collection of Capture Fishery Data. Fisheries Technical Paper 382. Rome: Food and Agriculture Organization of the United Nations.

Freeman, A. M. III. 2003 . *The Measurement of Environmental and Resource Values: Theory and Methods.* Washington, D.C.: Resources for the Future. Second edition.

GAO [General Accounting Office]. 1994. Ecosystem management: additional actions needed to adequately test a promising approach. GAO/RCED-94-111. Washington, DC: General Accounting Office.

Gordon, H.S. 1954. The economic theory of a common-property resource: the fishery. *Journal of Political Economy* 62:124-142.

Goulder, L.H. and D. Kennedy. 1997. Valuing Ecosystem Services: Philosophical Bases and Empirical Methods. In Daily, G.(ed.), *Nature's Services: Societal Dependence on Natural Ecosystems.* Washington, D.C.: Island Press. 23-47.

Grigalunas, T.A.. 1997. Benefit-Cost Framework for Marine Pollution Prevention and Management in the Malacca Straits. Manila: UNDP/IMO/GEF Regional Programme for Marine Pollution and Management in East Asian Seas.

Grigalunas, T.A. 1998. Practical Uses of Environmental Economics in Large Scale Marine Ecosystems. Paper presented to the meeting on the Benguela Current LME, Cape Town, South Africa (July).

Grigalunas, T., J. J. Opaluch, J. Diamantides. 1998. 'Estimating the Economic Cost of Oil Spills: Issues, Challenges and Examples', *Coastal Management* (June).

Grigalunas, T., J. Opaluch, D. French, and M. Reed. 1988. Measuring damages to marine natural resources from pollution incidents under CERCLA: Application of an integrated, ocean systems/economics model. *Marine Resource Econ.,* 5(1):1-21.

Grigalunas, T.A. and C.A. Ascari. 1982. 'Estimation of Income and Employment Multipliers for Marine-Related Activity in the Southern New England Marine Region', *J. of the Northeast Agric. Econ. Council* (XI) 1:25-34 (Spring).

Grigalunas, T.A. and J. Diamantides. 1996. The Peconic Estuary System: Perspectives on Use and Values. Peace Dale, RI: Economic Analysis, Inc.

Grigalunas, T.A. and J.J. Opaluch. 1988. 'Liability for Oil Spills: A New Approach for Providing Incentives for Pollution Control' *Natural Resources Journal* Vol. 28 (Summer).

Grigalunas, T.A. and J.J. Opaluch. 1989. 'Managing Contaminated Marine Sediments: Economic Considerations', *Mar. Policy.*

Grigalunas, T.A. and J.J. Opaluch. 1990. Social Costs. In Scott Farrow (ed.). *Managing Our Outer Continental Shelf Resources: Oceans of Controversy.* New York: Taylor and Francis Publ. Co.

Grigalunas, T.A., J.J. Opaluch, and M. Mazzotta. 1998. Liability for oil spill damages: Issues, methods, and examples. *Coastal Management* 26(2):61-98.

Grigalunas, T.A. and J.J. Opaluch. 1998a. Natural Resource Damage Assessment and the Straits of Malacca. Manila: UNDP/IMO/GEF Regional Programme for Marine Pollution and Management in East Asian Seas.

Grigalunas, T.A. and J.J. Opaluch. 1998b. Sustainable Financing for Ship-Based Pollution Prevention and Management in the Malacca Straits. Manila: UNDP/IMO/GEF Regional Programme for Marine Pollution and Management in East Asian Seas.

Hanna, S., C. Folke and K-G. Maler. 1966. *Rights to Nature: Ecological, Economic, Cultural, and Political Principles of Institutions for the Environment.* Island Press, Washington, D.C.

Hanna, S.S. 1998. Institutions for marine ecosystems: economic incentives and fishery management. *Ecological Applications* 8(Supplement):S170-S174.

Hanemann, M. 1994. Valuing the environment through contingent evaluation. *J. Econ. Perspect.* 8(4):19-43.

Hennessey, T. 1994, 'Governance and Adaptive Management for Estuarine Ecosystems: The Case of Chesapeake Bay. *Coastal Management* 22:119-145.

Hennessey, T.M. 1997. Ecosystems management. In Soden, D., B. Lamb, J. Tennert, eds. *Ecosystems Management: A Social Science Perspective.* Dubuque, IA:Kendal/Hunt.

Hunt, R. ed. 1997. *Property Rights. Monographs in Economic Anthropology.* Vol. 14. University Press of America. Lanham, MD.

Imperial, M.T. and T. Hennessey. 1993. The Evolution of Adaptive Management for Estuarine Ecosystems: the National Estuary Program and its Precursors,' *Ocean and Coastal Management 20:*147-180'

Imperial, M.T. and T. Hennessey. 1996. 'An ecosystem-based approach to managing estuaries: An assessment of the National Estuary Program. *Coastal Management* 24:115-139.

Interorganizational Committee on Guidelines and Principles (ICGP). 1993. Guidelines and Principles for Social Impact Assessment. IAIA, Box 70, Belhaven, NC 27810 (40p).

Isard, W. 1972. *Ecologic-Economic Analysis for Regional Development.* The Free Press, New York.

Juda, L. 1993. Ocean policy, multi-use management, and the cumulative impact of piecemeal change: The case of the United States outer Continental Shelf,' *Ocean Development and International Law* 24:355-376.

Juda, L. 1999. Considerations in the development of a functional approach to the governance of large marine ecosystems. *Ocean Development and International Law* 30: 89-125.

Juda, L. and T. Hennessey. 2001. Governance profiles and the management of the uses of large marine ecosystems. *Ocean Development and International Law* 32:41-67.

King, D.M. and D.A. Story. 1974. Use of An Economic-Environmental Input-Output Analysis for Coastal Planning With Illustration for the Cape Cod Region. Amherst: University of Massachusetts, Center for Water Resources Research.

Kirkley, J.E. and I.E. Strand. 1988. The technology and management of multi-species fisheries. *Applied Economics* 20:1279-1292.

Kitts, A.W., and S.R. Steinback. 1999. Data Needs for Economic Analysis of Fishery Management Regulations. Northeast Fisheries Science Center, Woods Hole, MA. NOAA Tech. Memo. NMFS-NE-119.

Kopp, R. and V.K. Smith. 1993. *Valuing Natural Assets: The Economics of Natural Resource Damage Assessment.* Washington, D.C.: Resources for the Future, 358 pp.

Larkin, P.A. 1996. Concepts and issues in marine ecosystem management. *Reviews in Fish Biology and Fisheries* 6:139-164.

Lourens, J., C. van Zwol, J. Kuperus. 1997. Indicators for environmental issues in the European Coastal Zone, *Intercoast Network.* 31(Fall):3-4.

Lee, K. 1993. *Compass and Gyroscope: Integrating Science and Politics for the Environment.*Washington, D.C.: Island Press.

Libecap, G.D. 1989. *Contracting for Property Rights.* Cambridge University Press, New York.

Marine Protection, Research, and Sanctuaries Act, Public Law 92-532.

MARPOL, International Convention for the Prevention of Pollution from Ships. 1973, as modified by the Protocol of 1978 (MARPOL73/78)

McCay, B. and J. Acheson (eds.). 1987. *The Question of the Commons.* Tucson: University of Arizona Press.

McGlade, J.M. 1989. Integrated fisheries management models: understanding the limits to marine resource exploitation. *American Fisheries Society Symposium* 6:139-165.

McGlade, Jacqueline. 1995 . 'Intelligent Knowledge Based Systems for the Analysis of Coastal Zones: Design Logic of SIMCOAST,' in ASEAN-EU Workshop Report, Interdisciplinary Scientific Methodologies for the Sustainable Use and Management of Coastal Resource Systems (University of Warwick).

McGoodwin, J.R. 1990. *Crisis in the World's Fisheries: People, Problems, and Policies.* Stanford: Stanford University Press.

Musgrave, Richard. 1969. Cost-benefit analysis and the theory of public finance. *Journal of Economic Literature* (7)3.

National Research Council. 1997. *Striking a Balance: Improving Stewardship of Marine Areas.* Washington, D.C.: National Academy Press.

NEPA, National Environmental Policy Act, Public Law 91-190.

North, D.C. 1992. *Transaction Costs, Institutions, and Economic Performance.* ICS Press, San Francisco, California.

Osuga, Hideo. 1997. International conventions on liability and compensation for oil pollution damages. In Ross, S. Adrian, C.S. Tejam, and R.M.P. Rosales (eds.). 1997. Sustainable Financing Mechanisms: Public sector- private sector partnership. MPP-EAS Conference Proceedings No. 6, GEF/UNDP/IMO Regional Programme for the Prevention and Management of Marine Pollution in the East Asian Seas, Quezon City, Philippines. 352 p.

Ostrom, E. 1990. *Governing the Commons: The Evolution of Institutions for Collective Action.* New York: Cambridge University Press.

Opaluch, J.A. and T. Grigalunas. 1984. Controlling stochastic pollution events through liability rules: Some evidence from OCS leasing. *Rand J. Econ.* 15(1):142-151.

Opaluch, J.A., T.A.Grigalunas, M. Maazzotta, R.J. Johnston, J. Diamantides. 1999. Recreational and Resource Economic Values for the Peconic Estuary. Report prepared for: U.S. Environmental Protection Agency, Peconic Estuary Program. Kingston, RI: University of Rhode Island.

Pauly, D. V. Christensen, J. Dalsgaard, R. Froese and F. Torres, Jr. 1998. Fishing down marine food webs. *Science* 279:860-863.

Pollnac, R.B. 1992. Multi-Use Conflicts in Aquaculture – Sociocultural Aspects. *World Aquaculture* 23 (2):16-19.

Pollnac, R.B. 1984. Investigating territorial use rights among fishermen. *Senri Ethnological Studies* 17: 285-300.

Portney, P.R. 1994. The contingent valuation debate: why economists should care. Journal of Economic Perspectives 8(4):3-17.

Rorholm, N., H.C. Lampe, N. Marshall, J.F. Farrell. 1967. Economic impact of marine-oriented activities: A study of the Southern New England marine region. Kingston, RI: University of Rhode Island, Agricultural Experiment Station.

Rosen, S. 1974. Hedonic prices and implicit markets: product ifferentiation in pure competition. J. Polit. Econ. 82(1):34-55.

Ross, S.A., C. Tejales, and R.M Rosales (eds.). 1997. Sustainable Financing Mechanisms: Public Sector-Private Sector Partnerships. Proceedings of the Regional Conference on Sustainable Financing Mechanisms for the Prevention of Marine Pollution. Manila, The Philippines: UNDP IMO Regional Programme for the Prevention and Management of Marine Pollution in East Asian Seas.

Rowley, C.K., R.D. Tollison, and G. Tullock (eds.). 1988. The political economy of rent-seeking. Kluwer Academic Publishers, Boston.

Russell, S. 1994. Institutionalizing opportunism: Cheating on baby purse seiners in Batangas Bay, Philippines. In   J. M. Acheson. *Anthropology and Institutional Economics.* Lanham, MD: University Press of America. 87-108.

Schultz, C. 1975. *Public Use of the Private Interest.* Washington, DC: Brookings Institution Press.

Sherman, K., M. Grosslein, D. Mountain, D. Busch, J. O'Reilly and R. Theroux. 1996. The northeast shelf ecosystem: An initial perspective. In Sherman, K., N.A. Jaworski and T.J. Smayda, eds. *The Northeast Shelf Ecosystem, Assessment, Sustainability and Management.* Blackwell Science, Cambridge, MA. 103-126.

Sherman, K. 1994. Sustainability, biomass yields, and health of coastal ecosystems: An ecological perspective,' *Marine Ecology Progress Series* 112: 277-301.

Sherman, K. 1997. Large Marine Ecosystems: Assessment and Management from Drainage Basin to Ocean,' Paper to the Joint Stockholm Water Symposium/EMECS Conference.

Sherman, K. and L.M. Alexander, eds. 1986. *Variability and Management of Large Marine Ecosystems.* Boulder: Westview (AAAS) 319p.

Sherman, K. and L.M. Alexander, eds.. 1989. *Biomass Yields and Geography of Large Marine Ecosystems.* Boulder: Westview Press (AAAS). 493p.

Sherman, K., L.M. Alexander and B. Gold, eds. 1990. *Large Marine Ecosystems: Patterns, Processes, and Yields.* Washington, D.C.: American Association for the Advancement of Science.

Sherman, K., L.M. Alexander, and B.D. Gold, eds. 1992. *Food Chains, Yields, Models, and the Management of Large Marine Ecosystems.* Boulder: Westview Press.

Sherman, K., L.M. Alexander, and B.D. Gold, eds. 1993. *Large Marine Ecosystems: Stress, Mitigation and Sustainability* Washington, D.C.: American Association for the Advancement of Science.

Sherman, K., E.N. Okemwa and M.J. Ntiba, eds. 1998. *Large Marine Ecosystems of the Indian Ocean: Assessment, Sustainability, and Management.* Malden, MA: Blackwell Science.

Sherman, K., N.A. Jaworski and T.J. Smayda, eds. 1996. *The Northeast Shelf Ecosystem: Assessment, Sustainability, and Management.* Cambridge, MA: Blackwell Science.

Slocombe, D. S. 1993. Implementing ecosystem-based management. *BioScience* 43:612-622.

Solow, R.M. 1991. Sustainability: an economist's perspective. Paper presented as the Eighteenth J. Seward Johnson Lecture to the Marine Policy Center, Woods Hole Oceanographic Institution, Woods Hole, MA., June 14, 1991.

Stratton Commission. 1969. (Commission on Marine Science, Engineering and Resources), Our Nation and the Sea: A Plan for National Action. Washington, D.C.: Government Printing Office.

Underdal, A. 1980. Integrated marine policy: What? Why? How? *Marine Policy* 4:159-169.

United States Congress. 1990. Oil Pollution Act of 1990. Pub.Law 101-380, 104 Stat. 484, Aug. 18.

UNEP (United Nations Environment Programme). 1998. Conference of the parties to the Convention on Biological Diversity. Report of the Workshop on the Ecosystem Approach. UNEP/CBD/COP/4/Inf. 9. Available from: Secretariat of the Convention on Biological Diversity, Montreal, PQ.

Vallega, A. 1991. The human geography of semi-enclosed seas: the Mediterranean case: A first approach. In Smith, H.D. and A. Vallega. *The Development of Integrated Sea-Use Management.* Routledge, New York. 238-259.

Vallega, A. 1992. *Sea Management: A Theoretical Approach.* London: Elsevier Applied Science.

van den Bergh, J.C.J.M. 1996. *Ecological Economics and Sustainable Development, Theory, Methods, and Applications.* Edward Elgar, Brookfield, US.

Victor, P.A. 1972. *Pollution Economy and the Environment.* George Allen and Unwin Ltd. London.

Vogt, W. 1948. *Road to Survival.* New York: William Sloan.

Von Moltke, K. 1997. Institutional interactions: The structure of regimes for trade and environment. In Young,O., *Global Governance.* Cambridge: MIT Press. 247-272.

Walters, C. 1986. *Adaptive Management and Renewable Resources.* New York: Macmillan.

World Bank. 1995. Monitoring and evaluation guidelines for World Bank – GEF international waters projects. Washington, DC: The World Bank, Environmental Department.

Zeckhauser, R. 1981. Preferred policies when there is a concern for probability of adoption. J. Envron. Econ. Manage. 8(3):215-237.

Large Marine Ecosystems, Vol. 13
T.M. Hennessey and J.G. Sutinen (Editors)
© 2005 Published by Elsevier B.V.

4

Governance Profiles and the Management of the Uses of Large Marine Ecosystems[1]

Lawrence Juda and Timothy Hennessey

The ecosystem paradigm is emerging as the dominant approach to managing the environment and its natural resources.[2] This shift from the treatment of individual resources to the broad perspective of the total ecosystem has taken hold in a number of fields such as forestry and fisheries and has also become an important management approach in the United States federal government and in international organizations. This paper considers the application of this approach to the governance of large marine ecosystems (LMEs).

The concept of large marine ecosystems has been developed by Sherman and Alexander and is used to refer to regions of ocean space encompassing coastal areas from river basins and estuaries on out to the seaward boundary of continental shelves and the seaward boundary of coastal current systems. They are relatively large regions on the order of 200 000 km² or larger, characterized by distinct bathymetry, hydrography, productivity, and trophically dependent populations.[3]

Utilizing such criteria, 64 LMEs have been identified.[4] Consideration of LMEs as management units follows from a substantial body of scientific investigation that examines

---

[1] Originally published as, Governance Profiles and the Management of the Uses of Large Marine Ecosystems, *Ocean Development and International Law* 32: 41-67 (2001) (co-author: Timothy Hennessey). Re-printed here with permission.

[2] See, for example, United Nations Development Programme, *et al.*, *A Guide to World Resources 2000-2001: People and Ecosystems, The Fraying Web of Life* (Washington, D.C.: World Resources Institute, 2000) summary online <www.wri.org/wri/wr2000/pdf/summary.pdf>; Global Environmental Facility, *Draft Operational Program #12: Integrated Ecosystem Management*, November 4, 1999 online at <www.gefweb.org/operprog/OP12rev8nov4.pdf>; Norman Christensen *et al.* "The Report of the Ecological Society of America Committee on the Scientific Basis for Ecosystem Management," 6 *Ecological Applications* 665-691 (1996); D. Scott Slocombe, "Implementing Ecosystem-based Management," 43 *BioScience* 612-622 (1993); and Timothy Hennessey and Dennis Soden, "Ecosystem Management: The Governance Approach," in Dennis Soden and Brent Steele, *Handbook of Global Environmental Policy and Administration* (New York: Marcel Decker, 1989) pp.29-48.

[3] Kenneth Sherman, "Sustainability, Biomass Yields, and Health of Coastal Ecosystems: an Ecological Perspective," 112 *Marine Ecology Progress Series* 277-301 (1994) and Lewis M. Alexander, "Large Marine Ecosystems: A New Focus for Marine Resources Management," 17 *Marine Policy* 186-198 (1993).

[4] A map showing these LMEs, together with information on them, is found online at www.edc.uri.edu/lme and at www.lme.noaa.gov. The Large Marine Ecosystems of the World Web Page is cosponsored by the National Oceanographic and Atmospheric Administration, the World Conservation Union (IUCN), University of Rhode Island (URI), the International Council for the Exploration of the Seas (ICES), and the International Oceanographic Commission (IOC) of the United Nations Educational, Scientific, and Cultural Organization (UNESCO).

---

the interaction of fish species with one another and with the physical environment that they inhabit, and also the effects on them of human activities.[5]

In his exposition of the LME concept, Sherman has outlined five linked modules with which to assess LME sustainability: productivity of the ecosystem, fish and fisheries, pollution and ecosystem health, socio-economic conditions, and governance.[6] The first three modules center on natural systems while the latter two focus on the human uses of the LME and its resources. To date, research has been devoted primarily to the first three modules but as attention turns from conceptualization of system dynamics to development of management strategies, it becomes increasingly clear that attention must be given to the human dimension of LMEs, represented by the socio-economic and governance modules.

Governance

The focus of the present study is on governance, by which we mean the formal and informal arrangements, institutions, and mores which structure:
> • how resources and environment are utilized;
> • how problems and opportunities are evaluated and analyzed;
> • what behavior is deemed acceptable or forbidden; and
> • what rules and sanctions are applied to affect the pattern of use.[7]

Human behavior and patterns of conduct critically affect the state of the natural world.[8] Governance is crucially important because attempts to manage resources and the environment are really about managing human behavior and encouraging patterns of conduct which accord with the operation of the natural world.[9]

It is important to observe that the concept of governance is not equivalent to government but includes also other mechanisms and institutions that serve to alter and influence human

---

[5] See the contributions of the many authors in: Kenneth Sherman, *et al.*, (eds.), *Variability and Management of Large Marine Ecosystems* (Boulder: Westview, 1986); *Biomass Yields and Geography of Large Marine Ecosystems* (Boulder: Westview Press, 1989); *Large Marine Ecosystems: Patterns, Processes, and Yields* (Washington, D.C.: American Association for the Advancement of Science, 1990); *Food Chains, Yields, Models, and the Management of Large Marine Ecosystems* (Boulder: Westview Press, 1992); *Large Marine Ecosystems: Stress, Mitigation and Sustainability* Washington, D.C.: American Association for the Advancement of Science, 1993); *The Northeast Shelf Ecosystem* (Cambridge, MA: Blackwell Science, 1996); *Large Marine Ecosystems of the Indian Ocean* (Malden, MA: Blackwell Science, 1998); Large Marine Ecosystems of the Pacific Rim (Malden, MA: Blackwell Science, 1999); and H. Kumpf *et al.* (eds.), *The Gulf of Mexico Large Marine Ecosystem* (Malden, MA., Blackwell Science, 1999).
[6] Kenneth Sherman, "Large Marine Ecosystems: Assessment and Management from Drainage Basin to Ocean," Paper to the Joint Stockholm Water Symposium/EMECS Conference (1997).
[7] Lawrence Juda, "Considerations in Developing a Functional Approach to the Governance of Large Marine Ecosystems," 30 *Ocean Development and International Law* 89-125 (1999).
[8] The Science Advisory Board of the Environmental Protection Agency observes that "the primary drivers of ecological change are anthropogenic factors" which include human population characteristics, consumption per capita, globalization of the economy, technology, education, and environmental laws and policies. Ecological Processes and Effects Committee, Science Advisory Board, *Ecosystem Management - Imperative for a Dynamic World,* EPA-SAB-EPEC-95-003 (March 1995).
[9] John C. Pernetta and Lawrence Mee, in *The Global International Water Assessment, GIWA*, note that "Clearly, the first step in promoting a response to complex environmental problems is to understand the causal chain between perceived problems and their societal root causes." p.4, <www.giwa.net/index2.html>.

behavior in particular directions.[10]   There are three key, general mechanisms of governance: the marketplace, the government, and non-governmental institutions and arrangements. These mechanisms interact with one another in an ongoing, continuing pattern of dynamic interrelationships. Through the pressures they generate, they individually and cumulatively impact use behaviors (Figure 4-1). Failure to heed the signals from these institutions may lead to sanctions that range from economic loss, to incarceration or monetary penalties, or to expulsion or alienation from the community.

The marketplace, through the profit incentive, for example, certainly affects how the environment is utilized, what resources are exploited, and the manner in which these resources are exploited. Indeed, the current controversy over the policies of the World Trade Organization and how its policies in support of trade liberalization affect the environment reflects this concern.[11]   Of course, consumers frequently have an array of choices as to what to buy. Should purchase decisions of a sufficient number of consumers incorporate not only considerations of inherent product quality but also the process which produces desired goods and should they be sensitive to associated eco-labeling, then marketplace outcomes may be more supportive of natural ecosystem protection than had been the case in earlier periods.[12] Additionally, efforts to attach monetary value to ecosystem services,[13] which have been

---

[10] On the distinction between government and governance see James N. Rosenau, "Governance, Order, and Change in World Politics," in Rosenau and Ernst-Otto Czempiel, (eds.), *Governance Without Government: Order and Change in World Politics* (Cambridge: Cambridge University Press, 1992) and Oran Young, "The Effectiveness of International Governance Systems," in Young *et al.*, (eds.), *Global Environmental Change and International Governance* (Hanover, N.H.: University Press of New England, 1996). See also Mark Sproule-Jones, *Governments at Work: Canadian Parliamentary Federalism and its Public Policy* (Toronto: University of Toronto Press, 1993). For another discussion of instruments to affect behavior in regard to marine resources, see R. Greiner, et al., "Incentive Instruments for the Sustainable Use of Marine Resources," 43 *Ocean & Coastal Management* 11-28 (2000).

[11] For background on this controversy, see Thomas J. Schoenbaum, "International Trade and Protection of the Environment: The Continuing Search for Reconciliation," 91 *American Journal of International Law* 268-313 (1997).

[12] Note that eco-labeling has been controversial, sometimes being criticized as a type of neo-protectionism. See Atsuko Okubo, "Environmental Labeling Programs and the GATT/WTO Regime," 11 *Georgetown International Environmental Law Review* 599-646 (1999), Elliot B. Staffin, "Trade Barrier or Trade Boon? A Critical Evaluation of Environmental Labeling and its Role in the 'Greening' of World Trade," 21 *Columbia Journal of Environmental Law* 205-286 (1996), and Samuel L. Lind, "Eco-Labels and International Trade Law: Avoiding Trade Violations While Regulating the Environment," 8 *International Legal Perspectives* 113-153 (1996). The Food and Agriculture Organization has become increasingly concerned with this subject as it relates to fisheries products and held a technical consultation on the subject in October of 1998. See, for example, FAO, Issues Related to the Feasibility and Practicability of Developing Globally Acceptable, Non-Discriminatory Technical Guidelines for Eco-Labelling [sic] of Products from Marine Capture Fisheries, FI:EMF/98/2 accessible on line at <www.fao.org/WAICENT/FAOINFO/FISHERY/FAOCONS/ecolab/fi-emf2f.htm>.

[13] See Robert Costanza et al., "The Value of the World's Ecosystem Services and Natural Capital," 387 *Nature* 253-260 (1997) and Frances Cairncross, *Costing the Earth: The Challenge for Governments, The Opportunities for Business* (Boston: Harvard Business School Press, 1992). For a non-technical introduction to the economics of ecosystem valuation see the Ecosystem Valuation website at <www.ecosystemvaluation.org>. In regard to the marine and coastal environment, see also, The Independent World Commission on the Oceans, *The Ocean Our Future* (Cambridge: Cambridge University Press, 1998), chapter 4, "Valuing the Oceans," pp.97-137.

Three Key Governance Mechanisms

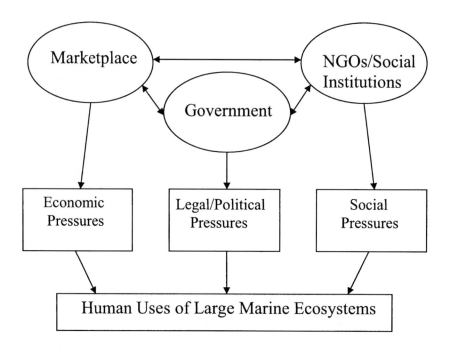

Figure 4-1. Three key governance mechanisms

regarded in the past as free, could serve to give a more concrete sense of value to those services and encourage the internalization of the costs of such services,[14] impelling more careful consideration of the natural environment. In a variety of ways the marketplace could make significant contributions to ecosystem protection.[15]

---

[14] In an examination of conditions under which international trade could contribute to ecologically sustainable development, Thomas Andersson, Carl Folke, and Stefan Nyström note that "Internalization means that the costs of environmental damage are included in the price of those goods and services which cause the damage." Among the methods they identify to internalize environmental costs are laws and regulations, markets for emission rights, environmental charges and taxes, and better defined property rights. *Trading with the Environment: Ecology, Economics, Institutions and Policy* (London: Earthscan Publications, 1995) pp.22-23. Note that Principle 16 of the 1992 Rio Declaration adopted at the United Nations Conference on Environment and Development calls on national authorities to promote the internalization of environmental costs. See also, Paul S. Kibel, "National Incentives to Protect Natural Resources: Preserving Their Place in International Trade," 29 *Environmental Law Reporter* 10411-10417 (1999).

[15] For consideration of a number of particulars in this regard, see Global Environment Facility, *Valuing the Global Environment* (Washington, D.C.: Global Environment Facility, 1998) pp. 43-65 and Robert Stavins and Bradley Whitehead, "Market-Based Environmental Policies," in Marian R. Chertow and Daniel C. Esty,

Government policy and regulation, at a local, regional, or national level are well recognized as mechanisms that can affect human behavior. Tax policies can provide incentives for particular types of conduct and, through government spending patterns, a substantial portion of society's resources may be directed so as to promote specific objectives.[16] Regulatory efforts, such as through zoning and permitting, can channel efforts along desired paths and, with their potential for unpleasant consequences in the form of fines or even imprisonment, can discourage undesired behavior. But in the long run, and perhaps most importantly, through education and outreach government may encourage environmental and ecosystem awareness.

Non-governmental organizations (NGOs) are becoming more evident in political activity at local, national, and international levels; there is a proliferation of NGOs that actively and purposefully seek to influence public policy relating to a very wide range of issues.[17] NGOs are a recognized force and play multiple roles in affecting behavior and public policy. They may serve as advocates of particular courses of action for government (*e.g.* stop licensing nuclear reactors) or of societal behavioral patterns (*e.g.* reject the use of furs) or seek to encourage or discourage particular pieces of legislation. In democratic and pluralistic societies, non-governmental groups play important constituency roles, affecting both governmental and marketing decisions with attendant ramifications for the natural environment. They also frequently harness scientific and technical expertise to monitor environmental impacts and change, as well as compliance with mandated expectations and, thereby, generate information relevant to decisions that will be made. In regard to environmental matters, many of them seek to educate the wider public on the workings of natural systems and on the particular issues at hand and, in doing so, help shape the framework in which problems and opportunities are analyzed and evaluated.

In traditional political usage, NGOs are exemplified by environmental organizations such as the Sierra Club and the World Wildlife Fund or trade associations such as the Chamber of Commerce or National Association of Manufacturers that have purposeful political agendas which, through explicit strategies, seek to influence public attitudes, governmental policy, and the marketplace so as to achieve particular goals. But for the purposes of the present analysis, NGOs should be thought of more broadly, and include bodies less overtly political in nature,

---

Thinking Ecologically: The Next Generation of Environmental Policy (New Haven: Yale University Press, 1997) pp.105-117.

[16] A classic examination underscoring the relationship of the government's budget to public policy efforts is found in Aaron Wildavsky, *The Politics of the Budgetary Process*, 4th edition (Boston: Little, Brown and Company, 1984). Wildavsky observes that "If one looks at politics as a process by which government mobilizes resources to meet pressing problems, then the budget is a focus of these efforts." p.4.

[17] Steve Charnovitz, "Two Centuries of Participation: NGOs and International Governance," 18 *Michigan Journal of International Law 183-286 (1997);* Thomas Princen and Matthias Finger, *Environmental NGOs in World Politics* (London: Routledge, 1994); Thomas G. Weiss and Leon Gordenker (eds.), *NGOs, the UN & Global Governance* (Boulder: Lynne Reinner, 1996); Lee A. Kimball, "Major Challenges of Ocean Governance: The Role of NGOs," in D. Vidas and W. Østreng (eds.) *Order for the Oceans at the Turn of the Century* (Kluwer Law International, 1999) pp.389-405; Grant J. Hewison, "The Role of Environmental Nongovernmental Organizations in Ocean Governance," in E.M. Borgese, Norton Ginsburg, and Joseph Morgan (eds.), *Ocean Yearbook 12* (Chicago: University of Chicago Press, 1996) pp.32-51.

from community associations to fraternal organizations to families and religious groups.[18] All of these may serve as agents of socialization and, thus, shape human perceptions, preferences, and attitudes. While they may not issue edicts which are legally binding in the way that law is in civil society or explicitly seek to change governmental or economic policy, they do influence ideas and patterns of thought and often generate meaningful social pressures that encourage adherence to particular norms of behavior. Accordingly, these non-governmental institutions and arrangements can affect ecosystem use patterns.[19]

On the natural system side, Sherman and his colleagues have identified three "driving forces" that impact LMEs: natural variability, overfishing, and pollution, with varying combinations of the three influencing each of the sixty-four identified LMEs.[20] The relative importance of each and the nature of the mix of driving forces need to be identified in order to develop responses appropriate to the problems identified in individual LMEs. Likewise, the above identified human aspect driving forces: the marketplace, government, and non-governmental organizations, arrangements, and practices also need to be understood in context.

The mix, character, and influence of these three mechanisms will vary from LME to LME and need to be understood empirically rather than assumed. Different governments, for example, have varying degrees of "reach," that is, effective control in both functional and spatial terms. Some governments try to directly influence a much wider range of human activities than others. And some have real control throughout the territory of the state while others have authority that may have limited impact once the capital city is left behind. Further, governments range from those that are highly centralized and unitary in nature to those much more decentralized and federal in nature, with significant power in the hands of local governments. In traditional societies poor communications networks and limited capability for monitoring and enforcement may be the order of the day and socially and culturally determined constraints may be of greater significance than government in influencing behavior.

Moreover, operations of the market and economic systems in different localities may well not correspond to or behave in ways that North Americans or West Europeans, for instance, take for granted. Economies that are subsistence or locally based in nature may have significantly different characteristics from those economies that are export oriented, and the social and ecological implications of economic actions may be weighted differently or even ignored.[21]

---

[18] One example of theological concerns with the environment is seen in Marlise Simons, "Eastern Orthodox Leader Preaches Environment," *The New York Times*, December 6, 1999, p.A10. The article focuses on Patriarch Bartholomew of Constantinople, the spiritual leader of some 200 million Orthodox Christians, who maintains that polluting is a "sin" and a "sacrilege."

[19] A thoughtful consideration of and analytical approach for assessment of what are termed "context variables," that is, socio-economic and political factors at community and individual and household levels that influence program outcomes is found in R. Pollnac and R. Pomeroy, *Evaluating Factors Contributing to the Success of Community Based Coastal Resource Management Projects: A Baseline Independent Method*, (Manila: International Center for Living Aquatic Resources Management: Anthropology Working Paper Number 54, January 1996).

[20] Kenneth Sherman, "Achieving Regional Cooperation in the Management of Marine Ecosystems: The Use of the Large Marine Ecosystem Approach,"29 *Ocean & Coastal Management* 165-185 (1995).

[21] Andersson et al., *Trading with the Environment*, supra note 11, observe that many local societies know how to utilize resources and ecosystems in a sustainable manner but globalization can undercut traditional institutions

Ecosystems and Governance

Ecosystems present special challenges for those concerned with management of natural resources and the wider environment. In most of the identified LMEs, governance involves institutions, interests, and people of more than one state since the boundaries of LMEs typically do not coincide with those of state jurisdiction. Yet the reality is that the ongoing availability of many desired goods and services significantly valued by stakeholders depends on the continued functioning of transboundary ecosystems.

The divergence between "ecologically defined space" and "politically defined space"[22] gives rise to a host of management problems and might provide either a rationale for international cooperation or, alternatively, in situations in which international cooperation is weak or has not been forthcoming, an abandonment of national efforts because if such efforts are undercut by the actions of those in other jurisdictions they will be rendered ineffective anyway. Accordingly, achievement of an appropriate level of regional cooperation to foster effective use management is an important objective.

Aside from this spatial incongruity, additional difficulties are often caused by the mismatch in time frames of policy makers and the need for long term efforts and commitment to the end of ecosystem protection.[23] A program to improve water quality, for example, may have to be pursued for many years before significant effect is demonstrated, yet office holders have to justify their actions and their use of public resources every election cycle. Education and greater public knowledge of ecosystem concepts and dynamics can contribute to a more consistent public policy approach and can increase public support for needed longer term efforts, but educational efforts may be time consuming.

Further, the areas encompassed by LMEs are subject to multiple human uses, each of which may be monitored and regulated by different government agencies with specific and limited functional areas of responsibility. Accordingly, no particular agency will have responsibility for the "big picture," the totality of what occurs within the spatial framework. Likewise, private stakeholders tend to focus on their particular uses rather than on that "big picture." The need for a system, as opposed to a use, perspective was recognized in the classic 1969 report of the Stratton Commission.[24] The governmental and stakeholder fragmentation seen

---

and property rights and reduce the significance of social restraints to sustainable use of local resources. p.43.

[22] "Ecologically defined space" is used to indicate the area over which natural ecosystems extend while "politically defined space" refers to the geographic area encompassed by particular governance systems. Juda, "A Functional Approach to the Governance of Large Marine Ecosystems," supra, note 6 p.93.

[23] Lawrence Juda and R.H. Burroughs, "The Prospects for Comprehensive Ocean Management," 14 *Marine Policy* 23-35 (1990); R.H. Burroughs and Virginia Lee, "Narragansett Bay Pollution Control: an Evaluation of Program Outcome," 16 *Coastal Management 363-377 (1988); Timothy Hennessey and Dennis Soden, "Ecosystem Management: The Governance Appro*ach," in Dennis Soden and Brent Steele (eds.), *Handbook of Global Environmental Policy and Administration* (New York: Marcel Decker, 1989) pp.29-48.

[24] Commission on Marine Science, Engineering and Resources (Stratton Commission), *Our Nation and the Sea: A Plan for National Action.* (Washington, D.C.: Government Printing Office, 1969).

in the United States, rather than being unique to the United States, is seen in other governments and in international organizations as well.[25]

Approaches to Ecosystem-Based Management

Despite the obstacles faced by ecosystem-based management approaches, a 1994 report by the General Accounting Office (GAO) notes strong and growing support among officials, scientists, and policy analysts to move away from management based on individual uses of land units or protection of individual natural resources in favor of an ecosystem management approach.[26]   The report notes that the four primary federal land management agencies (National Park Service, Bureau of Land Management, Fish and Wildlife Service, and the Forest Service) are already using or intend to use an ecosystem approach to managing lands and natural resources. But as indicated in that report, the ecosystem approach has different meanings to the disparate groups supporting its utilization.[27]

Based on a review of the scientific and policy studies on ecosystem management, the GAO report provides a flow chart for ecosystem management (Figure 4-2). While the indicated steps may be of guidance value to government agencies, they may have more limited utility in the wider context of efforts to achieve ecosystem-based management. In particular, the linking of the box of understanding ecosystems ecologies to that of making management choices fails to give needed attention to the fundamental importance of the governance structures that shape human behavior in relation to the environment; this is a significant point of deficiency. Further, the GAO model may be critiqued for its focus on government and governmental policy as the driving forces of ecosystem-based management. It is suggested that a wider consideration of governance as outlined above is more appropriate.

A recent study by the National Research Council, *Our Common Journey: A Transition Toward Sustainability*[28] uses a pressure-state-response model and clearly notes that human use of the environment has effects on that environment. But the reasons for and the factors that structure the uses of the environment are not considered.[29] To effect change in the pattern of human use it is necessary not only to consider **how** the environment is used and with what effects but also **why** it is used in that particular manner. Accordingly, it is necessary to consider explicitly governance mechanisms and issues which serve to structure the pattern of

---

[25] See, for example, The Independent World Commission on the Oceans, *The Ocean Our Future, supra note 10; Jean-Pierre Levy, "Towards an Integrated Marine Policy in Developing Countries," 12 Marine Policy* 326-342 (1988); United Nations Economic and Social Council, *Development of Marine Areas Under National Jurisdiction: Problems and Approaches in Policy-Making, Planning and Management,* E/1987/69 (8 May 1987). *Agenda 21* points to the need to strengthen the coordination among UN organizations with marine and coastal responsibilities. See chapter 17, para. 17.17 <sedac.ciesin.org/pidb/texts/a21/a21-17-oceans.html>. Note also the creation by the UN Administrative Committee on Coordination of a Subcommittee on Oceans and Coastal Areas. Online at <acc.unsystem.org/-subsidiary.bodies/accsoca.htm>.

[26] GAO, *Ecosystem Management: Additional Actions Needed to Adequately Test a Promising Approach,* GAO/RCED-94-111 (August 1994).

[27] Ibid. p.38. In this regard, the report dryly notes that "there is not enough agreement on the meaning of the concept to hinder its popularity."

[28] National Research Council, *Our Common Journey: A Transition Toward Sustainability* (Washington, D.C.: National Academy Press, 1999).

[29] Ibid., pp.235-236

behavior as the key link between ecosystem ecology and human use and subsequent management choices.

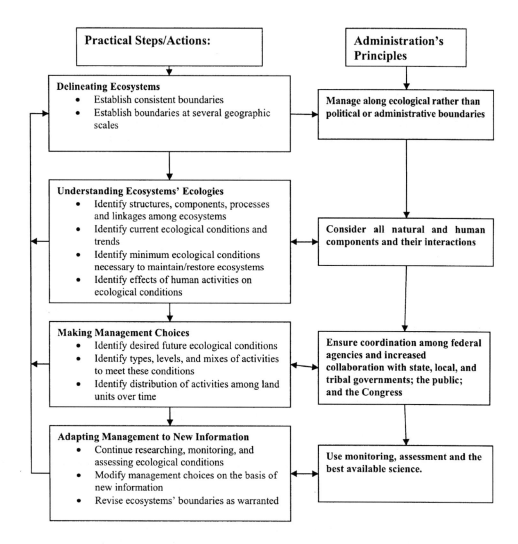

Figure 4-2. Relationships between practical implementation steps and ecosystem management principles [GAO/RCED-94-111 Ecosystem Management].

Governance Needs

As suggested above, a wide perspective of governance is required because values and expectations underlie human uses of the marine environment.[30]   Questions about sustainability, ecosystems, and ecosystem management are not simply questions about science: they are about values.[31]   Indeed, definitions of sustainability and ecosystems themselves reflect the values underlying them.[32]   And it is important to note that different environmental policies have varied distributive effects on societies that raise questions of equity.[33]   Management occurs within an institutional setting that, more or less, successfully recognizes the different values and expectations of a variety of user groups and those of the wider public.   Miller and Kirk have noted the difficulty or even impossibility of reconciling competing values.[34]   But values and expectations are subject to change over time with consequent modifications of behavior which brings us back to the mechanisms which shape those values and expectations.

Governance of large marine ecosystems requires consideration of a substantial amount of data as well as comprehension of a variety of relationships within the natural environment and also the effect of human uses on that environment.   Those who make decisions regarding the use of the natural environment and its resources need to be aware of and sensitive to the pattern of interaction resulting from their decisions if the sustainability of the environment that supports human needs is to be maintained.   In this regard there is a clear need, which has been frequently noted, for integrating science into public policy.[35]   While there is a need for greater awareness of what people want or expect from natural systems, there is also a requirement for

---

[30] Andersson *et al.*, *Trading with the Environment,* supra note 11, note in this regard "Our world view, values, knowledge and institutions influence to a great extent the way in which society relates to nature and the environment.   If people believe they rule and are separate from nature then a 'conquering' technology will develop which strives to create a society independent of nature.   If people regard themselves as part of nature and recognize their dependence on its support then a more collaborative type of technology, known as ecotechnology will develop." pp.40-41.   They further hold that "...experience suggests that ethical rules for sustainable behavior develop when people are directly confronted with the consequences of their actions." p.42

[31] On this point see Marc L. Miller and Jerome Kirk, "Marine Environmental Ethics," 17 *Ocean & Coastal Management* 237-251 (1992).   A report of the National Research Council observes that determining what to sustain is not only a biological question but one of values as well. National Research Council, Committee on Science and Policy for the Coastal Ocean, *Science, Policy, and the Coast: Improving Decisionmaking* (Washington, D.C.: National Academy Press, 1995), p.43.

[32] Valerie Normand and Debra Salazar, "Assessing the Meaning of Ecosystems Management in the North Cascade," in Dennis Soden, Berton Lee Lamb, and John Tennert (eds.), *Ecosystems Management: A Social Science Perspective* (Dubuque, Iowa: Kendall-Hunt, 1998).

[33] James F. Hammitt notes that "Concerns about social equity arise because environmental polices can alter the allocation of health, environmental quality, and financial costs across the citzenry.   The natural sciences cannot resolve these 'value' issues. They can be elucidated by the social sciences, but ultimately value choices must be resolved through the political system."   Hammitt, "Data, Risk, and Science," in Chertow and Esty, Thinking Ecologically: The Next Generation of Environmental Policy, supra note 12, pp.150-169 at p.151.

[34] Miller and Kirk, "Marine Environmental Ethics," supra note 30, p.246.

[35] See, for example, National Research Council, *Improving Interactions Between Coastal Science and Policy: Proceedings of the Gulf of Maine Symposium* (Washington, D.C.: National Academy Press, 1995) and Michael Healey and Timothy Hennessey, "The Utilization of Scientific Information in the Management of Estuarine Ecosystems, 23 **Ocean & Coastal Management** 167-191 (1994).

a better understanding of the capability of those systems to deliver desired goods and services and what is needed to maintain the viability of those underlying systems.[36]

Scientific understanding can assist in identifying internal natural system needs, the requirements of desired resources and conditions, and the limits to system productivity;[37] such considerations must be incorporated into decisions relating to human use of the environment. While productivity may sometimes be "tweaked" as through the use of fertilizers and pesticides in agriculture, there are still limits to productivity and, further, there are questions relating to the environmental, economic, and social costs as well as the benefits that are generated by the "tweaking" of nature.

In efforts to advance ecosystem-based management, contributions from both the natural and social sciences are needed and, further, these inputs need to be integrated. Fundamentally, the natural sciences can provide an understanding of the functioning of natural systems, the interrelationship and dynamics of system components, and the impacts of human use on the operation and changing states of those natural systems. They may also be able to suggest the human use implications of system changes.

On the social science side, the focus is on use management and efforts to modify use patterns to advance purposes such as system sustainability. Social scientists need to understand how the natural environment is utilized and why it is used as it is. The management goal is to affect human behavior vis-à-vis the ecosystem and to do this it is necessary to comprehend (1) the linkages between action and effect in the natural system, and (2) how the environment is perceived by users and what motivates particular behavioral patterns. What needs to be understood is not simply the reason for a use, such as fishing, but also the rationale for the use of specific methods of utilization. Such an understanding would enhance the potential for achieving needed behavioral change.

To integrate natural science and social science perspectives and to foster ecosystem sustainability, it is important to organize data on the interplay of human activities with natural processes to illuminate interrelationships and encourage thinking about what Pernetta and Mee term "causal chains."[38] In what follows we use several interaction matrices that can

---

[36] Lynton Caldwell has noted that "Not all human preferences are realizable in the real world; possibilities are not infinite, and basic relationships between man and nature are not negotiable. Nature does not bargain, and the biosphere is not a marketplace." Caldwell, *Between Two Worlds: Science, The Environmental Movement, and Policy Choice* (New York: Cambridge University Press, 1990) p.4.

[37] It has been noted that if sustainability of natural systems is the goal of ecosystem management, then it is not possible to set some arbitrary, desired level of commodity output since it is the natural system that determines what output levels are consistent with system sustainability. On this point, see Jerry F. Franklin, "Ecosystem Management: An Overview," in Mark Boyce and Alan Haney (eds.), *Applications for Sustainable Forest and Wildlife Resources* (New Haven: Yale University Press, 1997) and Norman Christensen et al., "The Report of the Ecological Society of America Committee on the Scientific Basis for Ecosystem Management," supra note 1.

[38] Pernetta and Mee, *The Global International Waters Assessment,* supra note 8 emphasize the importance of causal chain analysis. According to them:

"A causal chain is a series of statements that demonstrate and summarize, in a stepwise manner, the linkages between problems and their underlying or 'root' causes. Uncertainties accompanying each linkage should be clearly stated. The analysis also permits barriers to resolving the problems to be investigated.

serve as diagnostic tools and provide a framework for analysis and consideration of management problems and possibilities. The use of such matrices encourages the systematic and more holistic,[39] as opposed to purely sectoral,[40] consideration of actions.

The following analysis is not meant to provide an input/output model that predicts outcomes or a manner of automatically determining choices or policies. Rather, it seeks to promote understanding of relationships and to encourage the utilization of adaptive management approaches[41] that take full advantage of experience and learning.

Some introductory comments are needed on the matrices. First, the authors recognize that a variety of matrices have been used by others in conjunction with generic coastal or ocean use.[42] Perhaps more than these earlier perspectives, however, the current study considers their use in the context of ecosystem-based governance.

Second, we recognize that the nature of interaction among uses or the interplay between a particular use and the natural environment may often be site specific. That is, the same actions occurring in different natural settings may not have the same impacts on the physical environment or in the potential for creating conflict of use situations. The effects of the disposal of wastes in different marine environments, for example, may vary due to factors such as natural flushing action in open as opposed to enclosed sea areas. And conflict of use situations are affected by variables such as population densities, regional patterns of customary use, and degree and nature of economic and social dependence on ocean and coastal resources.

A causal chain presents the nature of the problem itself, including the effects and transboundary consequences, and then probes the linkages between problems and its societal causes. In its practical application, it can serve as a model into which regionally relevant information may be inserted."

[39] The authors are aware of the common overuse of this term. It has been noted that "holistic" came into vogue in the period of the UN Conference on Environment and Development in 1992 and is currently seen in statements on the environment "with the monotonous routine of applying vinaigrette dressing to salads." and Mee, *The Global International Waters Assessment,* supra note 8, p.2. The term remains useful, however, in distinguishing a more general, systemic viewpoint than that associated with sectoral approaches to problems.

[40] A recent report to World Commission on Water for the 21st Century notes the problem of sectoral approaches in relation to water resources observing that: "The current institutional system is fragmented, largely because water institutions have been driven by the concerns of other sectors. This led, for example, to creation of institutions managing water for agricultural use, for hydropower, for navigation, etc. Thus sectors may be planning on using the same water without coordinating with each other." M. Catley-Carson, *et al., Report of the Thematic Panel on Institutions, Society and the Economy and Its Implications for Water Resources,* p.2. Accessible through <www.watervision.org>. The problem of sectoralism is a major problem faced by efforts to promote ecosystem-based management.

[41] Kai N. Lee, "Appraising Adaptive Management," 3(2) *Conservation Ecology* (on line) <www.consecol.org/vol3/iss2/art3> (1999); Mark Imperial et. al., "The Evolution of Adaptive Management for Estuarine Ecosystems: the National Estuary Program and its Precursors," 20 *Ocean & Coastal Management* 147-180 (1993); Kai N. Lee, *Compass and Gyroscope: Integrating Science and Politics for the Environment* (Washington, D.C.: Island Press, 1993); C. Walters, *Adaptive Management and Renewable Resources.* (New York: Macmillan 1986).

[42] See, for example, Alistair Couper, *The Times Atlas of the Oceans.* (New York: Van Nostrand Reinhold Company, 1983); J.E. Halliday, J.E. and H.D. Smith, "The Integration of Coastal and Sea Use Management," in Smith, Hance D. (ed.), *Advances in the Science and Technology of Ocean Management* (London: Routledge, 1991) pp.165-178; and Adalberto Vallega, *Sea Management: A Theoretical Approach.* (London: Elsevier Applied Science, 1992); Porter Hoagland, *et al, Marine Area Governance and Management in the Gulf of Maine* (Woods Hole Oceanographic Institution, Marine Policy Center 1996).

Third, the matrices used in this study are rudimentary and are used solely to illustrate needed types of data and approaches. Clearly, it is necessary that broad categories of activity be subdivided appropriately: in the case of fishing, for example, operations are conducted in many different ways around the world and even among fishermen of a particular state. Commercial, industrial, artisanal, recreational fishing, and the use of different gear and techniques, while all coming under the general rubric of fishing, may each have varying impacts on the biomass, the physical environment, and other human uses, and, accordingly must be assessed differentially. While this study uses simple, two dimensional matrices, as noted below, actual utilization of coastal/ocean areas typically involves more than two uses and requires consideration of the cumulative, interactive effects of multiple uses. In this context it is necessary to evaluate:

(a) the compatibility of particular uses, given their inherent nature and requirements, in relation to other uses (compatible, conditionally compatible, or incompatible). Compatibility implies either that the uses do not interfere with one another or, possibly, that they may serve to enhance one or both of the uses through positive externalities. Incompatibility indicates detrimental effects of one use on another or both on each other through negative externalities. Conditional compatibility refers to situations in which potential negative externalities may be limited to acceptable levels through use limitations or restrictions. In many cases judgments regarding compatibility of use are situational rather than absolute in character, involving considerations of factors such as the amount of available space, the geographical layout of the area, and cultural and individual values.

(b) the effect of particular human uses on the natural environment and the operation of ecosystems. It is indisputable that human activities affect the workings of natural systems and, in some cases, can overwhelm and radically alter or completely destroy them. As in the case of use conflict, consideration must be given to the dynamics of particular ecosystems, taking into account existing stress and resilience. System specifics such as water currents, tidal flows, and natural flushing action may be of significance in evaluating behavioral patterns. Complicating matters is the fact that while sometimes effects are relatively quickly demonstrable, often they are of an insidious and cumulative nature, becoming manifest only after a substantial period of time. Yet judgments must be made as to the degree of impact (substantial, moderate, or inconsequential) of some use on the functioning of the natural environment, both short term and long term, and its consequent effect on sustainability.

Among the considerations which affect determinations of compatibility and degree of environmental impacts are:

(1) the nature of the use and the degree to which it puts stress on natural systems and limits future alternatives. Some uses are inherently more demanding of the natural environment, have more substantial and lasting effects than others (disposal of toxic wastes as opposed to recreational boating), and have greater potential to interfere with other human uses.

(2) the level of activity of a particular use (high, moderate, or low). The amount or frequency of activity needs to be considered as an independent variable since at low levels of use, uses may be compatible and environmental impacts may be insignificant, while this might not be the case with high levels of activity.

(3) the cumulative impact of varied uses requires consideration. Is there some type of synergy at work which magnifies the impacts of uses examined individually? The importance of cumulative impacts is increasingly recognized, as seen in the 1991 Protocol on

Environmental Protection to the Antarctic Treaty,[43] but the phenomenon presents substantial problems for analysts and policy makers.[44]

(4) the normative characterization (desirable, undesirable, or indifferent) of use interplay and environmental effect.   For example, is the by-catch of turtles in shrimp fishing normatively acceptable? Is the destruction of mangroves acceptable to the end of increasing shrimp production? Normative characterizations are determined largely in a cultural context, but local and regional ways of doing things are increasingly subject to outside pressures associated with trade, economic development, and imported technology and values.   The varied and changing cultural context is a factor which once more underscores the need for a site specific analysis of human interaction with the environment.   And there should be an awareness that difficult problems could be encountered should different states or communities share ecosystems but not traditions, values, and priorities.

The problems of evaluating and operationalizing the elements discussed above are substantial. As suggested by McGlade "fuzzy logic" may be of assistance in this regard.[45] But beyond the matter of assessing each of the four elements, the difficult question remains as to how the data will be aggregated.[46] Whatever device or procedure is used to organize and evaluate data, there can be no escape from a significant element of subjectivity. Moreover, values and preferences aside, the fact is that decisions will be made under conditions of imperfect knowledge and uncertainty and that fact raises questions regarding the controversial concept of precaution.[47]

With all these considerations in mind, there remain two basic reasons supporting the need for governance efforts: (1) incompatible human uses of LME space and resources that result in mutual interference, and (2) human uses of the LME environment that interfere with natural

---

[43] 30 *International Legal Materials* 1461-1486 (1991). Prior assessments of activities required under this protocol are to take "full account of…the cumulative impacts of the activity, both by itself and in combination with other activities in the Antarctic Treaty area". Article 3 (2)(c)(ii).

[44] On the problem of cumulative effects of multiple uses see William Odum, "Environmental Degradation and the Tyranny of Small Decisions," 32 *BioScience* 728-729 (1982); Peter M. Douglas et al., "Managing the Cumulative Impacts of Development: An Opportunity for Integration?," in National Research Council, *Improving Interactions between Coastal Science and Policy: Proceedings of the California Symposium* (Washington, D.C.: National Academy Press, 1995) pp.184-205; and Frances Irwin and Barbara Rodes, *Making Decisions on Cumulative Environmental Impacts: A Conceptual Framework* (Washington, D.C.: World Wildlife Fund, 1992).

[45] Jacqueline McGlade, "SimCoast: An Expert System for Integrated Coastal Zone Management and Decision-making," paper presented at the NOAA-NMFS Workshop on Biological and Physical Changes Within the Northeast Shelf Ecosystem of the USA, September 17-19, 1997 and "Intelligent Knowledge Based Systems for the Analysis of Coastal Zones: Design Logic of SIMCOAST," in ASEAN-EU Workshop Report, *Interdisciplinary Scientific Methodologies for the Sustainable Use and Management of Coastal Resource Systems* (University of Warwick, 1995).

[46] Arild Underdal, "Integrated Marine Policy: What? Why? How?," 4 *Marine Policy* 159-169 (1980).

[47] There is a growing body of literature on the subject of precaution and the precautionary principle. See, for example, David Freestone and Ellen Hey (eds.), *The Precautionary Principle and International Law: The Challenge of Implementation* (Boston: Kluwer Law International, 1996); Food and Agriculture Organization, *Precautionary Approach to Capture Fisheries and Species Introductions* (Rome: FAO Technical Guidelines for Responsible Fisheries, 2, 1996); Harald Hohman, *Precautionary Legal Duties and Principles of Modern International Environmental Law* (London: Graham & Trotman/Martinus Nijhoff, 1994); John M. Macdonald, "Appreciating the Precautionary Principle as an Ethical Evolution in Ocean Management," 26 *Ocean Development and International Law* 255-286 (1995); and S.M. Garcia, "The Precautionary Principle: Its Implications in Capture Fisheries Management 22 *Ocean & Coastal Management* 99-125 (1994).

processes and limit the potential for future use of that environment. The first two matrices directly address these matters.

Matrix 1
Use Interaction

Human Uses

Human Uses	Shipping/ ports	Fishing	Aqua-culture	Industrial siting	Re-creation	Waste Disposal	Housing	Military Uses	Agriculture	Forestry
Shipping/ ports										
Fishing										
Aquaculture										
Industrial siting										
Recreation										
Waste Disposal										
Housing										
Military Uses										
Agriculture										
Forestry										

scale:

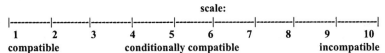

1	2	3	4	5	6	7	8	9	10
compatible			conditionally compatible					incompatible	

1. Use interaction matrix (Matrix 1)

The concept of conflict of use is basic to the fields of coastal zone management[48] and land use planning; as ocean uses increase and intensify, that concept has been similarly recognized as relevant in sea or ocean use management.[49] Often incompatibility is demonstrated in practice as sectoral-based decisions are implemented and negative externalities are generated. In the

---

[48] John Clark, *Coastal Zone Management Handbook* (Boca Raton: Lewis Publishers, 1996).

[49] Vallega, *Sea Management: A Theoretical Approach,* supra, note 41 Tundi S. Agardy, *Marine Protected Areas and. Ocean Conservation* (Austin: R.G. Landes Company, 1997) pp.65-69.

face of such experience, planners and coastal managers have resorted to devices such as zoning to keep apart activities that have significant incompatibilities.[50]

Clearly, those who make decisions must have some understanding of how different uses of the same area interplay and, not surprisingly, matrices have been employed in an effort to understand interactions among uses. The example of use interaction shown in Matrix 1 is extremely simplistic and in actual use would require much greater detail and sophistication. As noted earlier, a category such as fishing would have to be subdivided in a variety of ways, taking into account descriptors such as the scale of fishing, the gear used, and the time of the year. In his description of a "marine interaction model," Vallega[51] has provided a detailed breakdown of sea uses that could be modified and expanded for employment in an LME use interaction matrix. In developing such a matrix utilized data would have to be site specific if the matrix were to serve a useful purpose for local decision makers.

Matrix 2
Effects of Human Use on Ecosystems

Ecosystem alterations

Human Use	oxygen depletion	eutrophication	habitat destruction	turbidity increase	coastal erosion	change in biodiversity	pathogen contamination	introduction of alien species	change in water temperature
Fishing									
Aquaculture									
Dredging									
Navigation									
Military Uses									
Waste Disposal									
Recreation									
Industrial Siting									
Agriculture									
Forestry									
Off-Shore Oil									

scale:

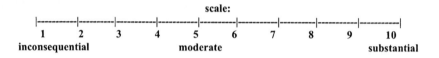

| 1 | 2 | 3 | 4 | 5 | 6 | 7 | 8 | 9 | 10 |
| inconsequential | | | | moderate | | | | | substantial |

---

[50] Jens C. Sorensen and Scott T. McCreary, *Coasts: Institutional Arrangements for Managing Coastal Resources and Environments* (Washington, D.C.: National Park Service, 1990) and Hilary Sargent, "Group Seeks Zoning Rules for Oceans," *Wall Street Journal,* May 10, 2000, pp.NE1,4.

[51] Vallega, *Sea Management: A Theoretical Approach,* supra, note 41.

2. Use/ecosystem effects interaction matrix (Matrix 2)

The notion that human use alters the natural environment is not new; what is relatively new is the degree to which that environment and its natural process may be affected by human actions. If sustainability of ecosystems is a matter of concern to decision makers, then it is necessary for those decision makers to consider the nature and character of the effects of human use on natural systems. The purpose of Matrix 2 is to highlight such impacts and to encourage an understanding of relationships between human behavior and ecosystem processes.

It may well be that the effects of human use are not well understood or, fully documented[52] and a degree of precaution may be called for to avoid irreversible damage or long term costs as decisions are made. Indeed, it would be useful for decision makers if some explicit assessment could be made as to data availability and the degree of understanding of natural processes that could be factored into decisions about the application of precaution. Consideration of interplay based on experience may be suggestive of priorities for future study where data or understanding is deemed insufficient.

To a considerable extent, human use of and effects on the ocean/coastal natural environment have been generally described. For instance, water quality has been monitored and evaluated, wetland loss has been studied, the introduction of alien species has been described, and coastal demographic changes have been documented. But in addition to studying changing conditions of the environment, greater consideration must be given to the practical consequences of those changes. The scientific community needs to highlight, in terms understandable to the layman, the consequences of those changes for human well being, a step that goes beyond observing the relationships of the type noted in Matrix 2.

A finding of depleted oxygen in coastal waters, for example, needs to be related to the practical, down-the-line potential consequences of fewer opportunities for commercial and recreational fishermen since such findings serve to motivate public concern and lead to action. Accordingly, we need a matrix which reflects the impact, that is the feedback implications, of ecosystem effects listed in Matrix 2 on outcomes of interest to stakeholders and the wider public. An expanded and more sophisticated version of Matrix 3 would encourage consideration of the impacts of observed ecosystem alterations on the potential for future human uses of the environment and its resources. The interdependent relationship between human use and environmental alteration is ongoing and needs to be monitored continuously. As noted elsewhere, site specificity is important in the relationships that Matrices 2 and 3 highlight. In regard to coastal management the need for a "vulnerability assessment" of specific environmental conditions has been noted[53] since variance in a number of natural conditions may alter the significance of possible threats. Indeed, the recognition of

---

[52] In its ongoing efforts to study ecosystems, the H. John Heinz III Center finds that of the three ecosystem types studied by it to date (croplands, forests, and coasts and oceans), "coasts and oceans suffer most from a lack of comprehensive and consistent information on key ecosystem goods, services, and properties" and the Center maintains that "there is no consistent information or solid scientific consensus on the key aspects of the arrangement and configuration of coastal and shoreline habitat areas." H. John Heinz III Center, *Designing a Report on the State of the Nation's Ecosystems*, <www.us-ecosystems.org/index.html>.

[53] Laurens *et al.*, "Indicators for Environmental Issues in the European Coastal Zone," *Intercoast Network*, Fall 1997, pp.3-4, 31.

special areas under the terms of the 1973 International Convention for the Prevention of Pollution

Matrix 3
Impacts of Ecosystem Alterations on Human Uses

Human Uses

Ecosystem alterations	Shipping/ ports	Fishing	Aquaculture	Industrial siting	Recreation	Waste Disposal	Housing	Military Uses	Agriculture	Forestry
oxygen depletion										
algae blooms										
eutrophication										
habitat destruction										
turbidity increase										
coastal erosion										
change in biodiversity										
pathogen contamination										
introduction of alien species										
change in water temperature										

scale:

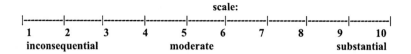

| 1 | 2 | 3 | 4 | 5 | 6 | 7 | 8 | 9 | 10 |
| inconsequential | | | | moderate | | | | | substantial |

from Ships[54] and designation by the International Maritime Organization of Particularly Sensitive Sea Areas,[55] the establishment of marine sanctuaries under the terms of the Marine

---

[54] 12 *International Legal Materials* 1319-1444 (1973).

[55] G. Peet, "Particularly Sensitive Sea Areas -- A Documentary History," 9 *International Journal of Marine and Coastal Law* 469-506 (1994) and A. Blanco-Bazan, "The IMO Guidelines on Particularly Sensitive Sea Areas (PSSAs)," 20 *Marine Policy* 343-349 (1996).

Protection, Research, and Sanctuaries Act,[56] and provision for special area management plans under the terms of the Coastal Zone Management Act[57] indicate recognition of vulnerabilities of particular areas.

Governance Profiles
As conflict of use and negative environmental consequences of human use become more obvious at a variety of levels, collective responses by society begin to emerge—in short, governance efforts evolve. This analysis urges the development of a baseline "governance profile" for each LME that describes the institutions, arrangements, and values at the core of existing governance in the area encompassed.

Just as natural ecosystems vary from one another, so, too do governance systems. The literature on natural ecosystems evidences problems with concepts of scale.[58] For analytical purposes and on the basis of designated scientific criteria, the authors of this study take as a given the LME, as described by Sherman and his colleagues, as the appropriate ecological level for consideration. Having done so it is necessary to recognize that governance has scale problems of its own: spatially, governance mechanisms may extend over very small geographical areas such as neighborhoods and range all the way to global arrangements. The appropriate scale of governance needs to be related to the particular ecosystem and its uses. But this need presents a challenge because smaller ecosystems are nested within larger ecosystems.

Governance arrangements already exist in areas encompassed by LMEs; they are not, however, presently organized around the concept of LMEs. Institutional, socio-cultural, and economic factors are of substantial significance in the use and management of the natural environment; as is the case with the natural environment, they are also site specific. In seeking to move toward governance arrangements that are more appropriate for ecosystem-based management, it is necessary to understand how existing institutional, economic, and cultural systems operate, their implications for the natural environment and its resources, and how needed change may emerge, given societal structures and norms.[59]

Development of governance baselines will allow for future comparisons and assessment of whether a "systems" perspective is advancing. In this context it is necessary to identify some indicators that would provide evidence of such a perspective: what might we expect to see in governance arrangements if such an ecosystem outlook were to increase in importance? What is being contemplated are indicators that serve, directly or indirectly, to signal transition toward an ecosystem-based orientation. Further, it is recognized that change is reflected in multiple indicators through their individual and their cumulative effects.[60] Indicators as used in relation to governance may relate to process or to result; the former refer to changes in the way in which decisions are made, institutions work, and values are prioritized while the latter

---

[56] P. L. 92-532 (October 13, 1972).
[57] 16 **U.S.C.** 1453.
[58] See, for example, Norman Christensen *et al.* "The Report of the Ecological Society of America Committee on the Scientific Basis for Ecosystem Management," supra note 3.
[59] Juda, "Functional Approach to Management of Large Marine Ecosystems," supra note 6.
[60] On the subject of indicators and their use in regard to sustainable development, see National Research Council, *Our Common Journey: A Transition Toward Sustainability*, supra note 27, pp.233-274.

focus on whether the process changes, in fact, have led to better results in terms of actual management.[61] The focus of this study is on process, since process change precedes result change. In regard to process, it is suggested that change may be anticipated in the following five interrelated areas:

1. perceptions and attitudes
2. institutions
3. processes and procedures
4. policies and programs
5. public participation

First, one would expect change in how situations are perceived and how problems are assessed and treated by governments, with indications of a shift away from sectoral to broader, more holistic approaches. Evidence of sensitivity of economic activity to ecological concerns would be apparent in terms of changes in product line, production processes, and marketing. Likewise, among non-governmental groups and the general public as well attitudinal change would be discerned. For example, greater awareness and understanding of the phenomena of negative spillovers or externalities would be seen with increasing comprehension that particular uses have impacts on other uses. Accordingly, one would expect to see a widening of the "stakeholder" community in regard to particular uses as other users increasingly recognize that their own interests are inter-related with and affected by the action of others. The great concern of fishermen with the impact of offshore oil development on living resources, seen in recent decades, was not apparent when the legal regime for offshore oil and gas development was established by the Congress of the United States in 1953; fishermen did not act as stakeholders concerned with the potential for oil spills in the early 1950s but they did later when use conflicts became more visible.[62]

Second, as externalities become increasingly manifest, one would expect the adaptation of existing governance arrangements and institutions so as to provide for "appropriate reach," for a better fit between jurisdiction and the extent of ecosystems. One would anticipate greater degrees of coordination and representation in relevant decision making. At the international level there might be efforts toward harmonization of law, participation in regional institutions and efforts, creation of treaty regimes, and provision for dispute settlement. Within governmental bodies interagency memoranda of understanding, interagency committees, task forces, and consultation will increase and expanded roles will emerge for non-governmental organizations, providing technical, scientific, and/or political advice. Negotiated rule-making with the intimate involvement of non-governmental groups and governmental agencies, as opposed to more traditional top-down, governmental-imposed

---

[61] For varying views of effectiveness and how it might be assessed, see Oran Young, "The Effectiveness of International Governance Systems," in Oran Young, George Demko, and Kilaparti Ramakrishna (eds.), *Global Environmental Change and International Governance, supra* note 17, pp.1-27; Oran Young (ed.), *The Effectiveness of International Environmental Regimes* (Cambridge: MIT Press, 1999); Olav Schram Stokke and Davor Vidas, "Effectiveness and Legitimacy of International Regimes," in Stokke and Vidas (eds.), *Governing the Antarctic* (Cambridge: Cambridge University Press, 1996) pp.13-31; and John Vogler, *The Global Commons: A Regime Analysis* (Chichester: John Wiley & Sons, 1995) pp.153-182.
[62] Lawrence Juda, "Ocean Policy, Multi-use Management, and the Cumulative Impact of Piecemeal Change: The Case of the United States Outer Continental Shelf," 24 *Ocean Development and International Law* 355-376 (1993).

rule-making may be in evidence.[63] The phenomenon of eco-labeling provides an example of how the changing attitude of consumers can influence the marketplace and through that influence alter patterns of ecosystem use.

Third, one would anticipate changes in processes and procedures so as to encourage fore-thought and precaution before actions are taken so as to minimize detrimental effects to the ecosystem and its resources. Included in this category would be requirements for data collection and analysis, environmental impact assessment, and notification and consultation. Such requirements would force consideration of broader system concerns.

Fourth, as the problems associated with sectoral approaches to problems become apparent, efforts are made to overcome them. One approach which may be utilized is the adoption of legislation and the development of governmental programs that reach across sectoral divides and force consideration of externalities. The National Environmental Policy Act[64] provides one such legislative example as the requirement for the use of an environmental impact statement mandates attention to the subject of externalities.

In the United States major federal, state, and local programs have the potential to impact LMEs. Such programs now encompass all of the coastal watershed leading to areas of fisheries and marine habitat. Watershed management emerged through the passage of section 6217 of the Coastal Zone Act Reauthorization Amendments of 1990 that mandate efforts to control non-point source pollution in coastal waters. Coastal states are required to use a watershed planning and control approach to deal with sources of pollution from agriculture, forestry, urban development, marinas, recreational boating, and hydromodifications. Plans must address the preservation and restoration of wetlands and riparian areas. States are to develop enforceable management measures to treat these sources of pollution.[65]

The National Estuary Program, established in 1987, complements the above efforts by providing funding to states to develop a comprehensive planning process to improve water quality and enhance living resources. There are currently 30 estuary programs in the United States, including four in the Gulf of Maine watershed.[66]

Coastal habitat issues have recently come to the forefront and have been addressed through the Sustainable Fisheries Act of 1996 that built on the Magnuson Fisheries Conservation and Management Act and required the National Marine Fisheries Service to specify "essential fish habitat" for all managed species and fisheries. Each fishery management council must amend its fishery management plans to: identify and describe the essential fish habitat for each managed species, identify the fishing and non-fishing related threats to the habitat, and develop management and conservation alternatives for that habitat.[67] It would be worthwhile to explore such legislative or programmatic mandates relevant to LME management to

---

[63] See, for example, the Negotiated Rulemaking Act of 1990, P.L.101-648 (November 29, 1990).
[64] National Environmental Policy Act, Public Law 91-190 (January 1, 1970).
[65] Mark T. Imperial and Timothy Hennessey, "An Ecosystem-based Approach to Managing Estuaries: An Assessment of the National Estuary Program," 24 *Coastal Management* 115-139 (1996).
[66] Ibid.
[67] P.L. 104-247.

understand how they may alter traditional agency activities and how they may serve to contribute to more holistic management approaches.

Fifth, one might expect provisions for greater public participation in the governance process, in regard to the shaping, implementing, and modifying of governance regimes and norms. Such participation is needed to obtain informational and attitudinal inputs, to allow for representation of views of different stakeholders, to secure needed understanding and support for decisions, and to provide assistance in implementation and assessment. For example, public meetings that allow for a two way flow of information between users and managers[68] and educational efforts aimed at enhancing public understanding of objectives and means are among the measures that could contribute to effective public participation and build support for needed efforts.

The next matrix directly considers governance aspects of LMEs and, in more sophisticated form, may contribute to the development of the suggested governance profile.

3. Governance/use matrix (Matrix 4)

Traditionally, governance arrangements have developed along sectoral lines on an *ad hoc*, piecemeal basis. As noted in the classic 1969 report of the Stratton Commission,[69] in governmental contexts, a problem is brought to light one way or another and some department or agency is given the responsibility to address that particular problem. Over time, responsibilities are spread among levels of government and among departments over a host of activities and areas. Eventually, interactional problems become evident since decisions are being taken without due regard to externalities: the lack of coordination leads to mutual interference, inefficiencies, and uncoordinated management.

While substantial attention has been given to mapping ecosystems, the mapping of governance systems, too, deserves attention. There is no question but that the nature and character of governance systems affect the pattern of use of coastal/ocean areas and, more generally, what has been termed "ecosystem health."[70] As long noted by political scientists and office holders, institutional arrangements can be instruments of delay and introduce the element of political and bureaucratic "turf" into all decisions.[71] But the interplay of different elements of government and governance, however, can also play a positive role by widening perspectives and forcing consideration of externalities.

---

[68] It is recognized that in a variety of situations users and managers may be the same people. Lee notes that those who harvest resources such as fishermen or farmers "are usually those who know most, in a day to day sense, about the conditions of the ecosystem. Their reports constitute much of the information that can be obtained at reasonable cost." Citing several studies Lee maintains that such harvesters "see themselves as stewards of the resources upon which they rely, a claim that frequently turns out to be well-founded." Lee, "Appraising Adaptive Management," supra note 40.

[69] Commission on Marine Science, Engineering and Resources (Stratton Commission), *Our Nation and the Sea: A Plan for National Action*, supra note 23.

[70] Robert Costanza et al., (eds.), *Ecosystem Health: New Goals for Environmental Management.* (Washington, D.C.: Island Press, 1992).

[71] Anthony Downs, *Inside Bureaucracy* (Boston: Little, Brown and Company, 1967).

Matrix 4
Governance/Use: **Example of the Gulf of Maine**

Governance Units	Uses											
	Shipping	Fishing	Aquaculture	Industry sites	Military uses	Recreation	Waste disposal	Mining	Housing	Agriculture	Forestry	Offshore oil/gas
General Purpose Regional Intergovernmental:												
a) International												
GOM Council												
b) National												
Canada Atlantic Action Plan												
Single Purpose Regional Intergovernmental:												
NAFO												
National Government												
a) U.S.												
NOAA												
b) Canada												
DFO												
State/ Provincial Governments:												
Maine, Massachusetts, New Hampshire, Nova Scotia, New Brunswick												
Local Authorities												
NGOs												
Market Place												

Scales:

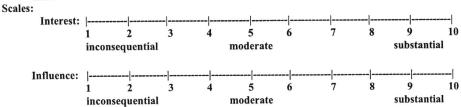

Interest: |----------|----------|----------|----------|----------|----------|----------|----------|----------|----------|
            1   2   3   4   5   6   7   8   9   10
            inconsequential   moderate   substantial

Influence: |----------|----------|----------|----------|----------|----------|----------|----------|----------|----------|
            1   2   3   4   5   6   7   8   9   10
            inconsequential   moderate   substantial

It is helpful (and sometimes horrifying) to know who is responsible for what and how the elements of governance, like those of ecosystems, interrelate and interact. They, too, are part of the "working environment" and must be taken into account as efforts are made to provide for effective use and protection of ecosystems. A fuller and more sophisticated version of matrix 4, which as shown utilizes examples from the Gulf of Maine for illustrative purposes, could yield a basic governance profile indicating networks of influence, jurisdiction, responsibilities, and interests. The dimensions of governance include (a) levels of governance (e.g. international, national, regional, or local), (b) sectoral areas (e.g. fisheries, offshore mining, waste disposal, recreation), and (c) stakeholders (e.g. fishermen, corporations, real estate interests, or port authorities). As is the case with particular LMEs, governance arrangements have site specific characteristics that need to be recognized and understood.

Relating to levels of governance, it is important to consider the level at which a problem should be addressed. The principle of subsidiarity suggests that authority belongs at the lowest level capable of effective action.[72] In fact, the European Union in its Integrated Coastal Zone Management Programme has adopted this principle and calls for problems to be addressed in the order of local, regional, national, and EU levels.[73] In this context, different levels of governance share responsibility and coordination is provided at higher levels. What is the appropriate level of governance to oversee particular uses? Clearly, problems exist in regard to the need for a considerable degree of both vertical (between levels) and horizontal (at the same level) consistency.[74]

It is apparent from our discussion of ecosystems that they present substantial challenges to resource managers. The most fundamental of these challenges is that ecosystem management must be able to cope with the uncertainty associated with the complexity of ecosystems as natural systems and the organizational and institutional complexity of management. We may think of these institutional structures and processes as the ecology of governance.[75] Adaptive management seems to be the most promising approach to coping with the uncertainty facing many decision makers as they try to manage such systems. Adaptive management involves learning by doing; that is, by treating programs and policies as experiments. By "linking science and human purpose, adaptive management provides reliable knowledge that serves as a compass for a sustainable future."[76] But the establishment of adaptive management is by no means easy in real world institutions; Lee notes the significant institutional constraints affecting the establishment and operation of ecosystem management (Table 4-1).

---

[72] Konrad Von Moltke, 1997 "Institutional Interactions: The Structure of Regimes for Trade and Environment," in Oran Young, *Global Governance* (Cambridge: MIT Press, 1997) pp. 247-272.

[73] European Union, "Overview of the Programme: Integrated Coastal Zone Management Demonstration Programme," <europa.eu.int/en/comm/dg11/iczm/overview.html>

[74] Underdal, "Integrated Marine Policy," supra note 45 and Biliana Cicin-Sain and Robert W. Knecht, *Integrated Coastal and Ocean Management* (Washington, D.C.: Island Press, 1998).

[75] This terminology is used by Timothy Hennessey in "Governance and Adaptive Management for Estuarine Ecosystems: The Case of Chesapeake Bay," 22 *Coastal Management* 119-145 (1994).

[76] Kai N. Lee, *Compass and Gyroscope: Integrating Science and Politics for the Environment*, supra note 40, p.9.

Table 4-1

Institutional Conditions Affecting Adaptive Management

Kai N. Lee, **Compass and Gyroscope: Integrating Science and Politics for the Environment**, p.85

There is a mandate to take action in the face of uncertainty. But experimentation and learning are at most secondary objectives in large marine ecosystems. Experimentation that conflicts with primary objectives will often be pushed aside or not proposed.

Decision makers are aware that they are experimenting anyway. But experimentation is an open admission that there may be no positive return. More generally, specifying hypotheses to be tested raises risk of perceived failure.

Decision makers care about improving outcomes over biological time scales. But the costs of monitoring, controls, and replication are substantial, and they will appear especially high at the outset when compared with the costs of unmonitored trial and error. Individual decision makers rarely stay in office over times of biological significance.

Preservation of pristine environments is no longer an option, and human intervention cannot produce desired outcomes predictably. And remedial action crosses jurisdictional boundaries and requires coordinated implementation over long periods.

Resources are sufficient to measure ecosystem-scale behavior. But data collection is vulnerable to external disruptions, such as budget cutbacks, changes in policy, and controversy. After changes in the leadership, decision makers may not be familiar with the purposes and value of an experimental program.

Theory, models, and field methods are available to estimate and infer ecosystem-scale behavior. But interim results may create panic or a realization that the experimental design was faulty. More generally, experimental findings will suggest changes in policy; controversial changes have the potential to disrupt the experimental program.

Hypotheses can be formulated. And accumulating knowledge may shift perceptions of what is worth examining via large-scale experimentation. For this reason, both policy actors and experimenters must adjust the trade-offs among experimental and other policy objectives during the implementation process.

Organization culture encourages learning from experience. But the advocates of adaptive management are likely to be staff, who have professional incentives to appreciate a complex process and a career situation in which long-term learning can be beneficial. When there is tension between staff and policy leadership, experimentation can become the focus of an internal struggle for control.

There is sufficient stability to measure long-term outcomes; institutional patience is essential. But stability is usually dependent of factors outside the control of experimenters and managers.

Perhaps the most fundamental observation about the institutional environment of ecosystem management is made by Lee who observes that social learning is most needed in large scale ecosystems the governance of which presents challenges for science, management, and politics. He notes the need to: *study how human institutions deal with the interdependence created when human boundaries cut across ecological continuities…what makes an ecosystem 'large' is not acreage but interdependent use: the large ecosystem is socially constructed.*[77] And just as ecosystems have a number of dynamic parts operating at a variety of levels so do the policy and institutional elements of the governance system reflecting a dynamic system of interdependence and complexity.

The issue of governance complexity is raised by Ostrom in her groundbreaking research on institutional analysis and design. She argues that "any governance system that is designed to regulate complex biological systems must have as much variety in the actions that it can take as there exists in the system being regulated."[78] In her research on the governance of natural resources around the world, she found that: the most notable similarity among the successful systems is the sheer perseverance of institutions which have the capacity to modify their rules over time according to a set of collective choice and constitutional choice rules in environments which are complex, uncertain, and interdependent.[79]

Ostrom found that all of the sustainable management institutions had clearly defined boundaries; a congruence between appropriation and provision of rules and local conditions; collective choice arrangements; monitoring; graduated sanctions; conflict resolution mechanism; minimum recognition of rights to organize; and nested enterprises.[80] She views what we have termed the ecology of governance in the following manner:

> The problem that we face is not pitting one level of government against another as a solitary source for authoritative decisions. Rather, the problem is developing institutional arrangements at multiple levels that enhance the likelihood that individual incentives lead participants toward sustainable uses of biodiversity rather than imprudent uses. Given the diversity of biological scales involved, Ashby's rule of requisite variety commends a variety of institutional arrangements at diverse scales. One key to understanding how to craft nested institutional arrangements at many levels is the analysis of how actions at one level change the incentives of actors at another level.[81]

Ostrom has developed an analytical framework for analyzing institutions that refers to a range of costs to be considered when designing such institutions. Coordination costs, the information costs of time and place and scientific information, and the strategic costs of free riding and rent seeking are fundamental factors in her approach to institutional analysis. She suggests that overall institutional performance be judged by the criteria of efficiency, fiscal equivalence, redistribution accountability, and adaptability.[82]

---

[77] Ibid., p.11.

[78] Elinor Ostrom, "Designing Complexity to Govern Complexity," in S. Hannah and M. Munusinghe (eds.), *Property Rights and the Environment: Social and Ecological Issues* (Washington, D.C.: The Beijer Institute of Ecological Economics and the World Bank, 1995) p.34.

[79] Ibid., p.34.

[80] Ibid., pp.35-40.

[81] Ibid., p.41. Ashby's Law of Requisite Variety states that diversity of scale must be matched by diversity of scale in self organizing systems. See Ross Ashby, "Principles of Self-Organizing Systems," in H. Van Foerster and Z.W. Zopf (eds.), *Principles of Self-Organization* (New York: Macmillan, 1962) pp.255-278.

[82] Elinor Ostrom, Larry Schroder, and Susan Wynne, *Institutional Incentives and Sustainable Development* (Boulder: Westview Press, 1993).

CONCLUSIONS

Much attention has been given to the natural science aspects of large marine ecosystems. If sustainability of those systems and their resources is to be enhanced, then greater systematic attention will have to be given to human interactions with those systems and the governance arrangements which shape the pattern of human uses.

Just as the natural features of LMEs are studied, so too, must the human use and governance arrangements be examined. Baseline studies have been conducted of natural systems that allow for later comparisons to evaluate the degree and nature of change over time. And various indicators are used or have been suggested to determine ecosystem health and stability. In addition, it would be useful to develop baseline studies of governance in each LME that could provide a benchmark with which to appraise change, allow assessment of progress over time in promoting sustainable uses of LMEs, and provide guidance for additional needed change. To that end, this paper suggests the use of a "governance profile" as a basic part of the strategy of advancing ecosystem-based management and as an indicator of "progress" toward an ecosystem orientation.

Governance profiles may be utilized to understand the human context in which use of resources and the natural environment proceeds. They recognize the unique systems of human use, institutions, values, culture, and priorities in each LME and provide needed understanding to encourage behavioral patterns that could accord with natural system sustainability. If we understand how human systems operate and the motivating forces behind them it is more likely, than in the absence of such understanding, that human activities may be modified to protect ecosystem integrity and, thereby, provide sustainable social benefits. The use of matrices, more sophisticated than those used in this paper, together with appropriate analyses can highlight significant relationships between human uses and ecosystem effects and may provide a useful tool to educate the wider public.

In looking to the future and considering how ecosystem-based management efforts may be improved, it is necessary to take the current governance system as a given and as the point of departure. Changes will be needed in terms of institutions, mores, and values if there is to be a shift away from sectoral approaches to management of natural systems and their resources. Movement toward ecosystem sensitive management does not have to be total and all at once but can result from cumulative, incremental change over time. Identification of incremental modifications would be desirable since such changes are easier to adopt and implement than more radical changes and, cumulatively, may still have substantial effects. As suggested by consideration of five identified areas: perceptions and attitudes; institutions; processes and procedures; policies and programs; and public participation, change is already occurring "on the ground" as high level discussions continue on the need for and theory of ecosystem-based management.

Large Marine Ecosystems, Vol. 13
T.M. Hennessey and J.G. Sutinen (Editors)

5

A Total Capital Approach to the Management of Large Marine Ecosystems: Case Studies of Two Natural Resource Disasters

Christopher L. Dyer and John J. Poggie

INTRODUCTION

Despite laudable efforts to integrate impacts on users, and user impacts on resources, into the management of marine resources in the U.S., the *status quo* remains one of conflict between managers and user populations, with users ultimately having little real input and control over final policy decisions. Origins of conflict, and the solutions to reduce it, are not well understood, but the outcomes have been disastrous. Despite the best efforts to manage user behavior or understand the user–resource interface and even with legal and political pressure from conservation groups and recreational fishing lobbyists dedicated to saving fish populations and habitats, declines and collapses continue to occur. In some cases, these declines are dramatic, sudden, and should be classified as "natural resource disasters" (NRDs) (Dyer, Poggie and Allee 2002).

We discuss the collapse of two Northeastern U.S. fisheries, the groundfishery of northern and central New England and the lobster fishery of western Long Island Sound, as natural resource disasters (Dyer, Poggie and Allee 2002). Natural resource disasters can be devastating to user populations. Further, they may be indicative of more ominous processes of global environmental change brought on by human actions and subsequent ecological perturbations (Dyer 2002b).

This chapter poses two basic questions: what happens when a natural resource disaster impacts a marine-linked human ecosystem and, how might a holistic human ecosystem approach, the total capital model, be applied to help prevent such disasters? We propose that integrating total capital forms—human, cultural, biophysical, social, and economic capital— into legislation aimed at achieving a more holistic approach to management of LMEs can improve the speed of recovery and lessen the overall negative impacts of regional natural resource disasters on user populations. Moreover, we suggest that the frequency and severity of natural resource disasters can be mitigated through a proactive total capital approach to resource management.

Disasters are primarily defined by their impacts on human populations, and our focus is on the human consequences of Natural Resource Disasters (NRDs) in the context of Large Marine

Ecosystems (LMEs) (Sherman *et al*. 1992; 1993; 1998). When such disasters occur, recovery may be possible, as with recent evidence for the recovery of groundfish stocks (New England Fisheries Management Council 2002), but the prolonged costs and losses born by the fishing-dependent families and communities from fishery NRDs can destroy significant amounts of total capital (Dyer and Poggie 2000), with loss of a way of life and a reduced ability of the fishery to support or recover the business and infrastructure of fishing. This is particularly severe if fishers experience protracted delays in renewal of catch effort in the face of stock recovery or, if no recovery is in sight, are not assisted to seek other diverse occupational options in the maritime or fishing industry or other related and appropriate occupations (Dyer, Poggie and Allee 2002).

The disaster recovery model presented here is designed to balance stock recovery with recovery of the other forms of total capital, with the unit of analysis being the fishing community, and networks of fishing communities within identified Natural Resource Regions (Dyer and Poggie 2000; Symes 2000). When natural resource disasters strike, they provide opportunities to understand weaknesses in management design thereby giving insights into how better to deal with the human-environment dynamic in a way that minimizes total capital loss. However, if such insights are realized but not applied, with the blame directed towards user populations, marine fisheries will continue to be vulnerable to future resource disasters, particularly as their frequency and geographic spread increases.

CONCEPTS AND UNITS OF ANALYSIS

Worldwide, there are 64 LMEs designated for marine bodies and their associated land masses and watersheds. Each LME is a potential ecosystem management zone where drainage basins and coastal areas are ecologically and hydrologically linked to continental shelves and dominant coastal currents. Explored in LME-GEF project strategic analyses (SAP) and transboundary diagnostic analyses and investigated in maritime anthropology studies, is the dynamic of how people and natural living resources act and interact, and the consequences to both of changes in this interaction (Dyer, Poggie and Allee 2002; Dyer 2002b; Dyer and McGoodwin 1994; Dyer and Poggie 2000). Unlike classical models of fisheries management, we do not posit a balanced system vacillating around a carrying capacity or defined by some catch effort paradigm such as Maximum Sustainable Yield. Rather, we propose an open system with fluctuations dependent on human migrations, techno-economic changes, political-ecological considerations, and temporal ecological variations which can shift interactions between human, cultural, social, economic and biophysical capital components. We also emphasize the importance of considering flows and transformations of capital components in the assessment of fisheries health.

Our conceptual framework emphasizes the connectivity of humans and the ecosystems in which they are embedded, and we define a fishing community as a type of *Natural Resource Community (NRC)* – a population of individuals whose primary cultural existence depends upon the use of renewable natural resources (Dyer *et al*. 1992; Dyer and McGoodwin 1994). *Capital* is defined as tangible or intangible resources that contribute to the long-term adaptation of a person, group, or population (Dyer and Poggie 2000:246). In fishing communities the NRC is broadly conceived to include human, social, cultural, economic, and biophysical elements. As previously noted:

An empirical question in regional studies is the extension of "capital" beyond the economic. In some cases, we suspect other forms or combinations of capital forms (social, human, cultural, and biophysical) may predominate, with economic capital being one of a complex mix with others (Dyer and Poggie 2000: 246).

Residents of a natural resource community frequently display a relatively common worldview and way of life that result in interdependent flows between economic actions and outputs (economic capital), the residents' patterns of values and traditions (cultural capital), networks of social and occupational relationships (social capital), the population and held skill sets of dedicated resource users and marketers (human capital), and the fish populations and supporting habitats in which the fish exist and from which they are harvested (biophysical capital).

Variations and patterns of capital flows define regions of interacting communities. In a comprehensive study of the fishing communities of New England (Hall-Arber, Dyer and Poggie 2002), eleven distinct networks of total capital flows were identified which, taken together, comprise the New England Natural Resource Region (Dyer and Poggie 2000).

Total capital, then, is conceptualized as the sum of all the component units of capital and their interactive states within a region (Dyer and Poggie 2000:248), and a Natural Resource Region (NRR) interfaces with the regional networks of natural resource communities (NRCs), their capital flows, and the associated Large Marine Ecosystem.

Using a total capital approach, the praxis of fisheries management is shifted towards consideration of the impact of legislated policy and regulations on communities of users, with the maximum good, or service to commerce becoming a focus on the well-being of user populations and their supporting community and regional networks. This does not ignore the importance of maintaining viable fishery stocks. Rather, a total capital approach emphasizes that understanding flow and interactions of total capital forms is the single best way to sustain a human ecosystem and minimize the potential for catastrophic collapse – the natural resource disaster. Moreover, it presents legislators and fishery resource managers with a central holistic framework that should allow them to provide the best possible management strategies to sustain biophysical capital without compromising other capital forms.

Recognition of natural resource disasters as increasingly frequent events has been recently observed by Dyer, Poggie, and Allee (2002), and institutionalized response mechanisms to deal with such events are few. Coastal populations experiencing marine resource disasters are dependent on government responses dominated by 'scientific' (biological, biochemical, biophysical) research. In fact, the realization that human consequences are worthy of attention is only now changing in the eyes of agencies such as the National Marine Fisheries Service and related state agencies. Still, in the culture of maritime research, mainstream scientists and resource managers seem perplexed by those who insist that fisheries should be about the interactions of people and natural resources, and not simply a series of independent "scientific" investigations of non-human environmental factors. For example, a keynote speaker at a recent lobster symposium on Long Island Sound remarked is his presentation: "I frankly don't know why there is a section on socioeconomic impacts in this forum" (Dyer, Poggie and Allee 2002: 1).

The preceding statement was made in the face of a natural resource disaster directly and indirectly threatening the livelihood of 1200 lobster fishing families. The fisheries scientist quoted above assessed this disaster excluding socioeconomic dimensions. When given to those making fisheries management decisions, such limited commentary cannot present a sufficiently realistic view of the dynamics of the human ecosystem that must be managed. It should be noted that historically, the human, social, cultural, and economic capital elements of any particular natural resource disaster management crisis have received major critical attention only *post hoc*, under the threat of litigation by stakeholders experiencing the brunt of legislative mandates.

Human, cultural, economic, and social capital use the biophysical resources of the adjacent marine environment to define the character of the region in which communities interact. **Economic capital** refers to formal and informal exchanges of goods and services, with the primary source derived through production and transformation of biophysical capital (*e.g.* marine resources). For example, lobster fishing in Long Island Sound includes exchange of labor (human capital such as helping offload lobsters) and information (cultural capital such as letting folks know where the creatures are) among community residents without any formal monetary exchanges. Such reciprocity results in the flow of long-term sustainable benefits that carry much more than narrowly conceived economic value.

The concept of **social capital**, the configuration and functions of people's personal ties, was explicitly articulated by the late James Coleman (1988) but earlier versions have appeared in sociological and anthropological theory. Like other forms of capital, social capital is productive, making possible the achievement of certain ends that would not be attainable in its absence. Like physical capital and human capital, social capital is not completely fungible, but is fungible with respect to certain activities. Social capital in the lobster fishery is virtually non-fungible, in that it is linked to the very specific activities of lobster fishing.

Cultural capital is less well known and less widely recognized by the general public, yet most potential employers inadvertently consider cultural capital in selecting employees. Cultural capital consists of specific behaviors, values, and skills transmitted among and between members of a population, including across generations, applied to their adaptation to specific environments including the transformation and utilization of natural, human, and social resources in those environments.

Biophysical capital is the sum total of natural resources as they are used and transformed by human activity. Remaining ecosystem components also function to support the system, as the air, water, soil, and other biophysical features allow both lobsters and humans to exist and thrive. Interactions among the various forms of total capital define natural resource regions, and disruption or loss of capital components by disaster can lead to significant disruptions, or even collapse if disruptions are severe enough, of the human ecosystem (Dyer 2002b).

Table 5-1 summarizes the components of total capital and the benefits accrued through community and regional-based capital exchanges and transformations. Forces external to the Natural Resource Region of a fishery and its Large Marine Ecosystem, influence the flow of total capital within the ecosystem, and can change the direction and magnitude of flow between capital components.

Table 5-1. Total Capital Interactions with External Forces in a Natural Resource Region

Total Capital Forms	Major Externalities				
	Technology Change	Market Change	Governance Change	Environmental Change	Population Change
Economic Capital Formal and informal exchanges of goods and services, with the primary source derived through production and transformation of biophysical capital	Slight Moderate Extreme	Slight Moderate Extreme	Slight Moderate Extreme	Slight Moderate Extreme	Slight Moderate Extreme
Human Capital A human population and its individual formal and informal occupational roles	Slight Moderate Extreme	Slight Moderate Extreme	Slight Moderate Extreme	Slight Moderate Extreme	Slight Moderate Extreme
Social Capital The sum total of people's personal ties and the networks within and between communities	Slight Moderate Extreme	Slight Moderate Extreme	Slight Moderate Extreme	Slight Moderate Extreme	Slight Moderate Extreme
Cultural Capital Behaviors, values and knowledge transmitted among and between members of a community	Slight Moderate Extreme	Slight Moderate Extreme	Slight Moderate Extreme	Slight Moderate Extreme	Slight Moderate Extreme
Biophysical Capital Resources derived from the dynamic between human action and the environment	Slight Moderate Extreme	Slight Moderate Extreme	Slight Moderate Extreme	Slight Moderate Extreme	Slight Moderate Extreme

For example, the herring fishing of New England supports populations of workers in the herring processing plants of coastal Maine. Dyer, Poggie and Hall-Arber (1998) determined that a proposed management policy allowing for processing of herring by offshore factory vessels in the Gulf of Maine would have devastated the onshore processing workers (human capital) and their families (social capital). Understanding the various capital forms and potential interactions and transformations is central to our analysis of impacts of marine resource disasters and models for disaster recovery.

The formula for any sustained total capital system is an adaptive and changing dynamic that varies from one society to the next. Any insider in a total capital system is embedded in a human environment filled with traditional patterns of behavior as well as the people who make up the social networks through which these patterns are created and recreated across space and time. Empirical measurement of these patterns is a difficult matter and to some degree a controversial one but, at a higher level of abstraction, most would agree that the patterns are enormously important to the life ways of people and communities everywhere and deserve to be measured over time.

In summation, the total capital model provides an overarching framework that can:
- Reveal the interconnections between externalities and human, social, economic, cultural and biophysical capital flows—an understanding which brings a holistic unity to the process of sustainable fisheries management;
- Encourage development of useful quantitative and qualitative methods to measure outcomes of the human-environment interface, both on the individual fisheries level and as part of a holistic ecosystems approach to fisheries management; and
- Engage and empower legislators, users and resource managers in Natural Resource Regions and associated LMEs, resulting in improved fisheries policy, and more effective communications and community participation in the management process.

We turn now to our comparative case studies of natural resource disasters within the framework of our Natural Resource Region model, and their intrinsic total capital flows.

THE WESTERN LONG ISLAND SOUND LOBSTER FISHERY AND THE 1999-2000 DIE-OFF

Brief Historical Overview of the Fishery

Commercial lobster fishers came to Long Island within the last thirty years from Connecticut and Rhode Island, or drifted in from the direction of New York City. The earliest fishers claim to have settled in areas east of Northport and west of Orient Point, and the early commercial fishing of lobsters has been likened to agricultural activity such as dairy or wheat production. In fact, the response to natural or economic disasters impacting diary farming has comparative relevance to this case. Recent research (New York Seafood Council 2003) established that bait used to trap lobster in effect is a major source of food for all age cohorts of lobsters. Because of the reliance of lobster on trap feed such as herring bait, skate and rack (what remains of a fish after filleting), some fishers describe lobstering as "sea ranching."

These early commercial ventures were characterized by modest efforts, using 200-800 traps. A key respondent indicated that fishers on the south shore did not come over to Long Island Sound because people thought "...there were no lobster off the northern shore." The last decade saw an expansion of the lobster fishing population out from the west and extending across the length of the Sound, with some fishers setting up to 2500 traps in an area.

The economic value of lobster to the commercial sector exceeds that of all other commercial fishery landings in the region. For example, the 1999 lobster landings in New York State put them at number one in dollar value at $27,332,599, accounting for 35.9% of total catch, with the next highest value being quahogs (a shellfish) at $17,777,034, or 23%. In 1969, the lobster fishery was a significantly lower percentage (10.4%) of the overall landings, with most of the catch (58.3%) being quahog clams. By numbers, the 1969 catch was 1,416,300 compared to 7,062,678 for 1999. This rise in catch numbers is consistent with increasing effort in this young fishery, and it confirms the lack of a high historical catch effort before about twenty years ago, which is when many of the current full-time lobster fishers came to the area.

For 1999, there were 288 registered lobster fishermen in Connecticut and 657 registered lobster fishers in New York. In Connecticut, approximately 114 (39%) of lobster fishers were affected by the natural resource disaster on the western end of the Sound in Connecticut and approximately 300 (46%) of lobster fishers on the New York side were affected (from Bridgeport west in Connecticut and Northport west on Long Island).

The western Long Island Sound lobster fishery is a young fishery, only about a generation old, with the current catch effort a result of the recent (1969-1999) expansion of the fishing population. Because of the highly specialized nature of the fishery and the lack of diversity in fishing options, the fishery is highly vulnerable to a collapse of biophysical capital (the lobster stocks). Furthermore, no adaptive management strategies or community and household strategies were in place to deal with a natural resource disaster. Such a disaster occurred with the catastrophic die-off of lobsters in the fall of 1999. The 1999-2000 declines in participation in the fishery are directly related to the die-off. The social, economic, and psychological consequences of the natural resource disaster are enormous as fishing families are economically displaced and require significant social service support to survive.

The Die-Off as a Natural Resource Disaster

Long Island Sound has seen fluctuations in catch (biophysical capital) over the years, with the last occurring around 1989. Despite such fluctuations, the lobster die-off of 1999 represents a true Natural Resource Disaster, because of the totality, rapidity and centrality of the lobster population collapse. The recovery of the lobster stocks has been very slow, leading to pernicious and long-term impacts on the fishing families and related industry networks on both sides of the western Long Island Sound. A survey recorded the observations of lobster fishermen in the fall of 1999 when the die-off struck (Table 2), noting potentially key events such as hypoxia, water color changes, and Malathion spraying for West Nile virus-carrying mosquitoes. Malathion is a chemical fatal to lobsters at very low concentrations. Following the passing through the watershed of Hurricane Floyd (after the widespread Malathion spraying), a massive die-off of lobsters was reported throughout the western Long Island Sound marine area.

A potentially important issue here is that there was reportedly no consideration of potential impacts to the ecosystem before the insecticide was applied. Those who used the resources of the Sound were not consulted or advised. A total-capital based ecosystem approach which involved fishers might have alerted authorities to the possible consequences of spraying. Moreover, a total capital response would have better alleviated the disaster impacts that followed.

Table 2. Time Line of Observations Surrounding the Lobster Die Off

August 1999	3	Hypoxia events peaked in WLIS; low oxygen levels reported near Hempstead Harbor. Hypoxia zone extends from Hempstead Harbor/Greenwich/ Mattituck /Guilford. Thunderstorms occurred intermittently throughout the entire month
	15	Lobsters appear to disappear from Stamford & Greenwich, after thunderstorms
September 1999		
	3	Malathion spraying started in NYC and continued use intermittently for the rest of the month
	6	Water changed to milky gray coloration and then unusually clear in some areas
	13	Deep trenches devoid of lobsters off Norwalk Very few live lobsters are reported from inshore areas
	14	Numerous dead lobsters reported from both deep and shallow waters off Northport and other western areas of Long Island Sound
	15	Tropical Storm Floyd passed through region
	20	Traps reported to be devoid of fouling organisms and grass Sulphur odor being emitted from traps 80% reduction in landings off of Lloyd's Neck
	21	Complete absence of live lobsters off Greenwich and Stamford
	22	Many dead lobsters brought up in traps Sound wide, including sub-legal molting stages Numerous dead spider crabs and no lobsters observed in deep holes
	26	Branford lobster run does not appear, and lobster fishers begin delaying setting of traps
	27	Lobsters absent off Greenwich and Stamford but still being caught off Darien
	28	Reports of new-shell lobsters appearing off Darien, but mortality rate of trapped lobsters 1-in-6 after 3-day trap sets
	29	Hard-shelled lobsters weakened and lethargic, eyes cloudy- with lobsters dying within a few hours of being landed
	30	Dead spider crabs widely reported
October	1	Noticeable absence of lobsters off Greenwich and Stamford
	8	Wholesalers report lobsters dying in tanks within a few days of receipt
	9	Dead lobsters observed in their burrows off New Rochelle by divers
	19	Molting berried females (sign of physiological stress) found in large numbers off Bramford
	29	Very few live lobsters caught off Bramford, Bridgeport, Norwalk and Darien

Natural Resource Disaster Impacts to the Human Ecosystem

Impacts of the disaster to the human ecosystem are most intense in the western end of the Sound and diminish as you travel east. The area, from Northport west in Long Island and about Bridgeport west on the Connecticut side, experienced an almost complete collapse of lobster stocks and this corresponds to the area of highest water temperature in the Sound.

Recovery in this region has not yet occurred, and those few lobsters caught are often weak. Many die in holding tanks before being sold. The secondary economic effects, or secondary disasters, go beyond the lobster-fishing household to include the marketing, transportation, and restaurant sectors. Disruption of the flow of biophysical capital and public fears concerning the safety of the lobster product together have a ripple effect that impacts other families and households outside the fishery.

Table 5-3. Age Distribution of Surveyed Lobster Fishers

Age Cohort (Connecticut)	Distribution CN	NY
70-79	1	0
60-69	3	6
50-59	7	8
40-49	12	12
30-39	9	6
20-29	1	1

With 69% of the Connecticut sample and 78% of the New York above 40 years in age, and only one below 30 years for both population samples, it is likely to be more difficult for these fishermen to move into alternate occupations or go through retraining, thus worsening the potential long-term human impact of the fishery disaster. Dedication to an occupation is in most cases indicated by the number of years doing it. The average number of years fishing for the Connecticut sample was 23 years, and ranged from 7 to 43 years. For New York, the average was 20 years, ranging from 9 to 45 years, indicating that lobster fishers in western Long Island Sound represent a stable population with a long-term commitment to fishing.

Economic capital generated from lobster fishing is critical to the maintenance of families and households. For example, 20 out of 31 (64%) (2 no answer) of the survey are married, with 7 (22%) single and only four (13%) divorced. These fishermen support from 2 to 5 individuals per household, with the average being 2.8 individuals. When asked how many households in the community are "directly and indirectly supported by lobster fishing", the responses ranged from 1 to 150, with an average response of 31 households per community. These supported households include primary and secondary businesses such as diesel fuel operators, marine supply retailers, fish marketers, welders, trucking operations, and seafood restaurants.

Assessing their commitment to fishing (Table 5-4), all (100% -33 out of 33) Connecticut respondents in the survey sample considered themselves full-time lobstermen before the disaster, while only 63% consider themselves full-time today. In New York the pattern was similar (93% and 57% respectively). This decline certainly stems from the disaster. It is accompanied by considerable despair over being idle from work. One of those from Connecticut who answered, "No," to this question (36%) also remarked, "I'm out of business." Two of those who responded, "Yes," to full time status also remarked "but out of business," and "with no income." Thus, they want to be considered as full-time but have been idled by the disaster. All (100%) of those sampled (Table 5-4) would like to continue lobster-fishing if able, and none (0%) in Connecticut and only one in New York was considering quitting lobster fishing before the disaster. This supports the interpretation that the lobster fishers are committed to their industry and want to remain participants if at all possible.

Table 5-4. Assessment of Dedication to Fishing.

	Yes CN % / N	Yes NY % / N	No CN % / N	No NY %/ N
Before the disaster, did you consider yourself A full-time fisherman?	100 / (33)	93 / (31)	0 / (0)	7 / (2)
Do you consider yourself A full-time fisherman today?	63 / (20)	57 / (19)	36 / (12)	43 / (21)
If you were able to, would you like to continue lobster-fishing?	100 / (33)	100 / (33)	0 / (0)	0 / (0)
Were you considering quitting lobster-fishing before the disaster?	0 / (0)	4/(1)	100 / (33)	96 / (32)

Besides the many years of vested experience on the job, there are shared sets of cultural capital (behavioral characteristics) that can influence fishers' world view and make it difficult for them to shift into alternate job roles. The term fungibility refers to the ability to move to other jobs, places, and lifestyles. Not being directly dependent on the harvest of natural resources in a particular place to make a living, most American workers can move into other jobs if necessary, even if some training is needed, and are thus highly fungible. However, it is well-documented that fishermen are not so adaptable.

The vulnerability of those involved in the fishing business in the Sound stems also from the specialization in lobstering, the equipment and fishing technique, the lack of alternative uses to which gear could be switched, and the high costs of purchasing and maintaining gear, boats, and docking access. Fishers' loss of income from lobstering can lead to insurmountable bills, loss of boat and home, car repossessions, and loss of gear. Business losses are made most severe by the (1) high degree of economic specialization and lack of economic diversity in the population, and (2) the rapidity and severity of the event itself, which did not allow for any flexibility in response by those being impacted, nor by managers and other government representatives charged with mitigating the impacts.

Many of those pursuing lobster in this part of the Sound had invested significant economic and human capital in their means of production, and maintained production relationships through networks of social capital both on and offshore. The losses they experienced were made most severe because of the pattern of economic specialization geared to lobster fishing. For example, from a sample survey of lobster fishers from the Connecticut side, overall earnings ranged from 40% – 100%, with only 18% (6 out of 33) indicating other sources of income. Thus 82% of those surveyed (27 out of 33) indicated they earned 100% of their salary from lobster-fishing only. On the New York side, overall earnings ranged from 50-100%, with only 12% (4 out of 32) indicating other sources of income. Of those surveyed, 88% (28 out of 32) indicated they earned 100% of their living from lobstering.

Table 5-5. Economic Capital Losses for Lobster Fishers.

Loss of gear/other business capital
Loss in gear value
Down-grading operation – fewer traps/ smaller boat
Selling out all traps and boat – out of business
Lack of resources for proper boat/gear maintenance
Lack of resources to pay overhead expenses/boat loans
Legal threats/ Repossession of business and household property
Total depletion of economic capital
Loss of home
Loss of good credit standing
Working outside minimum wage job; must give up occupation
Fishing alone- inability to pay deck hands

The average reported loss from the fishery between 1998 and 1999 was $31,566, ranging from a high of $156, 876 to a low of $30. A total of 138 respondents showed losses in 1999 as compared to 1998. Many of those who showed gains or stayed even in 1999 reported that their losses occurred during the 2000 fishing season. Unfortunately, losses suffered in 1999 continued into 2000 as the lobsters stocks and fishery failed to recover. Key respondents indicated that the compensation program, since it does not cover continuing loss and non-recovery, was nothing more than a stop-gap measure. Compensation came very late after impacts began, with significant losses in total capital continuing to occur as fishers attempted to cope with the loss of their livelihood, total capital, and way of life. Without sustained support, and without recovery of the fish stocks, little opportunity for recovery presented itself to the full- time, heavily invested core fishers.

From key respondent interviews, from dockside intercept interviews, and from a standardized survey of 33 lobster fishers impacted by the die-off, those in the far western end of the Sound had the greatest losses, and also typically fished more traps than those in the middle to eastern areas of the Sound. As we move east from Northport, the number of traps used declines and the number of surviving lobsters rises. As of August 2001, lobster fishers in the mid-to-eastern Sound reported little lingering evidence of the 1999 die-off effect beyond a decrease in the size of lobsters caught. By contrast, fishers west of Northport, Long Island and Bridgeport, CN, had seen no significant recovery of the native stocks.

A summary of critical total capital impacts is found in Table 5-6. The categories of responses are organized as production impacts, economic impacts, and lifestyle impacts. Under lifestyle impacts we subsume social, cultural, human and psychological variables. The loss of fishing productivity, including important access to infrastructure such as docks and moorings, can cripple a commercial industry that is already squeezed for space by competition from recreational vessels. The inability to sustain productivity includes loss of boats and gear and can also wipe out what was a thriving fishery.

Table 5-6.   Summary of Total Capital Impacts from Disaster.

Productivity Impacts	Economic Impacts	Lifestyle Impacts
Dropping boat insurance	Loss of customer base	Loss of way of life
Deterioration of equipment, unable to pay for boat maintenance	Public fear of contaminated lobsters	Loss of contact with fishing networks and friends
Loss of experienced boat crew	No steady work	Loss of contact with marine environment
Loss of dock area	Cannot pay for daily costs to go lobstering	Loss of identity as fishermen
Sale of equipment	Out of business income, unable to pay business bills	Gained new respect for environment
Boats for sale	Business halted, no lobsters	Fighting with wife and family
Unable to pay winter storage- Boat kept on moorings	Loans needed to sustain business	Not being able to provide for family by fishing
Loss in value of bait, traps and boats	Used up savings from previous years' work to sustain business	Loss of family vacations

Loss of the fundamental economic basis of the fishery has had devastating effects on fishing households and other support industries. The quality of life is thus directly impacted by the loss of personal total capital flexibility.   For example, advantages in childcare, social networking and family life are lost.   Many fisher households urgently suffered from lack of resources to pay bills, forcing heads of households to look for new jobs outside of fishing and forcing other family members to work. Unfortunately, most of the new jobs paid far less than lobstering, which led to the inability to pay the bills associated with the fishing business. Wives who until then were able to stay at home to take care of the household and children, had to find work outside the home.

We now turn to another case in the same LME, that of the New England groundfish fishery, which suffered a similar collapse.

THE GROUNDFISH FISHERY COLLAPSE OF NEW ENGLAND

A Brief Historical Overview of the Fishery

Groundfishing in New England dates back to the establishment of the Massachusetts Bay Colony in 1623.   Prior to that time, vessels would come in the summer months and fish off Cape Ann, the Grand Banks, Gulf of Maine, and Cape Cod and then return to England. Poor soil in many parts of New England encouraged settlers to focus on a mixed fishing-farming strategy, and ports such as Gloucester, Portland and New Bedford soon specialized in fishing. Through the centuries waves of immigrants came to take advantage of the wealth of fish stocks of New England. Canadians, Scots, Yankees and Portuguese were early entrants into the fishery.

In the early 20th century, many fishers came from Sicily and from several southern Italian fishing ports. For example, the Sicilians came to Gloucester in two waves. The first, in the 1920s, consisted of 343 Italian fishermen, who mostly came without their families. When the fishing grounds played out in the Mediterranean, fishers left for New England and the groundfish fishery, to stalk the abundant populations of cod, haddock, and halibut, as well as secondary fisheries of redfish, herring, and mackerel. After World War II, a second wave of Sicilian immigrants arrived, mainly from Terrasini, Italy. They swelled the ranks of the fishery at a time when the fishing economy was growing. Unlike the relatively young and highly specialized (single species) lobster fishery of Long Island Sound, the New England fishery is one of old rich heritage and a diversity of fishing techniques and ethnic persuasions. Migration to the ports of New England was based on extensive social capital networks, where close friends and family would establish an economic and occupational bridgehead for others to follow (Dyer 2002a). This is often referred to as "chain migration."

Today, long-established fishing kin networks use gillnets, longlines, and dragging gear to fish for the diversity of species making up the groundfish stocks. A typical groundfishing boat includes a captain, first mate, cook, engineer, and crew. The key to a successful boat is the ability of the crew to get along and work well together. Offshore trips can last a week to ten days, whereas inshore boats with crews of three will go out for a day or two. Vessels can be stern or side trawlers (Eastern Rigs), gillnetters, or long-liners. Long-liners fishers will bait a series of lines, with each line containing thousands of hooks. Dragging gear are purse-shaped shaped nets that are pulled (dragged) by the boat just above the bottom. Traditional dragger fleets are composed of three size categories of vessels. Vessels over seventy feet in length with crews of eight to ten fish offshore from seven to ten days at a time. Mid-sized vessels ranging from fifty to sixty-nine feet are manned by a crew of two or three and fish three to five days with dragging gear. The smallest vessels are forty-five feet or less. These "day boats" have crew of one to two, and usually fish using gill nets, long lines or otter trawls (small stern purse seines).

The Natural Resource Disaster of the Groundfish Stock Collapse

Until 1976, the domestic fishing communities of the US competed with foreign fishing vessels in all waters beyond the ten-mile limit. In order to stop the perceived loss of fish stocks to foreign fleets, and to boost the domestic fishing economy, legislative action was taken. The 1976 Magnuson Fisheries and Conservation Act, [reauthorized (1996) as the Magnuson-Stevens Act] was instituted to protect the marine resources of the United States through the establishment of a 200-mile Exclusive Economic Zone (EEZ). To revitalize the domestic fleet and develop the now protected marine resources of the EEZ, the National Marine Fisheries Service provided low-interest loans to build up the domestic fleet. Unfortunately, this loan program was not designed with an understanding of the total capital dynamics of coastal fishing communities. There was no assessment of the historical needs and human-ecological interactions of these communities – their total capital flows (cf. Hennessey and Healy 1998 and 2000 for an additional analysis of the groundfish collapse).

External investors recognized the opportunity afforded by the loan program, and quickly increased the fleet size particularly in the thriving fisheries of the New England region. Just about every major east coast port including Gloucester, Boston, and New Bedford in

Massachusetts; Portland and Rockland in Maine; and Newport and Point Judith in Rhode Island, built up their fleets and increased the number of those invested and participating in the fishery with powerful stern trawlers, under this Fishing Vessel Obligation Guarantee Program. Besides the buildup of the large dragger fleet, many small and medium-sized vessels were built, increasing fishing pressure on inshore, nearshore, and offshore stocks. Some of these smaller boats even ventured offshore, facing hazardous conditions which they were not designed for, to reach the rich fishing areas of Georges Bank, offshore fishing grounds of the Gulf of Maine and other prime spots such as Cashes Ledge, Franklin Sewell, Three Dory Ridge and Plats Bank. Key respondents in affected fishing communities claimed that many of those taking advantage of the fishing vessel loan program were newcomers to the fishery (Dyer and Griffith 1996). Many of these outsiders had no prior connections to the social, cultural or human capitals invested in the fishing communities and were thus not bound by the responsibilities and reciprocal exchanges of total capital that marked traditional fishing families, households, and social capital networks of embedded groups such as the Gloucester Fishermen's Wives Association, the Offshore Mariners Association of New Bedford, Massachusetts, or the Point Judith Fishermen's Cooperative in Rhode Island.

Build-up of the economic capital of the fishery, as well as the accompanying social, cultural and human capital changes, resulted in intense fishing pressure on stocks, from inshore to offshore. As competition for groundfish resource increased with new fishing crews coming in —new boats, new technologies, and expanded markets to carry the volume of increased catches—acrimony increased both within and between fishing-dependent communities, and between fleets of different gear types.

Increased fishing capacity and pressure from newcomers to realize a "return on investment" resulted in overfishing of stocks. The traditional fishing families found themselves caught up in a spiral of increasing effort, declining catch, and increasing costs. Such a negative spiral might have been avoided if management had considered what was already in place and worked for slow growth controlled by established community stakeholders. Rather than infusing outside capital to expand the fishery, the goal could have been a sustainable fishery in balance with existing economies of scale and their sustaining total capital flows.

In 1995-1997, the NMFS established that the primary groundfish stocks in the management area of the New England Fishery Management Council were seriously depleted. A 1998 National Research Council review of the ill-fated policies of expansion without community input concluded that recruitment of harvesters and increased investments in the industry occurred, *"evidently in excess of what the fishery could sustain,"* (NRC 1998).

The policy response to this crisis was the institution of two amendments—numbers five (5) and seven (7)—which drastically cut back on the number of allowable days at sea (DAS) for the groundfishing vessels, from an existing 260 DAS to 88 DAS. It also eliminated significant exceptions from current effort control regulations and broadened area closures to protect juvenile and spawning fish. This was undertaken with little understanding of the community costs and consequences of these measures.

Losses in human capital in the harvesting and processing sectors were dramatic. In 1980, there were 3500 finfish harvesters and 5700 workers in the processing sector in Massachusetts alone. By 1992, finfish harvesters had decreased to 1500 and processing workers to 2700 (a

58% and 53% decline, respectively, in 12 years). Related losses of social capital were apparent in declines in attendance of, and participation in, local fishermen organizations. This included out-migration as some of the monolingual older fishers with limited education returned, unsuccessful, to Sicily and Portugal as their jobs were lost from crew reductions and boat buybacks.

Losses in social networks led to fierce conflict between draggers, gillnetters and longliners, many of whom were previously friends both on and off the water. Further increasing the social stress were financial declines in catch profits and subsequent withdrawal of loan support and health insurance coverage among fishing households.

Social and *cultural capital* impacts at the household level included increased domestic strife, avoidance behavior, alcoholism, suicide, and also avoidance of community ritual celebrations.

> We used to go out to the club and go to church, but I do not do that anymore. What is the point? There is nothing good to talk about. We just go from the boat to the house. Sometimes we go to church, but it's now only on Easter or other holidays.

Such losses in cultural capital, expressed through non-participation in specific cultural groups and declining celebration of culturally associated rituals (*e.g.* the St. Peters festival- patron saint of fishermen) were also apparent. Indicators of decline in social and cultural capital can be seen in declining participation in other social activities or groups. In Gloucester, Massachusetts, participation in the local fishing association of *Societa Siciliana* decreased from 304 in 1991 to 89 in 1995 (a 70% decline), and *Sons of Italy* from 200 in 1991 to 79 in 1995 (a 60% decline). By comparison, in the same period membership in culturally "neutral" (non-fishing) associations such as the *Gloucester Elks,* which consisted mostly of newly-arrived Boston suburbanites, increased from 76 in 1991 to 185 by 1995, a 143% increase. A disturbing indicator of *cultural capital* loss was expressed by the actions of a newly appointed religious leader in Gloucester who, during this time of community crisis, disparaged the celebration of local Sicilian saint days as contrary to the wishes of the Catholic Church. Subsequent declines in parish attendance were in part attributed to such attacks of ignorance against local forms of cultural capital.

Ports throughout New England exhibited a gentrification trend for this important fishing festival. More and more boats going through the annual "blessing of the fleet" ritual were no longer commercial fishing vessels but recreational craft. In fact, a marked gentrification of the entire region, in ports large and small, occurred simultaneously as the fortunes of the commercial fishing sectors declined in each community (Hall-Arber *et al.* 2002). This made it even more difficult to maintain the infrastructure and dock space necessary to support commercial fishing enclaves. Valuable fishing infrastructure (ice houses, docking space, and marine railways) was lost from many sites as gentrification took over. One positive adaptive consequence of this was an increase of capital flows among communities as they came to rely more on one another's infrastructure to keep the commercial fishing industry alive.

Initial government reaction to the natural resource disaster included a $25 million dollar buy-back program of groundfishing vessels and retraining programs for fishermen. Bought out vessels were either scrapped or sunk at sea, or transferred to non-fishery uses. Ironically, the buyback program represented an attempt to decrease the over capacity in economic capital (fishing boats) originally created by the vessel loan program.

In summary, the natural resource disaster of the groundfishery collapse had complex and far reaching impacts on the total capital of the Natural Resource Region. A cascade of multiplier effects from the impact on fish stocks (biophysical capital) of the original (1976) loan programs, and subsequent severe reductions in allowable fishing effort, included dissolved credit relationships, created job and infrastructure losses, decreased sharing and cooperation, on shore and at sea, loss of cultural capital and increased social problems ranging from domestic violence and suicide to declining community participation and plummeting job satisfaction.

A TOTAL CAPITAL RESPONSE APPROACH TO NATURAL RESOURCE DISASTER MITIGATION

Basic Components

The Long Island Sound lobster die-off and New England groundfish fishery collapse were unanticipated natural resource disasters that had complex and far-reaching impacts on the total capital of their associated human ecosystems. Consequently they provide an opportunity to explore how a total capital approach could mitigate losses in the post-disaster environment:

> When hazards and disasters occur, they both reveal and become an expression of the complex interactions of physical, biological and socio-cultural systems. Hazards and disasters not only manifest the interconnections of these three factors but also expose their operations in the material and cultural worlds (Oliver-Smith and Hoffman 2002:5).

Resource managers and stakeholder communities must respond to the interactions revealed by natural resource disasters in a manner that is both rapid and comprehensive in its scope. Natural Resource Communities, under normative conditions of resource ebb and flow, can flexibly adapt to these changes by strategies such as shifting catch effort to different species in the LME, modifying gear efficiency to decrease catch of undesirable sizes, or limiting effort. These shifts are all part of the natural cycles of NRCs embedded in NRRs and their associated LMEs, and can be present in folk, co-managed, or state managed systems (Dyer and McGoodwin 1994). The problem arises when the total capital balance is disrupted by some external force, or poor management decisions. Sometimes, the balance is so disrupted that recovery is not possible, or can only occur at great costs to (or even loss of) the participating communities and stakeholder groups. In the worst of cases, natural resource disasters induced by pollution events, such as the suspected effect of insecticide spraying in the Long Island Sound ecosystem, or the known effect of the Valdez oil spill in Prince William Sound, Alaska, can permanently destroy the NRR-LME link.

In what follows, we review the five forms of capital, impacts from NRDs, and potential responses to mitigate capital losses. The total capital response model for natural resource disasters is represented in Figure 5-1.

Social Capital Component of Response Model

Social capital is defined as the interactive network of human social relationships among natural resource communities in an NRR and is essential to the maintenance of occupational viability in relation to LME resource extraction (fishing). Social capital is key to the flow of other forms of capital, as well as central to the dynamics of governance and resource utilization. Social support services deal with the ebb and flow of social capital. Social capital flows reflect the activities of families, the exchange of ideas and communication among acquaintances and colleagues, and insure the continued aspects of social life in a community.

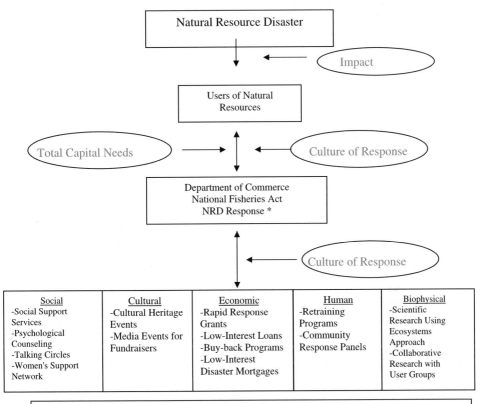

Figure 5-1. Total Capital Response Model for a Natural Resource Disaster

The breakdown of social capital has devastating effects on families. As was illustrated in these case studies, outcomes can include economic disruptions, isolation, depression, loss of social relationships, increased stress within the family unit, non-communication with family members, alcohol and drug abuse, and domestic violence. Loss of ties with other fisher families means a corresponding loss of community networking that usually provides needed social and psychological support in times of stress (Dyer and Griffith 1996; Dyer 1993). In the work environment, loss of social relationships can mean loss of economic ties, breakdown of work contracts and crew relationships, loss of marketer-producer ties, including credit relationships, and the decline in communication between fishers, marketers, and others in the network about prices, fishery conditions and other fishery-related topics. Participation by fishers in organizations such as the Western Long Island Sound Lobstermen's Association (now moribund) and other groupings has also declined, weakening the effectiveness of such organizations and decreasing the overall communication between fishers and managers.

Assessment of the social capital outcomes of disaster response is crucial to cutting recovery costs, improving efficiency, and effectively mitigating impacts to disaster victims. Community, state, and federal agencies find themselves more frequently pulled together in responding to these disasters. Yet, because there is no formal mechanism to assess outcomes, it is difficult to integrate lessons learned from one case to the next, and to coordinate resources between local and external support agencies to better prepare for future events.

For the New York and New England fishing communities, we recommend a community FISHNET Program that links fishing families with social service agencies and managers in times of economic crisis. FISHNET would establish a listing of fishing households, collecting baseline and longitudinal social, economic, and family data that gives social service providers and managers a snapshot over time of the social-economic status of each household and its potential vulnerability to disasters.

Cultural Capital Component of Response Model

Cultural capital includes the values, knowledge, and culturally-transmitted behavior and ideas of a population, applied to the transformation and utilization of natural resources. The cultural capital of lobster fishing defines the role between lobster fishers and their environment, and gives identity to those who are active participants in the occupation of lobster fishing. In the face of a natural resource disaster, the fabric of the fishing community tends to unravel, and efforts to re-weave that fabric may improve understanding, but cannot restore the dignity and self-worth lost when fishermen can no longer fish for a living. Celebrations of cultural heritage can bring the community closer together, for example, and might help forge new alliances with individuals or groups interested in conservation or historical preservation. People band together when something well-defined can be accomplished; solutions for fishing communities are seldom clear.

Economic Capital Component of Response Model

The term economic capital means the exchange of goods and services and the production activities linked to this exchange. Understanding economic capital flows is critical to the restructuring of communities after a disaster. The economic capital maintained and created by these fishers is now lost from the system, and they are in a desperate condition of trying to recreate other forms of economic capital. Strategies include:

- Seeking alternative jobs on land (non-marine related)
- Seeking alternative marine-related jobs
- Selling out their production capital (boats, gear)
- Moving to other regions to continue their fishing lifestyle
- Attempting to hang on using a variety of economic activities in hopes that there will be a natural recovery of stocks, or alternate means of recovery (e.g. restocking) to hasten recovery.

Unfortunately, the loss of so many from the fishery has not resulted in the recovery of economic capital for the remaining fishing households, and there is no evidence that the fishery aid program, meant to cover losses from the 1999 lobster die-off, has had any substantial impact on the economic well being of those affected by the disaster. The funds were well-intended but were inadequate to cover the ongoing economic losses experienced by households with the non-recovery of the stocks. With no recovery in sight for the lobster fishery in this area, the aid program could not be considered a viable stopgap measure allowing the fishers to get back on their feet. Without the recovery of the biophysical capital that fuels the system, there can be no successful rebound of the families having so intensively and singularly invested their economic capital in lobster fishing.

Because of the devastation and non-recovery of the present fishery, promoting economic capital recovery must entail a broad future-looking approach that seeks to prepare the region for future impacts. Economic capital flows are most sustainable when their integrative effects on the ecosystem as a whole have been considered during the development process. Short-term solutions to promote economic capital flows that further compromise the health and viability of the human-marine ecosystems should be discarded over options that promote sustainability and low environmental impact.

Rapid economic capital assistance such as loans and grants that cover operating and capital losses should be available within the critical first couple of months after the disaster. It should become a part of the institutional knowledge that a strong economic capital response is necessary in order to prevent the severe collapse seen in this case.

Economic diversification is key to strengthening the overall fishing culture of the western Long Island Sound, where fishers can link up to alternatives rather than being highly committed to the one resource. This can best be accomplished if there are fewer traps and other forms of economic capital committed to the endeavor, with a diversity of other activities included in household strategies to adapt to life near the water. Some alternatives are covered in the next section on human capital responses through retraining.

To promote an ecosystem response model requires regional responsibility to respond to disaster. Agencies such as FEMA are working to create a culture of '*disaster resistance*' by divesting responsibility to states, counties and communities for disaster response planning and mitigation [see the MOU between FEMA and NOAA of 2000, for example at www.nws.noaa.gov/om/wcm/share.NOAA-FEMA-MOU-2000]. Divesting responsibility to locals for managing local disasters requires the development of infrastructure and policy that allows a response flexible enough to suit the uniqueness of local environments, cultures, and economies.

An example of a regional response model is FEMA's *Project Impact* [http://www.app1.fema.gov/impact/]. *Project Impact* encourages community action by providing federal funding to selected communities to focus their energy on ways to make themselves more disaster resistant. This includes outreach efforts to the diversity of businesses, non-profits, volunteer and state organizations that collaborate in the event of a natural disaster. For natural resource disasters, increasing disaster resistance that mitigates the loss of economic capital can include the following aspects. It is important to note that several of these address sociological and cultural attributes as well:

- Characteristics of natural resource base (biophysical capital);
- Linguistic and cultural background of fishing populations;
- Number and kinds of voluntary associations associated with fishing;
- Transportation infrastructure and access;
- Degree of educational attainment of resource users;
- Complexity/diversity of local economies/income patterns (IMPLAN modeling);
- Number and diversity of church organizations;
- Nature and complexity of familial networks;
- Degree of cultural isolation (community enclaves or community networks);
- Number and diversity of institutions of higher learning;
- Frequency of use of external assistance programs and resources.

Human Capital Component of Response Model

Human capital in a fishery refers to the human population and the knowledge and skills acquired by that population from formal and informal education associated with the occupational roles of natural resource extraction. Thus human capital—the lobster fishing families and their unique histories and talents—is being lost from the system as households are forced into other means of livelihood or even forced to leave the area or state to seek opportunity elsewhere. Retaining and maintaining human capital can be accomplished through two strategies: (1) retraining programs focused on keeping individuals working in marine related professions, and (2) community response panels to expand the participation and integration of the commercial fishing sector with the wider community and lay a foundation for local human ecosystem management (folk management) of vital coastal and marine resources.

Within many American communities, the ability of people to work together towards common goals through the public policy process has been weakened by a lack of participation in governance and by distracting social problems that threaten the health, education and safety of children and families, particularly those of 'minority' occupational status such as commercial fishers. To develop a regional based initiative to develop and diversify human capital, we suggest the creation of community response panels.

A community response panel consists of a cross-section of community participants selected by community organizations and neighborhood leaders whose function is to communicate concerns and assess issues of importance to both leaders and the public. The panel is selected to represent the social, economic, and occupational diversity within a target population, with an emphasis on those who are linked to the marine environment , such as the tourist, marina, restaurant, commercial fishing, and recreational fishing businesses. Panels are drawn from community workshops designed to educate and engage the variety of stakeholders interested in the management of the regional environment. It should be noted that such a multisectoral panel is completely consistent with the LME approach to large marine ecosystem assessment and management, a supra-national approach currently used globally whereby, with up-to-date information from 5 modules—productivity, fish and fisheries, pollution and ecosystem health, socioeconomics and governance—all representative participants (government and non-government) and stakeholders meet over time to identify and prioritize threats to the (shared) large marine ecosystem and to implement agreed-upon actions to solve those problems within specific time frames.

Community panels are conceived to consist of eight individuals, with five alternates. The panel is selected to represent the range of diversity within a target population by social, economic, and ethnic status. Panels will be convened twice a month to review and develop the economic plans coming out of the workshops as related to the ecosystem approach of total capital. Panel members will also be asked to address issues of concern to themselves and to other members of neighborhoods and their informal social networks. Significant project benefits to be realized from workshops and the subsequent establishment of response panels are:

- engage the commercial fishery sector and other under-represented groups in public discourse on economic development;
- increase development capacity through culturally appropriate training of grassroots stakeholders to leverage additional funding;
- create sustained communication links among state development agencies, public leaders and the target community that will improve understanding of the critical economic development needs and concerns of these constituents;
- foster better understanding of the economic development process in the target community;
- promote sustained collaboration through participating agencies and the public.

A significant outcome would be the creation of planning models that surpass traditional efforts by including all aspects of capital (social, human, cultural, economic, biophysical) in the planning process, and where the focus is on development of economic capital diversity and inclusion rather than maximization of economic capital benefits for selected sectors. Note that these community goals and results are also entirely in harmony with the LME concepts and strategies as they are realized in multi-national, multi-sectoral agreements and specific actions to preserve and restore the marine ecosystems and their watersheds, so vital to the economic interests of the countries involved.

The creation of economic diversity is a key to increasing the adaptive flexibility of any human ecosystem. Strengthening the human capital component of the fishery can occur through creating new opportunities to develop diversity in work skills, both for those who choose to remain in the marine context and those who cannot. One strategy to increase diversity in the face of recent declines of fishery stocks and restrictive government fishing regulations has been retraining programs for commercial fishers.

The Corporation for Business, Work and Learning of the State of Massachusetts has been running a very successful retraining program in three centers. Programs like the Gloucester Fishermen and Family Assistance Center (GF&FAC) have trained hundreds of fishers and their wives and many are now augmenting their fishing income using their new job skills, or have stepped out of the fishery altogether to pursue alternate occupations.

Key to the success of the Massachusetts fishermen centers is the encouraged collaboration between the centers and the community. The mission statement of the center describes the community link: to "provide a gateway to the fishing community to transition from a fishing-dependent life to other marketable occupations by working in partnership between the individual and the community resources to provide innovative, quality Employment and Training services."

The most popular jobs, those noted to have the highest satisfaction rating, were those in which fishers could use their skills to put themselves back out on or at least near the water. Jobs such as captain of a private yacht, able seaman, tugboat operator, charter boat captain, aquaculture and marine construction were said by the coordinators of the GF&FAC to be preferred positions. Particularly prized were charter boat captain and tug boat operator. When classes were given for these positions, hundreds of applicants attended them.

This has not occurred without some considerable grief on the part of participants, many of whom represent trans-generational families rooted in an old world fishing culture. This program has been successful in buffering the impact of downturns in the fishery. However, recent regulations have even more severely depressed the fishery to the point where protests are occurring because reductions in fishing grounds and days at sea (DAS) quotas are making it extremely difficult to continue fishing.

The greatest job satisfaction in this program remains with those activities, such as tugboat or charter boat captain, that keep former fishers near or on the water (David Bergeron, personal communication). Within a particular region, however, the number of these jobs would likely be fewer than the number of applicants. Additionally, many women, whose families were once supported by a single income from fisheries, have taken advantage of the retraining program as a way to get themselves into the work force to provide needed and, in some cases, the sole family income. These changes from traditional occupations in fishing communities herald cultural changes as well.

Biophysical Capital Component of Response Model

Improving the state of the biophysical capital is best done in a collaborative effort, with the recognition between users and managers that biophysical capital affects all other forms of capital, and that the fishery should be managed in an ecosystem context that fully integrates the participants in the management process, as they are already integrated as part of a human ecosystem which makes up the fishery. Ideally, "…fisheries management should proceed as a collaborative effort between professional managers and fishers. Fisheries managers who have an intimate understanding of the people they are charged with managing, who know what management strategies these people are likely to invent, and what strategies they typically prefer, will have a far better chance of getting cooperation than will managers who keep their distance and remain focused mainly on conservation. Likewise, fishers who are made aware of the culture of fisheries managers, who are given examples of successfully managed fisheries and who are encouraged to actively engage managers in a cooperative dialogue can empower themselves to achieve greater control and sustainability of the natural resources they "exploit " (Dyer and McGoodwin 1994:10).

In Natural Resource Communities, such as that of the lobster fishers, the dependence on biophysical capital creates a cultural pattern that locks users into a particular lifestyle. It is important that we understand this connection, for it is at the heart of the non-fungibility of those practicing fishing. Giving more responsibility to those who actually use the resources, including the monitoring of stocks, is one solution. Folk management of local resources has been demonstrated to lead to sustainable systems that minimize long-term negative impacts to human users, while retaining long-term viability of biophysical capital (Dyer and McGoodwin 1994).

The New England Fisheries Management Council and New York and Connecticut have already engaged fishers in collaborative research programs to track stock recovery and test gear modifications and other technological innovations designed to improve catch efficiency and reduce bycatch. This laudable program gives fishers a chance to be directly involved in the collection of scientific data, while building rapport with fishery scientists responsible for the monitoring of stocks.

Expansion and continuation of such programs, ecosystem wide, and linking other forms of research whenever possible with the commercial sector, could be of help. It could also help rebuild lost confidence among the fishing community in the management sector, and lead to building a foundation for a human ecosystem approach to the management of Long Island Sound and New England coast fisheries.

A RETROSPECTIVE AND CONCLUSIONS

In the Long Island Sound lobster fishery and New England groundfish fishery, we have seen two examples of what can happen when a natural resource disaster strikes LMEs, and the outcomes to the human actors (fishers and their families and communities) in the NRRs interfaced with these LMEs. Our contention is that LME management using a total capital approach can help prevent or mitigate the negative consequences of such disasters, and can even help prevent their occurrence. The natural resource disaster that wiped out the lobster population in Long Island Sound was an unexpected event that impacted a highly vulnerable and specialized population having little adaptive flexibility. The natural resource disaster of the New England groundfish fishery affected a more diverse and widespread population with more opportunities to adapt and change, but the consequences to the fishing sector have been profound. In both cases, there are opportunities to learn from the mistakes made, and to improve the ways in which such systems are managed in the future.

LMEs around the globe experience periodic fluctuations and perturbations in the dynamics of energy flows, fish population numbers, and other factors, as influenced by human actions such as fishing and coastal pollution events but also by the impacts to systems of wider environmental changes such as global warming. When LME fluctuations affect human populations reliant on the marine resources extracted from the system, there is a variety of possible responses. Prior to worldwide enactment of legislated management by state entities, human populations developed their own local solutions to fluctuations, including limits on effort through the folk management of user behaviors (Dyer and McGoodwin 1994).

As fishery management developed its own scientific momentum, mostly based on models derived from North Atlantic fisheries, management infrastructure and "scientific" legislated strategies spread across the globe. Unfortunately, the expertise guiding the development of these strategies emphasized understanding only one form of total capital-- the biophysical. Thus, any infusion of human, social, or cultural capital components was lost in the development of policies narrowly focused on managing fishery stock levels. The new legislated biophysical capital-centered fisheries models replaced folk-managed strategies, such as the system of limiting effort in the New Bedford, Massachusetts scallop fishery by voluntary "tie up days" (days when no boats fished). As bureaucratic momentum took over with the legislation of catch effort under the Magnuson Act in the U.S., consequences to and inputs from fishing communities were brushed aside as unscientific and irrelevant. The outcomes have been natural resource disasters in many of our domestic fisheries and equally pernicious declines in LME resources worldwide.

The Large Marine Ecosystem (LME) approach is multidimensional and can improve our response to natural resource disasters. The approach requires a recognition of the inter-connectivity of the various forms of capital. The LME approach, for example, involves socio-economic and governance modules that include adaptive management. Shifts, increases or decreases of any one capital form can have impacts on other forms within a human ecosystem. A decrease in one form , such as reduction in crew size through the introduction of new, more efficient fishing technologies, may not be disastrous. However, anything that forces the base biophysical capital to collapse has the potential to disrupt all the other capital forms, including the established patterns of capital flows that hold the system together– creating a natural resource disaster.

A total capital approach should be seen as an important component informing the socioeconomic and governance modules. This change must begin with legislated mandates that break from past systems and truly give voice to community stakeholders, for systems locked into a bureaucracy that puts biophysical capital first do not have the flexibility to integrate aspects of total capital in the decision-making process.

For systems that are small-scale, artisanal fisheries, such as the fishery of the Kerala coast of India, the best strategy would be to take steps to prevent technological externalities – industrial-scale fishing – to disrupt and overwhelm already sustainable total capital systems. For systems such as the New England groundfish fishery that are losing ground to externalities such as gentrification, what infrastructure remains should be protected by legislation, particularly as remaining infrastructure components (ice houses, docking space, fish auction facilities) become unique pieces of a network of inter-community total capital flows.

The total capital approach must be *a priori* in any management scheme—as a starting point for analysis of specific proposals to manage fishery issues, and as the best way to avoid Natural Resource Disaster scenarios in LMEs. For example, if Individual Transferable Quotas (ITQs) were seen as a solution to a particular fishery issue, then an assessment could be made of potential total capital impacts and consequences of initiating ITQs to an existing total capital system. Finally, the total capital configuration of each and every fishery under a new system of management should be scrutinized and, for each, an integrated approach devised that addresses all forms of capital, and analyzes their potential interactions in the face of new policies, issues, and regulations.

REFERENCES

Coleman, James. 1988. Social capital in the creation of human capital. *American Journal of Sociology* 94: 95-121.

Dyer, Christopher L., Duane A. Gill and John S. Picou. 1992. Social disruption and the *Valdez* oil spill: Alaskan Natives in a natural resource community. *Sociological Spectrum* 12(2):105-126.

Dyer, Christopher L. 1993. Tradition loss as secondary disaster: Long-term cultural impacts of the *Exxon Valdez* Oil Spill. *Sociological Spectrum* 13: 65-88.

Dyer, Christopher L. and James R. McGoodwin (eds.). 1994. *Folk Management in the World's Fisheries: Lessons for Modern Fisheries Management.* University Press of Colorado, Niwot, Colorado.

Dyer, Christopher L. 2002a. The Sicilian fishing families of Gloucester, Massachusetts. 233-257. In: Greaves, Tom (editor). *Endangered Peoples of North America: Struggles to Survive and Thrive.* The Greenwood Press, Endangered Peoples of the World Series. Greenwood Press, Westport, Connecticut.

Dyer, Christopher L. 2002b. Punctuated entropy as culture-induced change: The case of the *Exxon Valdez* oil spill. 159-186. In: Hoffman, S. and A. Oliver-Smith (eds). *Catastrophe and Culture: TheAnthropology of Disasters.* School of American Research Advanced Seminar Series. School of American Research Press. Santa Fe, NM.

Dyer, Christopher L., John J. Poggie and David Allee. 2002. *Assessment of the impact on lobster fishers and their families from the 1999-2000 lobster die-off in western Long Island Sound.* Human Ecology Associates and Cornell University Cooperative Extension. Economic Development Administration., U.S. Department of Commerce. Washington, D.C.

Dyer, Christopher L. and John J. Poggie. 2000. The natural resource region and marine policy: A case study from the New England groundfish fishery. *Marine Policy* 24:245-255.

Dyer, Christopher L. and David Griffith. 1996. *An Appraisal of the Social and Cultural Aspects of the Multispecies Groundfish Fishery in New England and the mid-Atlantic Region.* Aguirre International, Bethesda, MD.

Dyer, Christopher L., John J. Poggie and Madeleine Hall-Arber. 1998. *A social impact assessment of the New England Atlantic herring fishery.* New England Fisheries Management Council. Newburyport, MA.

Hall-Arber, Madeleine, Christopher L. Dyer and John J. Poggie. 2002. *New England's fishing communities.* MIT Sea Grant. Boston, MA.

Hennessey, T. and M. Healy. 2000. Ludwig's ratchet and the collapse of New England groundfish stocks. *Coastal Management* 28:187-213.

Hennessey, T. and M. Healy. 1998. The paradox of fairness: The impact of escalating complexity on fishery management. *Marine Policy* 22:109-118.

National Research Council. 1998. *Review of Northeast Fishery Stock Assessments.* National Academy Press. Washington, D.C.

New England Fishery Management Council. 2002. *Heading toward recovery: Rebuilding New England's fisheries.* Newburyport, MA.

Oliver-Smith, Anthony and Susanna Hoffman. 2002. Introduction: Why Anthropologists Should Study Disasters. 3-22 In: S. M. Hoffman and A. Oliver-Smith, eds., *Catastrophe and Culture: The Anthropology of Disaster.* School of American Research Advanced Seminar Series. School of American Research Press, Santa Fe, New Mexico.

Symes, David. (ed). 2002. *Fisheries Dependent Regions.* Fishing News Books. Blackwell Science Ltd. Oxford. Blackwell Science. Malden, MA.

Sherman, Kenneth, Ezekiel N. Okemwa and Micheni J. Ntiba. 1998. *Large Marine Ecosystems in the Indian Ocean: Assessment, Sustainability and Management.* Blackwell Science Ltd. Oxford. Blackwell Science. Malden, MA.

Sherman, Kenneth, Lewis M. Alexander and Barry D. Gold (eds). 1993. *Large Marine Ecosystems: Stress, Mitigation, and Sustainability.* American Association for the Advancement of Science Press. Washington, D.C.

Sherman, K., L.M. Alexander and B.D. Gold (eds). 1992. *Food Chains, Yields, Models, and the Management of Large Marine Ecosystems.* Westview Press. Boulder, CO.

Large Marine Ecosystems, Vol. 13
T.M. Hennessey and J.G. Sutinen (Editors)

6

Ownership of Multi-Attribute Fishery Resources in Large Marine Ecosystems[1]

Steven F. Edwards

ABSTRACT

LMEs encompass many living and non-renewable resources, each having multiple physical or relational attributes. For example, fish stocks can be differentiated by species, age structure, growth rates, distribution, gene pool, diet, and habitat. Each resource attribute is potentially valuable, but attributes are also costly to manage. Property rights arrangements that bundle fishery resource attributes are recommended. Bundled property rights could evolve from a comprehensive assignment of usufruct rights in fisheries, such as IFQs (individual fishing quotas) or harvest cooperatives, which reduce the transaction costs of gathering information on unspecified stock attributes and of internalizing interactions due to ecology (*e.g.*, predation), unspecialized fishing technologies *(e.g.*, bycatch in mixed-species fisheries), and impacts of fishing gear on habitat and non-commercial species.

INTRODUCTION

Large Marine Ecosystems (LMEs) supply humans with natural resource commodities, services, and amenities (Sherman and Alexander 1986), such as seafood, waste disposal, and recreation, that have proven difficult for government agencies to manage. The world's fishery resources, in particular, are harvested unsustainably (Pauly *et al.* 1998) and uneconomically (Wilen 2003) by fleets with considerable excess capacity. Fishing gear also impacts non-target species (Gislason 2001; NMFS 1998) and the sea floor (Kaiser *et al.* 2001). Sewage effluent, industrial pollution, oil spills, and coastal land-use can be detrimental to fisheries as well as the environment (Rosenberg 2001; Sherman and Duda 1999).

This mix of environmental problems has renewed interest in ecosystem approaches to fisheries management (EPAP 1999; NRC 1999; Sherman and Duda 1999). The United Nations' "Code of Conduct for Responsible Fisheries" urges governments to manage the stock and environmental attributes of fishery resources, including age structure, genetic diversity, recruitment variability, predation, and the habitat elements of carrying capacity (see http://www.fao.org). Similarly, the 1996 amendments to the Magnuson-Stevens Fishery Conservation and Management Act (Magnuson-Stevens Act) require fishery managers in the United States to prevent biological overfishing in the EEZ (*i.e.*, extended economic zone), to

---

[1] Most of this work was previously published in the Elsevier journal, *Ecological Economics* (Edwards 2003).

minimize the incidental harvest of undersized, non-target, and protected species (*i.e.*, bycatch), and to conserve and enhance fish habitat. This more detailed attention to the complex attributes of marine fisheries and ecosystems is not accompanied, however, with advice on how to meet the new demands and promote economic growth at the same time. Working within an out-of-date regulatory environment which arose to manage stock biomass alone, managers are artificially dividing fishery resource attributes with single-species bycatch limits and habitat area closures as expedient ways to satisfy laws.

If societies hope to gain economically from fishery and other marine resources, new institutional arrangements are needed which reflect the complexity, scale, and uncertainty of ecosystems (Hanna 1998). Elsewhere in this volume, Larry Juda and Tim Hennessey discuss decision-making infrastructures and responsibilities for resource management in LMEs. In contrast, my interest is with the property rights foundation of economic production. In particular, is it better (from an economics standpoint) to continue to practice divided ownership of marine resource rights or to accommodate interrelationships among resource attributes with attribute-bundling?

By way of contrast, this focus on the physical and relational attributes of fishery resources differs markedly from the recent investigation in the fisheries management literature of the quality characteristics of a property right. Scott (1988) explained that fishing rights, such as a license or an IFQ, can be differentiated according to the content of six characteristics: duration, flexibility, exclusivity, quality of title, transferability, and divisibility. Divisibility is no longer considered crucial in this literature (*e.g.*, Arnason 2000), and is generally thought of as a scaling factor, such as the divisibility of harvest quotas or fishing territories. This perspective contrasts with a major theme of the mainstream property rights literature which investigates the conditions that favor either the division or combination of property rights to resources with multiple attributes and uses. This line of thought can be traced back at least to the seminal work of Coase (1960) who carefully analyzed, for example, the competing interests of a railroad that emitted sparks which set fire to adjacent farmland. This type of situation is ubiquitous in fisheries, including the incidence of bycatch and habitat impacts. The chapter continues with some necessary background information on the property rights theory of multi-attribute resources.

PROPERTY RIGHTS THEORY AND MULTI-ATTRIBUTE RESOURCES

To appreciate why property rights theory is important to management of complex resources in LMEs, we need to review what it says about multi-attribute resources. Two phenomena concern us. One is why there are always some attributes of a resource that are exposed in the public domain where their value is dissipated. The second concerns spillovers that result from attribute interactions.

Transaction costs and unspecified attributes
At its core, economics is the study of property rights over scarce resources (Furubotn and Pejevich 1972). Property rights are entitlements to the income (or utility) that can be derived from resources (natural or manufactured) which are sanctioned, or at least condoned, by society and protected by a higher authority (De Alessi 1983; Demsetz 1998). Protection is provided by "the force of etiquette, social custom, ostracism, and formal legally enacted laws

supported by the states' power of violence or punishment" (Alchian 1977, p. 129). Legal status is not a necessary condition of property rights, however (Barzel 1989). Often, property rights, such as fishing rights, are de facto, having developed under rules of first possession (Lueck 1998a).[2]

Since Coase's (1960, p. 44) seminal work on the subject, it has become meaningful to think about the ownership and exchange of property rights, not physical entities: "[w]e may speak of a person owning land and using it as a factor of production but what the land-owner in fact possesses is the right to carry out a circumscribed list of actions ... [including] the right to do something which has a harmful effect (such as the creation of smoke, noise, smells, etc.) ..." In other words, factors of production, manufactured goods and services, and the natural environment are categories that can be used to bundle property rights. Property rights are exchanged through markets, contracts, gifting, and involuntary ways, including "harmful effects," or what the literature refers to as spillovers or externalities. Spillovers benefit the causal party at the expense of others' entitlements, although the respective property rights often are incomplete and ill-defined.

This economics perspective on ownership and exchange provides an analytical basis for understanding several aspects of property rights arrangements. First, "ownership" of, say, a car or fishing vessel, is rarely, if ever, absolute. All property rights are restricted by laws, regulations, or custom to protect peoples' property (*e.g.*, criminal laws), health and safety (*e.g.*, emission standards and speed limits), or sensibilities (*e.g.*, cruelty to animals). Second, property rights are incomplete, ill-defined, or even missing because the cost of information, definition, and protection for all attributes is too high (Cheung 1970). Non-exclusive resource attributes are, therefore, ubiquitous, including quality characteristics of goods and services and environmental spillovers such as pollution.[3] The costs of property rights need to be compared to resultant gains in the value of production or utility (Coase 1960; North 1990). Finally, the economics perspective of property rights also deviates from conventional notions of ownership and exchange when more than one party owns rights to use different attributes of the same resource. Here, too, there are many examples, particularly where the environment is concerned, such as private ownership of grazing or farming rights and government ownership of wildlife rights on the same land (Lueck 1995a,c), and common ownership of the right to beaver meat for personal consumption shared by Indians in eastern Canada separated from the private, territorial right of smaller bands to sell the meat and pelts (McManus 1972).

Transaction costs—*i.e.*, "the cost of establishing and maintaining property rights" (Allen 1991: 1)—are integral to the evolution of property rights. Transaction costs include the (opportunity) costs of gathering information on the quantity, quality, and relational attributes of resources in order to stipulate a private or social contract, as well as those used to negotiate, monitor, and enforce the contract (Barzel 1989; De Alessi 1983; Eggertsson 1990). Transaction costs are key to an understanding of property rights development (Libecap 1989b), spillovers (Cheung 1970; Coase 1960), law (Lueck 1998a), and the way production is

---

[2] First possession is the process by which most property rights have been established in practice and in law (Lueck 1998a). People who are first to take the flows from a resource or to capture the asset generally are assigned property rights.

[3] By non-exclusive, Cheung (1970) is referring to situations when users do not have the ability or property right to exclude others from the asset. For example, IFQs bestow harvest rights but not the right to decide who has access to the fishery resource.

organized (De Alessi 1983; Williamson 1979, 1984), making it central to economic growth (North 1990). Institutions, including property rights regimes, evolve, in part, to economize on both production (input transformation) and transaction costs (North 1990). In terms of sheer magnitude, transactions costs reportedly comprise nearly half of national income in the United States (North 1990).

Coase (1960) explained that, aside from income distribution, the initial partitioning of property rights does not affect the total value of an economy's production if transaction costs are zero. In this make-believe world, property rights to all resources are privately-owned and costlessly exchanged in markets (at a price) until combinations that maximize the total value of production are realized. In reality, however, transaction costs can preclude exchange and new arrangements of property rights when they exceed the potential net gains of exchange. This provides governments with opportunities to promote economic growth by assigning property rights in ways that lower transaction costs.

A profound teaching of property rights theory is that it is prohibitively costly to measure all attributes of a resource and capture their potential value in property rights (Barzel 1989; Cheung 1970; North 1984). Also, each attribute may require several stipulations which affect its use, benefits, and exchange (Cheung 1970). For example, fishery management plans stipulate gear type and attributes, catch limits, and fishing seasons and areas. As a result, some portion of all manufactured and natural resources end up in the public domain where its rent is dissipated along unspecified, incompletely specified, or unenforced margins (Cheung 1970). North (1990) gives several commonplace examples to underscore the pervasive incidence of this phenomenon, including fresh orange juice and pricing of theater tickets. Specifically, you might value fresh orange juice for its flavor, volume, and vitamin C content, but oranges are sold by number or weight. Customers and produce managers will expend resources to capture the benefits of unpriced attributes. Similarly, theater seats are valued for their location relative to the screen, but they sell at a uniform price, causing movie goers to arrive early for choice seats. A more pertinent example is highgrading (low value species or sizes are discarded) in fisheries where annual or trip limit quotas are allocated to individual vessels.

Spillovers
Spillovers (externalities) reveal the absence of a contract between parties, a contract with incomplete stipulations to attributes, or stipulations that are somehow inconsistent with a party's objectives (Cheung 1970). Coase (1960) explored how, by assigning property rights and reducing transaction costs, it is possible to resolve spillovers. The result is property rights arrangements that either divide or bundle resource attributes.

Alchian (1977, p.132) illustrated divided ownership of property rights as follows: "What are the effects of various partitionings of use rights? By this I refer to the fact that at the same time several people may each possess some portion of the rights to use the land. A may possess the right to grow wheat on it. B may possess the right to walk across it. C may possess the right to dump ashes and smoke on it. D may posses the right to fly an airplane over it. E may have the right to subject it to vibrations consequent to the use of some neighboring equipment." A marine analogy could be commercial fishing, whale watching, municipal waste disposal, shipping, and naval bombing experiments, respectively.

A compelling reason to divide ownership of resource attributes is when production technologies (or people's preferences, in the case of preservation) are highly specialized (Barzel 1989), such as harpooning Atlantic bluefin tuna (*Thunnus thynnus*). Combined production can be enhanced by separate ownership in these situations (perhaps oil production and fisheries), but special efforts are necessary to control spillovers when the transaction costs of doing so do not exceed the spillover damages. The respective parties may police each others' activities and use the courts to resolve disputes. Or the parties may negotiate private contracts that stipulate restrictions, infringements, and compensations. When both of these approaches are too costly to negotiate and enforce privately or collectively, government may restrict either activity. As mentioned, government restrictions of property rights are widespread, including in capitalistic economies. Restrictions may either attenuate the scope of a property right (such as air pollution standards for cars or mesh size in a fishery), or they may completely exclude some activities (such as land use zoning, Marine Protected Areas (MPAs), and killing endangered species).

Government restrictions can reduce total income to a greater extent than they save on spillover damages or transaction costs, however (Alchian 1977; Coase 1960; Demsetz 1967; Eggertsson 1990). Coase (1960, p. 2) used the fishery to illustrate this caution as follows: "[i]f we assume that the harmful effect of pollution is that it kills fish, the question to be decided is: is the value of the fish lost greater or less than the value of the product which the contamination of the stream makes possible? It goes almost without saying that this problem has to be looked at in total *and* at the margin." Regulations that preclude tradeoffs between uses reduce total income. Further, regulatory environments worsen losses by encouraging competitive rent-seeking behavior by parties who lobby government officials for favorable decisions (*e.g.*, Stigler 1971; Buchanan *et al.* 1980) and by promoting the principal-agent problem in large bureaucracies such as government when managers' interests diverge from those of owners (*i.e.*, principals) (*e.g.*, De Alessi 1980; Niskanen 1968).[4]   Instead of regulations, then, government could assign property rights in ways that decrease transaction costs and thereby facilitate exchanges which repackage, or bundle, property rights to joint attributes in beneficial ways.

Combining property rights into bundles of related uses could follow either of two strategies. Demsetz (1964) discusses "tie-ins", or the joint supply of outputs, illustrated by shopping malls that provide free parking, and apple orchards that produce both apples and bee honey. Alternatively, one party could expand its output mix and thereby internalize the spillover as Coase (1960, p. 23), once again, illustrated: "[a]fter the railroad purchases title to enough land to make it worthwhile, it could take into account the effect of its output of sparks on land values and profitably bring about an adjustment of this output to the socially optimal amount - that which maximizes the joint value of railroading and landowning". Lueck (1995a,c) examined similar situations when high contracting costs of division warrant combining wildlife hunting rights and agriculture or grazing rights on farms. In LMEs, we can imagine property rights arrangements that combine rights to harvest fish and to alter habitat, including in ways that promote fish productivity. In general, bundling is favored when resource

---

[4] Rent-seeking occurs when governments are lobbied for favorable permissions or property rights. This behavior is generally thought to be a net drain on an economy because the scarce resources used up in rent-seeking (*e.g.*, people's time) do not increase economic output. The principal-agent problem can arise when agents (managers) are able to pursue personal gain at the expense of principals (owners) due to the costs of overseeing the agents. Although first observed in the private sector (*e.g.*, large corporations), it can arise in any bureaucracy, including government agencies.

complexity results in high transaction costs (Alchian 1977; Demsetz 1967) and total income can be increased by an administrative structure which internalizes spillovers (Coase 1960) and makes owners the residual claimants to the value of unspecified attributes (North 1990).

Technology plays an important role in bundled ownership of property rights, too. Citing wild capture fisheries and aquaculture, Cheung (1970) pointed out that rents from non-exclusive resources are dissipated by the choice of an inferior technology as well as outputs. Once spillovers are internalized by a rearrangement of property rights, however, technology can be modified, replaced, or innovated in order to increase combined income.

Property rights arrangements and the organization of production also reflect uncertainty in attribute levels and the frequency of property right exchanges because these factors increase transactions costs. Williamson's (1979, 1984, 1990) authoritative research on the subject identifies three forms of contractual relations that are influenced in part by uncertain information and frequent exchanges—classical, neoclassical, and relational contracts. Although he concerned himself with intermediate products, his work applies to natural resource problems as well.

Standardized commodities which are familiar to all parties and relatively inexpensive (such as movie tickets or seafood) lend themselves to divided ownership and classical contracts with market exchange. In contrast, neoclassical contracts, such as those covering athletes, apply to resources that are too complex for a simple market exchange, and occasional and unpredictable exogenous events alter resource value. Ownership can be divided in these circumstances because exchanges are infrequent, but a third party may be needed to resolve new (or exacerbated) spillovers or attribute values as they appear in the public domain. Finally, a high degree of uncertainty and frequent exchanges favor long-term, relational contracts such as found in industries with a high degree of vertical integration. In this case, production is unified within a single organization that has internalized the costs of exchanges. A regulatory government agency, a commons or cooperative with decision rules which are designed to resolve use conflicts (Ostrom 1990; Runge 1986), or a corporation that centralizes production decisions (Coase 1960; Demsetz 1964) are general forms of relational contracts and unified governance arrangements.

FISHERY RESOURCE ATTRIBUTES AND MANAGEMENT

The above discussion of property rights theory and multi-attribute resources provides an analytical framework to characterize fishery resources and management in the United States. A brief, but useful, extension to other marine resources, such as petroleum, is also provided.

Fishery resources
Fishery resources are stocks of target species and their environment. Gordon (1954, p. 129), who advanced an elegant theory of rent dissipation along only the biomass and productivity margins of a fish stock, actually described multiple attributes of fishery resources in some detail: "Demersal ... fishes ... live and feed on shallow continental shelves where the continual mixing of cold water maintains the availability of those nutrient salts which form the fundamental basis of marine-food chains. ... [I]n some cases the fish of different banks can be differentiated morphologically... The significance of this fact is that each fishing ground can

be treated as unique, in the same sense as can a piece of land, possessing, at the very lea. characteristic not shared by any other piece: that is, location." Cheung (1970), whose sen. work was cited extensively above, also referred to fishery resources collectively as the fis. sea floor, and water.

Fishery resources are comprised of myriad biological (*e.g.*, plankton, fish, and mammal species), chemical (*e.g.*, salinity, oxygen concentration), and physical (*e.g.*, sediment type, oil and gas reserve, currents, space) attributes which can be further differentiated by quantity, quality, and relational attributes. For example, a species of fish could be defined by stocks, biomass, population, structure (age, size, sex), geographic location and scale, gene pool, disease, dynamics (life cycle, migration, fecundity, recruitment, growth rate, natural mortality), coexistence with other species, diet, habitat requirements, and so forth. Many of the same attributes can also be used to describe different life stages, cohorts, or even individual fish. Expand this list by the large number of species of finfish and shellfish landed by fishermen—*e.g.*, more than 200 in the Northeast Region of the United States—and one begins to appreciate the scope of attributes associated with fishery resources.

The property rights literature comments on the high costs of enforcing property rights to large scale fish stocks, especially fugitive species, such as tunas, that migrate extensively (Eggertsson 1990; Lueck 1995a). However, the information and other transaction costs required to initially completely define each attribute of a fish stock and its environment for property rights are incalculable. No matter what type of governance arrangement is in place, we should expect to find some fishery resource attributes in the public domain where their value might be dissipated.

The high costs of defining fishery resources for property rights is compounded by relational attributes that manifest as spillovers in several ways, including gear conflicts, bycatch, habitat disturbance, and predation. Gear conflicts result from competition for the non-exclusive space attribute of fishery resources. Fixed gear in particular, such as pots, gillnets, and longlines, are vulnerable to mobile trawl and dredge gear. This conflict dates back to at least the late 19th century in New England (Baird 1873) and the Pacific Northwest (Higgs 1982) when the comparative advantage of fish wheels and in-river traps was reversed in favor of coastal trap and weir fishermen.

Bycatch refers to "the unintended capture or mortality of living marine resources as a result of a direct encounter with fishing gear" (NMFS 1998, p. 1). It is a difficult concept because it combines dissimilar problems, including inefficient choices of technology or gear (discarding undersized commercial and recreational species), de facto claims to harvest rights in competitive fisheries (discarding alleged non-target species as required by regulations), and purely accidental catches of uneconomic "trash" fish (*e.g.,* sea robins) and protected species, such as marine mammals (*e.g.*, whales, seals, dolphins), threatened or endangered species (*e.g.*, all sea turtles), and seabirds. Alverson *et al.* (1994) estimate that discards amount to one-quarter of the world's total commercial catch (by weight).

With some exceptions, such as speargun and harpoon, the major trawl, gillnet, dredge, and seine fishing technologies are only moderately specialized (at best) to catch the desired sizes and quantities of target species. In the Northeastern United States, for example, only about half of the catches by groundfish trawl, sink gillnet, and scallop dredge vessels that carried observers on trips to Georges Bank during the 1990s was comprised of traditional target species (Edwards *et al.* 2001). Several fisheries in the region also take protected species, including harbor porpoise (*Phocoena phocoena*) and the endangered right whale (*Eubalaena glacialis*) in the Gulf of Maine sink gillnet fishery, and leatherback (*Dermochelys coriacea*) and green (*Chelonia mydas*) sea turtles in the Atlantic swordfish (*Xiphias gladius*) and tuna (*Thunnus* spp.) pelagic longline fishery (NMFS 1998).

Vessels (vis-a-vis gear) are also only moderately specialized because they can be converted to use different gear depending on relative prices, costs, and biomass levels. For example, many Gulf of Maine vessels participate in groundfish, Northern shrimp (*Pandalus borealis*), and Atlantic sea scallop (*Placopecten magellanicus*) fisheries seasonally.

Fishing gear may also disturb the environment when it contacts the sea floor (Kaiser *et al.* 2001). Impacts on the geologic and biogenic structures caused by dredge and trawl gear are often cited for possibly affecting the survivorship of juveniles of valued species, and for possible impacts on marine biodiversity (Gray 1997).

Spillovers also stem from trophic relationships when the species claimed by one fishery eats the species claimed by others, or when a fishery harvests the prey or predator claimed by another fishery. Having co-evolved over many years, trophic interactions are coupled, continual, and difficult to measure and predict quantitatively, however. For example, Link (1999) reported that finfish, invertebrate, mammal, and bird species which inhabit the Northeast Continental Shelf are part of a highly generalized food web, including omnivory and extensive dietary overlap. American goosefish (*Lophius americanus)* and spiny dogfish (*Squalus acanthias)* are top predators in this food web, consuming many other commercial species which also prey on each other (*e.g.,* gadids and flounders). Other predators in the system, particularly silver hake (*Merluccius bilinearis)*, may be more responsible than fishermen for mortality of some commercial species.

Finally, fishery resource dynamics take place at different spatial and temporal scales, and are highly uncertain (Sherman and Alexander 1986). Larval survival is an enigma, and recruitment of species to a fishery is non-normal and quite variable. The possible influence of habitat disturbance on the productivity of fishery resources is also unknown, even qualitatively.

Marine fisheries management in the United States
Together, the Magnuson-Stevens Act and President Reagan's 1983 Proclamation on the EEZ made the Federal government (as agent for U.S. citizens) the presumptive owner of property rights to fishery and other marine resources (*e.g.*, oil and gas, hard minerals) within 200 miles of the shoreline. The U.S. Congress sanctioned the de facto common harvest rights of fishermen, and it transferred management rights to regional councils and the National Marine Fisheries Service (NMFS). Council members from the commercial fishing industry, state and federal fishery or wildlife agencies, and recreation, conservation, and consumer groups are appointed for short terms. Management plans are contracts between the federal government and fishermen which stipulate producers (*e.g.*, permits), outputs (*e.g.*, total allowable catch (TAC), vessel harvest quotas, minimum fish size, bycatch limits), inputs (*e.g.*, mesh size, number of crew, days-at-sea quotas, vessel size), technologies (*e.g.*, gear type, vessel monitoring electronics), and, increasingly, time and area closures. The Magnuson-Stevens Act

clearly denies the legal status of harvest rights, including in IFQ fisheries where annual quotas are allocated to individual fishermen or vessels (see Section 303(d)(3) of Public Law 94-265; 16 U.S.C. 1853). Fishing is more than a privilege, however, judging from how historical use determines membership in a limited access fishery and from federal programs that buy, not cancel, permits in fisheries suffering from excessive harvest capacity (*e.g.*, Northeast groundfish fishery). These actions seemingly recognize entitlements owned by fishermen.

Government agencies attempt to manage only a few attributes of fishery resources, however. Numerous regulations address only the measurable biomass, age structure, productivity, and, through closures, location attributes of target species or stocks. Management objectives are described primarily in terms of achieving maximum sustainable yield (MSY) from individual species, but there are an infinite number of MSYs depending on the mixed harvest characteristics of fishing gear, the size of predator and prey populations, and attributes of the environment, including habitat characteristics (*e.g.*, Larkin 1977). Furthermore, it is very expensive for NMFS and the Coast Guard to thoroughly monitor and enforce regulations on thousands of fishermen who are motivated to harvest fish without regard for opportunity costs or spillovers in the prevailing rule-of-capture management regimes.[5] Age-structure management is compromised by discarding of undersized individuals and highgrading of catches for the highest priced sizes. Spatial management of large year classes, such as in the Atlantic sea scallop fishery, is impeded by high transaction costs, including with other fisheries who share the fishing grounds and with environmental organizations who sue for restrictions upon re-opening.

The 1996 amendments to the Magnuson-Stevens Act also require management councils and NMFS to minimize all forms of bycatch, and to protect and conserve the waters and substrate necessary to fish for spawning, breeding, feeding, or growth to maturity—*i.e.*, "essential fish habitat" (EFH). As for overfishing, bycatch regulations, such as catch limits and gear modifications, address the biomass and age structure attributes of stocks. EFH currently is defined broadly in terms of large scale geographic locations of individual species, sometimes in association with sediment type. Information on bycatch levels and EFH attributes is very costly to gather and then develop into regulations, including the contribution of EFH to fish productivity. Nevertheless, NMFS currently faces numerous lawsuits by environmental organizations charging it for poorly executing the law.

As mentioned, harvest rights are predominantly divided according to gear and traditional target species. Regulatory bycatch rules are now widely used to restrict landings of species that "belong" to other fisheries. For example, vessels in the Northeast groundfish fishery are limited to 40 pounds of sea scallop meats, 300 pounds on goosefish tails, and 100 American lobsters (*Homarus americanus*), all per trip. Similarly, trip limits on vessels in the sea scallop fishery include 300 pounds of groundfish, 200 pounds of goosefish tails, and 100 lobsters.

Separate management plans for species that coexist, coupled with legal requirements to manage each species or stock at MSY-based target levels, also lead to divided ownership when specific locations are closed to rebuild one or a few stocks. For example, in 1994 three large areas of the U.S. Georges Bank and nearby waters were closed to all fisheries capable of

---

[5] The rule-of-capture says that you do not own rights to the income or other benefit of an asset until you extract and physically possess it.

catching significant amounts of depleted stocks of Atlantic cod (*Gadus morhua*), haddock (*Melanogrammus aeglefinus*), and yellowtail flounder (*Limanda ferruginea*), including the sea scallop, monkfish, and spiny dogfish fisheries. Short-term access to parts of these areas was granted to the sea scallop fishery in 1999 and 2000, but not since then even though about half of the sea scallop biomass is inside these closed areas (Dvora Hart, personal communication, 2002). Further, access ended early in 1999 when only two-thirds of the scallop quota was caught because the TAC of yellowtail flounder bycatch was reached.

Divided ownership and exclusivity are also applied to safeguard protected species of marine mammals and sea turtles. Gear might be modified to reduce the number or impacts of incidental contact (such as break-away lines in lobster pot fisheries or pingers on gillnets that ward off marine mammals), or prohibited in closed areas. For example, lobster traps and anchored gillnets are excluded from areas of the Gulf of Maine where endangered right whales are observed. Similarly, pound nets are being restricted in parts of the Chesapeake Bay in order to minimize sea turtle takings.

EFH policy is evolving towards exclusive closed areas. In particular, "habitat areas of particular concern" (HAPCs) are being designated for the protection of individual species and life stages, such as the juvenile Atlantic cod HAPC on Georges Bank. Similarly, Executive Order 13158—signed by President Clinton and reauthorized by President Bush—directs federal agencies to strengthen and expand the nation's system of MPAs and to prevent adverse impacts in these areas. HAPCs and MPAs are widely viewed as no-fishing zones.

The extensive use of divided ownership of property rights to fishery resources in rule-of-capture management regimes raises concern about the consequences for total fishery income. Regulatory bycatch limits and related gear restrictions do not create incentives for fishermen to voluntarily reduce discarding of undersized commercial and recreational species or uneconomic species that comprise the food web. Closed area policies that are designed to rebuild selected species or to preserve marine mammals, sea turtles, or habitat might be less costly to monitor and enforce than input or output regulations in the current governance arrangement, but they also create opportunity costs from lost production by excluded activities as just illustrated for the groundfish closures.

Other marine resources
A brief comment about property rights to other marine resources will suffice to expand this discussion to the ocean. Property rights to the use and management of fishery resources and protected species exist side-by-side and interact with those to oil and gas production, hard minerals mining (*e.g.*, sand), municipal waste disposal, shipping, national defense, and recreation (*e.g.*, cruises), to name a few. Property rights to these activities can cause spillovers that affect fisheries, protected species, and habitat as well (*e.g.*, Exxon Valdez oil spill in Alaska; collisions between merchant or naval ships and right whales), although not necessarily continuously. The technologies of these activities are highly specialized relative to each other and fisheries. Property rights to these other marine resources are divided among industries and different regulatory and other divisions in the Federal government, including the Minerals Management Service, Army Corps of Engineers, Environmental Protection Agency, Coast Guard, and Navy.

ALTERNATIVE PROPERTY RIGHTS AND GOVERNANCE ARRANGEMENTS

For reasons just discussed, it is very doubtful that the stock management and habitat policies which divide ownership of property rights to fishery resources promote income and economic growth. This section therefore suggests general alternatives to the present property rights and governance arrangements in U.S. fisheries that should contain transaction costs and match property rights structures to marine ecosystem attributes—both cited by Hanna and Munasinghe (1995) as general requirements to effectively sustain beneficial uses of environmental resources. A set of four guidelines from the review of property rights theory are used to develop the proposals: (1) the economic value of production should be looked at in aggregate and net of transaction costs (Coase 1960); (2) transaction costs increase with the number of resource attributes (Cheung 1970) and their jointness, uncertainty, and frequency of exchange (Williamson 1979); (3) government is pivotal to the evolution of property rights and governance arrangements because it has the authority to assign property rights and thereby reduce the transaction costs of subsequent contracts (Coase 1960); and (4) making those who can most easily influence the variability of particular attributes the residual claimants increases net resource value (North 1990).

Usufruct rights, such as IFQs and harvest cooperatives, significantly improve rents in fisheries because harvesters economize on input use and improve the timing, quality, and mix of catches (Leal 2003). They should also reduce the transactions costs of contracting uses to some of the remaining, non-exclusive attributes of fishery resources because the initial allocation of fishing rights and population of claimants are resolved. Gaining the rights of management and exclusion (Schlager and Ostrom 1995), in particular, would make those with harvest rights the residual claimants to rents from other attributes, including the age and genetic composition of fish populations, bycatch, predator and prey control, and environmental protection, as Scott (1986, 1989) noted. This evolution toward more complete ownership rights is taking place in New Zealand where IFQs are a national policy. For example, the Challenger Scallop Enhancement Company Limited is an organization of scallop IFQ owners who cooperatively finance dockside monitoring, scientific research, and seeding projects, and who negotiate with other fisheries which compete for the same fishing grounds (*e.g.,* clam, groundfish, recreation; Arbuckle and Drummond 2001).

The New Zealand experience suggests that a comprehensive assignment of usufruct rights in all fisheries—IFQs, harvest cooperatives, or something else—is a sound way to begin to address spillovers and rent dissipation associated with non-exclusive resource attributes. Regulatory bycatch conflicts between fisheries could be readily handled with simple classical contracts for quota exchanges, thereby increasing the overall economic value of production from fishing grounds. Whether harvest rights are bundled on a vessel depends, in part, on whether the bycatch rights can be standardized for frequent market exchange and on the degree of specialized technology in fisheries. Mixed output vessels (trip or seasonal) are a realistic option if income from specialization is small relative to the costs of frequent market exchanges.

Predation spillovers ("your fish ate my fish," or "you're harvesting the species that my fish eat") present a more difficult, common pool problem,[6] and they implicate the substantial

---

[6] The common pool problem involves resources with multiple users whose activities subtract from resource

discards of uneconomic "trash" fish which are the prey or predators of species which are sought by commercial and recreational harvesters (Gislason 2001). Information costs are extremely high because predation is poorly understood from a quantitative scientific standpoint—especially indirect effects across several trophic levels—and variable and difficult to predict due to irregular recruitment and spatial patterns. Predation is also continuous, making the costs of repeatedly negotiating, monitoring, and enforcing thousands of individual contracts impractical and vulnerable to free-riders. Yet, even with restrictions due to ecological requirements (*e.g.*, maintain trophic levels) and social sensibilities (*e.g.*, forbid extinctions), there is potential to increase the economic value of fishery resources by manipulating the size of predator and prey populations.

The frequency and uncertainty of predation spillovers makes them most suitable to the unified, relational form of share contract which integrates production decisions across activities in order to avoid the transaction costs of continual negotiations. In the fisheries literature, this contractual arrangement is known as sole ownership by a corporation or collective (Scott 1955, 1989). It would most likely be too costly for a government agency to impose changes to harvest rights that were gained through first possession rules and designed to produce single-species MSYs, and then to monitor and enforce regulations on predation attributes. Instead, fishermen's organizations or communities could negotiate share contracts that stipulate either new distributions for harvest quotas and, where it applies, compensation formulae (cooperative model) or income shares in a unitized industry that pools its physical capital and harvest strategy (corporate model). Either model maintains the residual claimant status of fishermen and should induce endogenous changes in fishing technology that reflect management of a portfolio of fishery resource assets.

Of course, there are significant barriers to the development of such unified, portfolio harvesting arrangements.[7] First, contracts between heterogeneous organizations of fishermen would be costly to negotiate, monitor, and enforce (Johnson and Libecap 1982; Libecap 1989b; Ostrom 1990). The cooperative option would face these costs each time new information requires renegotiations, but the unitized corporate arrangement might under-produce due to shirking and agency loss (Lueck 1995b). Second, political opposition from other parties, such as environmental organizations and regulators, will swell contracting costs (Christy 1996; Libecap 1989a). For example, resource management by fishermen would require a transfer of rights from government regulatory agencies to industry. It would further entail a significant departure from the institutionalized single-species definition of overfishing and similarly narrow notions of EFH that would allow reductions in some populations below those corresponding to estimates of MSY. Third, it could be costly to monitor and enforce exclusive use of large scale assemblages, especially for highly migratory species that cross political jurisdictions. Scale economies afforded by new enforcement technologies (*e.g.*, air flights, vessel monitoring systems) might offer relief, however (Eggertsson 1990; Huppert and Knapp 2001). The classic example of a new technology that reduces the costs of exclusion is barbed wire and grazing land (Anderson and Hill 1975). Finally, ownership of management rights to large scale fishery resources may require a favorable antitrust review.

---

quantity or quality. Fish stocks, oil pools, groundwater, and the atmosphere are common examples of common pools that can be degraded as a result of the cumulative effects in individual behavior.

[7] Elsewhere in this volume Jason Link, Barbara Rountree and I discuss a portfolio approach to multi-species management.

Together, the transaction costs of collective or corporate management arrangements are substantial, but perhaps not overwhelming. Contracting might still be induced by large expected gains from lower transaction costs and a more complete specification of attribute rights, provided payoffs for losers (*e.g.*, those organizations that need to reduce harvest rates to increase portfolio income) can be reached (Libecap 1989a). Also, some of the costs might be eased if the initial system of comprehensive usufruct rights is area-specific (*i.e.*, reduces the number of claimants and their differences) and conforms to fishermen's preferred locations as well as the geographies of species assemblages.

The foregoing discussion about relational share contracts to attributes of species assemblages applies equally to the uncertain dependency of productivity on habitat attributes. Introducing habitat attributes to the bundle of property rights requires the right to exclude others from your territory. Similar to predation spillovers, the relationship between habitat disturbance that is caused by fishing gear and the productivity of fishery resources is continuous and poorly understood. Government allocation of area-specific rights has precedents in TURFs (*i.e.*, territorial use rights in fisheries; Christy (1996)), homesteading policies used by the U.S. government to settle the West (Allen 1991), and auctions now used to lease large tracts of the continental shelf to oil companies (Libecap 1989b). As Allen (1991) argued for Western land, a marine homesteading policy might replace the high costs of monitoring and enforcing government regulations with benefit incentives for residual claimant fishermen.

A path to the preservation attributes of public goods is less clear. Technically, the general public is claimant to preservation benefits, but it relies on the government, as its agent, to produce habitat and species preservation because it is prohibitively costly for individual citizens to define and enforce property rights to these attributes. Environmental organizations, who earn income from public donations and foundation grants, can influence environmental policy by suing and lobbying management agencies and Congress. Fishermen do not benefit from preservation (notwithstanding unclear positive spillovers between habitat and fish productivity) unless they can improve their joint income from producing fish and preservation. This arrangement is analogous to ranchers and farmers who also produce hunting opportunities on their land (Lueck 1995a).

There are at least two conceivable paths towards preservation of habitat and species biodiversity. One path, being taken in the United States and elsewhere, involves continuous rent-seeking by competing fisheries and environmental groups, and government regulation of fishing gear and closed area zoning. This path does not incorporate incentives to control transaction costs or to promote the combined benefits of fishery production and preservation.

A second and more productive path to preservation could use neoclassical contracts (see above discussion of Williamson's work) afforded by a comprehensive assignment of usufruct rights. One option is co-management (Jentoft 1989; Townsend and Pooley 1995) whereby fishing organizations negotiate contracts with government which stipulate restrictions, monitoring techniques, violations, and penalties. Ownership is divided because the government continues to be an active guardian of habitat and protected species. The contract will require revisions, however, because it is impossible to stipulate all conceivable circumstances in advance given poor information and uncertainties. The power-sharing characteristic of co-management may require arbitrators to conclude negotiations.

A second option—probably not available when species or unique types of environment are threatened or endangered—is for government agencies and/or environmental organizations to buy harvest rights (possibly area-specific) and become shareholders in a cooperative or corporate management arrangement. Their influence on habitat and species preservation would depend on their holdings, but being residual claimants to fishery income would force them to take into account the opportunity costs of preservation. Alternatively, government and environmental organizations might contract out of the unified governance arrangement for their own preservation zones if fishing technology and preservation preferences are too specialized to stay joined. For example, the Audubon Society has leased grazing and oil production rights on its Rainey Wildlife Sanctuary in Louisiana (Baden and Stroup 1981). The Nature Conservancy is taking a similar approach in its program to lease subtidal lands from state governments and form partnerships with the community and economic interests (Marsh *et al.* 2002). This behavior contrasts markedly with current demands and lawsuits by environmental organizations for government designation of exclusive HAPCs and MPAs.

Finally, the argument that relatively unspecialized fishing technology lends itself to bundled property rights arrangements and unified governance does not automatically extend to other marine industries that interact with fisheries, such as oil production, sediment mining, and marine transportation. For example, oil companies with Outer Continental Shelf leases could negotiate neoclassical contracts with fishing organizations and the government which divide ownership rights to fishery and energy resources. Or they could try to buy the usufruct rights to fishery resources as a way to control rent-seeking and other transaction costs in politicized competitions. Buying fishing rights could result in a mixed-output industry in which the impacts of oil production on fish yields are internalized. This possibility is not farfetched despite the disparate production technologies. For example, corporations that are identified with energy (*e.g.*, British Petroleum) and timber (*e.g.*, Weyerhauser) production have experimented with salmon ranching on the West coast even though the wild, common pool fishery intercepted ranched salmon upon their return from sea (Anderson 1985). Likewise, Baird (1873) reported testimony that accused the Cape Cod Railroad Company of monopolizing coastal fisheries more than a century ago in New England. Alternatively, fishing rights could be re-sold or leased under contracts that stipulate compensation for fish kills and habitat damage. Discussions of such disparate uses of resource attributes—e.g., wildlife, farmland, railroads, minerals, air quality, etc.—are already found in the property rights literature reviewed above (*e.g.*, Coase 1960), so there is no need to repeat them here.

CONCLUSIONS

A significant hurdle facing management of fishery and other resources in LMEs is the evolution of property rights and governance arrangements that promote economic growth. Marine fisheries governance in the United States, as well as in many other coastal nations, is primarily a federal and state government ownership regime, with access and harvest rights divided among fisheries and severed from the governments' rights to manage the resources and to determine who can harvest fish. Rents are dissipated along many attribute margins in addition to those usually associated with excess capacity (Gordon 1954) and product choice and quality (Cheung 1970; Wilen 2003). A poorly developed system of common law (Lueck 1998a,b) results in high transaction costs because of competitive rent-seeking, protracted rule-making, and, increasingly, lawsuits. Total fishery income is compromised because the values of fishing (and other) activities which are linked by ecology and moderately specialized fishing technologies are not adequately compared when regulations, such as bycatch rules and area closures, are designed. The high costs of information limit managers to only a few, measurable attributes of fishery resources, exposing many others to rent dissipation in the public domain. Compliance is poor because fishery resource attributes are non-exclusive to regulated fishermen; therefore, monitoring and enforcement costs are prohibitive.

The present governance arrangement in U.S. marine fisheries is probably incapable of administering the multi-attribute management of fishery resources as required by law and supporting economic growth at the same time. A compatible governance arrangement probably requires government to alienate some of its rights of management and exclusion to the fishing industry (or other interested parties) who would become the residual claimants of their harvest decisions. This proposal has little in common with the present arrangement whereby relatively few people from industry serve for short terms as members of regional fisheries management councils but are not personally accountable for management decisions. Instead, production would most likely organize as corporations and collectives which reflect the frequency and uncertainty of complex and mingled fishery resource attributes such as predation and habitat requirements. It seems unlikely, however, that unified forms of property rights and governance arrangements would evolve without first passing through a comprehensive system of usufruct rights in fisheries that share the same fishing grounds. This extraordinarily difficult accomplishment would reduce the transaction costs of contracting with government for management and exclusion rights and facilitate the creation of rents from more and more attributes of fishery resources.

REFERENCES

Alchian, A.A. 1977. *Economic Forces at Work.* Liberty Press, Indianapolis, Indiana.
Allen, D.W. 1991. Homesteading and property rights; or "How the West was really Won." *Journal of Law and Economics* 34:1-23.
Alverson, D.L., M.H. Freeberg, S.A. Murawski and J.G. Pope. 1994. A Global Assessment of Fisheries Bycatch and Discards. Food and Agriculture Organization Fisheries Technical Paper No. 339. Rome, Italy.
Anderson, J.L. 1985. Market interactions between aquaculture and the common-property commercial fishery. *Marine Resource Economics* 2:1-25.
Anderson, T.L. and P.J. Hill. 1975. The evolution of property rights: a study of the American West. Journal of Law and Economics 18:163-179.
Arbuckle, M. and K. Drummond. 2001. Evolution of self-governance within a harvesting system governed by individual transferable quota. Proceedings of the Tenth Biennial Conference of the International Institute of Fisheries Economics and Trade, Oregon State University, Corvallis, OR.
Arnason, R. 2000. Property rights as a means of economic organization. In R. Shotton (ed.), Use of Property Rights in Fisheries Management. FAO, Rome. Vol. 1, 14-25.
Baden, J. and R. Stroup.1981. Saving the wilderness: a radical proposal. *Reason* 13:28-36.
Baird, S. 1873. Report on the Condition of the Sea Fisheries of the South Coast of New England in 1871 and 1872. Government Printing Office, Washington, D.C.
Barzel, Y. 1989. *Economic Analysis of Property Rights*. Cambridge University Press,

NY.

Buchanan, J.M., R.D. Tollison, and G. Tullock (Editors).1980.*Toward a Theory of the Rent-Seeking Society*. Texas A&M University Press, College Station, Texas.

Cheung, S.N.S. 1970. The structure of a contract and the theory of a non-exclusive resource. *Journal of Law and Economics* 13:49-70.

Christy, F.T. 1996. The death rattle of open access and the advent of property rights in fisheries. *Marine Resource Economics* 11:287-304.

Coase, R.H. 1960. The problem of social cost. *The Journal of Law and Economics* 3:1-44.

De Alessi, L. 1980. The economics of property rights: a review of the evidence. *Research in Law and Economics* 2:1-47.

De Alessi, L. 1983. Property rights and transaction costs: a new perspective in economic theory. *The Social Science Journal* 20:59-69.

Demsetz, H. 1964. The exchange and enforcement of property rights. *Journal of Law and Economics* 7:11-26.

Demsetz, H. 1967. Toward a theory of property rights. *American Economic Review* 57:347-359.

Demsetz, H. 1998. Property rights. In Newman, P.(Editor), *The New Palgrave Dictionary of Economics and the Law*, Stockton Press, New York, volume 3; 144-155.

Edwards, S.F. 2003. Property rights to multi-attribute fishery resources. *Ecological Economics* 44: 309-323.

Edwards, S.F., B. Rountree, J.W. Walden and D.D. Sheehan. 2001. An inquiry into ecosystem-based management of fishery resources on Georges Bank. In Nishida,T., P.J. Kailola and C.E. Hollingworth (Editors), Proceedings of the First International Symposium on Geographic Information Systems (GIS) in Fishery Science. Fishery GIS Research Group, Saitama, Japan. 202-214.

EPAP (Ecosystems Principles Advisory Board). 1999. Ecosystem-Based Fishery Management. A Report to Congress by the EPAP. National Marine Fisheries Service, Silver Spring, Maryland.

Eggertsson, T. 1990. *Economic Behavior and Institutions*. Cambridge University Press, New York.

Furubotn, E.G. and Pejovich, S. 1972. Property rights and economic theory: a survey of the literature. *Journal of Economic Literature* 10:1137-1162.

Gislason, H. 2001. The effects of fishing on non-target species and ecosystem structure and function. Paper presented at the FAO Conference on Responsible Fisheries in the Marine Ecosystem, Reykjavik, Iceland. FAO, Rome.

Gordon, H. Scott. 1954. The economic theory of a common-property resource: the fishery. *Journal of Political Economy* 62:124-142.

Gray, J.S. 1997. Marine biodiversity: patterns, threats, and conservation needs. *Biodiversity Conservation* 6:153-175.

Hanna, S.S. 1998. Institutions for marine ecosystems: economic incentives and fishery management. *Ecological Applications 8(Supplement)*:S165-S169.

Hanna, S. and M. Munasinghe. 1995. An introduction to property rights in a social and ecological context. In Hanna, S. and M. Munasinghe (Editors). *Property Rights in a Social and Ecological Context: Case Studies and Design Applications*. Beijer International Institute of Ecological Economics and the World Bank, Washington, D.C. 3-11.

Higgs, R. 1982. Legally induced technical regress in the Washington salmon fishery. *Research in Economic History* 7:55-86.

Huppert, D. and G. Knapp. 2001. Technology and property rights in marine fisheries. In Anderson, T.L. and P.J. Hill (Editors), *The Technology of Property Rights.* Rowman and Littlefield Publishers, Inc., New York. 79-100.

Jentoft, S. 1989. Fisheries co-management: delegating government responsibility to fishermen's organizations. *Marine Policy* 13:137-154.

Johnson, R.N.and G.D. Libecap. 1982. Contracting problems and regulations: the case of the fishery. *American Economic Review* 72:1005-1022.

Kaiser, M.J., J.S. Collie, S.J. Hall, S. Jennings and I.R. Poiner. 2001. Impacts of fishing gear on marine benthic habitats. Paper presented at the FAO Conference on Responsible Fisheries in the Marine Ecosystem, Reykjavik, Iceland. FAO, Rome.

Larkin, P.A. 1977. An epitaph for the concept of maximum sustainable yield. *Transactions of the American Fisheries Society* 106:1-11.

Leal, D.R. 2003. *Evolving Property Rights in Marine Fisheries.* Rowman and Littlefield Publishers, Inc., Lanham, Maryland.

Libecap, G.D. 1989a. Comments on Anthony D. Scott's 'conceptual origins of rights based fishing'. In Neher, P.A., R. Arnason and N. Mollett (Editors), *Rights Based Fishing.* Kluwer Academic Publishers, Boston. 39-45.

Libecap, G.D. 1989b. Contracting for Property Rights. Cambridge University Press, New York.

Link, J. 1999. (Re)constructing food webs and managing fisheries. Ecosystem Approaches for Fisheries Management. AK-SG-99-01. Alaska Sea Grant College Program, Anchorage, Alaska, USA. 571-588.

Lueck, D.L. 1995a. Property rights and the economic logic of wildlife institutions. *Natural Resources Journal* 35:625-670.

Lueck, D.L. 1995b. Contracting into the commons. In Anderson T.L. and R.T. Simmon (Editors), *The Political Economy of Custom and Culture: Informal Solutions to the Commons Problem.* Rowman and Littlefield Publishers, Inc., Lanham, Maryland, pp. 43-59.

Lueck, D.L. 1995c. The economic organization of wildlife institutions. In Anderson, T.L.and P.J. Hill (Editors), *Wildlife in the Marketplace.* Rowan and Littlefield Publishers, Inc., Boston, Massachusetts.1-24.

Lueck, D.L. 1998a. First possession. In P. Newman (Editor), *The New Palgrave Dictionary of Economics and the Law.* Stockton Press, New York, volume 2, 132-144.

Lueck, D.L. 1998b. Wildlife law. In Newman, P. (Editor), *The New Palgrave Dictionary of Economics and the Law.* Stockton Press, New York, volume 3, 696-701.

Marsh, T.D., M.W. Beck and S.E. Reisewitz. 2002. *Leasing and Restoration of Submerged Lands: Strategies for Community-Based, Watershed-Scale Conservation.* The Nature Conservancy, Arlington, Virginia.

McManus, J.C. 1972. An economic analysis of Indian behavior in the North American fur trade. *Journal of Economic History* 32:36-53.

NMFS (National Marine Fisheries Service). 1998. *Managing the Nation's Bycatch.* National Oceanic and Atmospheric Administration, U.S. Department of Commerce, Washington, D.C.

NRC (National Research Council). 1999. *Sustaining Marine Fisheries.* Committee on Ecosystem Management for Sustainable Marine Fisheries, Ocean Studies Board, NRC. National Academy Press, Washington, D.C.

Niskanen, W.A. 1968. Nonmarket decision-making: the peculiar economics of
    bureaucracy. *American Economic Review* 58:293-305.
North, D.C. 1984. Transaction costs, institutions, and economic history. *Journal of
    Institutional and Theoretical Economics* 140:7-17.
North, D.C. 1990. Institutions, Institutional Change and Economic Performance.
    Cambridge University Press, New York.
Ostrom, E. 1990. *Governing the Commons: The Evolution of Institutions for Collective
    Action.* Cambridge University Press, New York.
Pauly, D., V. Christensen, J. Dalsgaard, R. Froese and F. Torres Jr. 1998. Fishing
    down marine food webs. *Science* 279:860-863.
Rosenberg, A.A. 2001. Multiple uses of marine ecosystems. Paper presented at the
    FAO Conference on Responsible Fisheries in the Marine Ecosystem, Reykjavik,
    Iceland. FAO, Rome.
Runge, C.F. 1986. Common property and collective action in economic development.
    World Development 14:623-635.
Schlager, E. and E. Ostrom. 1992. Property rights regimes and natural resources: a
    conceptual analysis. *Land Economics* 68:249-262.
Scott, A. 1955. The fishery: the objectives of sole ownership. *Journal of Political
    Economy* 63:116-124.
Scott, A. 1986. Catch quotas and shares in the fishstock as property rights. In
    Miles, E., R. Pealy and R. Stokes (Editors), *Natural Resource Economics and Policy
    Applications: Essays in Honor of James A. Crutchfield.* University of Seattle Press,
    Seattle, Washington, pp. 61-96.
Scott, A. 1988. Development of property in the fishery. *Marine Resource Economics* 5:
    289-311.
Scott, A. D. 1989. Conceptual origins of rights based fishing. In Neher, P.A.,
    R. Arnason and N. Mollett (Editors). *Rights Based Fishing.* Kluwer Academic
    Publishers, Boston, pp. 11-38.
Sherman, K. and L.M. Alexander (Editors). 1986. *Variability and Management of Large
    Marine Ecosystems.* Westview Press, Boulder, CO.
Sherman, K. and A.M. Duda. 1999. Large Marine Ecosystems: an emerging paradigm
    for fishery sustainability. *Fisheries* 24: 15-26.
Stigler, G.J. 1971. The theory of economic regulation. *Bell Journal of Economics* 2:3-21.
Townsend, R.E. and S.G. Pooley. 1995. Distributed governance in fisheries. In
    Hanna, S. and M. Munasinghe (Editors). *Property Rights and the Environment: Social
    and Ecological Issues.* Beijer International Institute of Ecological Economics and the
    World Bank, Washington, D.C., pp. 47-58.
Wilen, J. 2004. Property rights and the texture of rents in fisheries. In Leal, D.R.
    (Editor). *Evolving Property Rights in Marine Fisheries.* Rowman and Littlefield
    Publishers, Inc., Lanham, Maryland [in press as of preparation of the present volume]
Williamson, O.E. 1979. Transaction cost economics: the governance of contractual
    relations. *Journal of Law and Economics* 22:233-261.
Williamson, O.E. 1984. The economics of governance: framework and implications.
    Journal of Institutional and Theoretical Economics 140:195-223.
Williamson, O.E. 1990. A comparison of alternative approaches to economic
    organization. *Journal of Institutional and Theoretical Economics* 146:61-71.

Part II:
Economic Activity and the Cost of Ownership

Large Marine Ecosystems, Vol. 13
T.M. Hennessey and J.G. Sutinen (Editors)
© 2005 Elsevier B.V. All rights reserved.

7

Economic Activity Associated with the Northeast Shelf Large Marine Ecosystem: Application of an Input-Output Approach

P. Hoagland, D. Jin, E. Thunberg, and S. Steinback

ABSTRACT

The industries linked to the uses of a large marine ecosystem (LME) have a substantial influence on contiguous coastal economies. We estimate the economic activity of U.S. marine sectors associated with the Northeast Shelf LME. Our best *upper bound* estimate of total output impact is $339 billion, including a total "value-added" impact of $209 billion. Total employment impacts are estimated on the order of 3.6 million persons. The estimate of total value-added impact is approximately 10% of the $2.2 trillion total gross state product for the region. In the future, critical interactions between industrial sectors and the ecological health of the Northeast Shelf will affect economic activity in opposing ways.

INTRODUCTION

Ecosystem Valuation

Measuring the economic value[1] of a large marine ecosystem (LME) is straightforward at a conceptual level. Where an LME is defined on the basis of its relevant ecological features, its economic value is equivalent to the net present value of goods and services that flow from uses and "non-uses" of the resources and the environment. A calculation of this kind is only descriptive, and it is necessarily anthropocentric. To undertake such a calculation, one estimates the sum of consumer and producer surpluses associated with identifiable uses of the ocean, such as recreation, commercial fishing, marine transportation, or plausible non-uses, such as preservation or species protection. These surpluses must be forecasted into the future and discounted back to the present.

Although the estimation of economic value is descriptive, its purpose may be normative. Ideally, we would compare such an estimate with the economic value that obtains when uses and resources of the ecosystem are allocated differently. A comparison of the values associated with alternative feasible allocations would measure the opportunity costs of policy interventions or could be used to characterize the most economically efficient allocation.

---

[1] We refer here to "social value," or the value to society--not to specific firms or individuals.

In practice, ecosystem valuation can be very problematic.[2] Few studies include calculations of surpluses from specific uses of the ocean. The results of studies that make such calculations may not be readily transferable to other areas where no studies have been conducted.[3] Resource depletion, pollution, ecological interactions, or irreversibilities further complicate the valuation process. If the effects of these phenomena are lagged, it may be difficult to forecast surpluses into the future. Some species or ecosystem characteristics may be difficult to value because their services are not traded in established markets. If user costs, externalities, or non-use values are ignored, then valuations become incomplete and less useful for normative comparisons. Finally, there may be substantial sources of uncertainty about uses, their markets, and even wholesale data gaps that limit the usefulness of valuation exercises.

The Input-Output Approach

A different analytical approach to understanding the economic characteristics of an LME exists. This approach involves the use of an economic input-output model to estimate the economic activity (or "impact") of marine sectors in coastal economies. The input-output approach was developed by economists to provide a snapshot of the universe of linkages between the economic sectors of an economy. The input-output approach estimates the value of goods and services produced (*i.e.*, gross revenues) in different economic sectors that are linked to a marine sector, such as commercial fishing.

It is important to understand that the input-output approach is not a substitute for the calculation of surpluses. In particular, it does not provide an estimate of net benefits. As such, input-output analysis cannot be used as a normative tool to determine an efficient pattern of resource allocations in a large marine ecosystem (*cf.* Probst and Gavrilis 1987). Moreover, the conventional input-output approach does not capture the effects of resource depletion and environmental degradation in a way that would fully reflect the costs of such phenomena to society.[4]

Although the input-output approach is not useful in making normative decisions, it does have several useful features. First, and most importantly, an input-output model gives us an understanding of the direct and indirect effects of activity in a particular sector on all other sectors from which it purchases and to which it sells goods or services. Thus we can use the model to identify patterns of transactions and to understand the economic "influence" of a large marine ecosystem on all sectors of the relevant economy to which it is linked. Second, the model quantifies this influence in terms of sectoral outputs (in dollars), employment, and other economic measures. The employment measure is important because the level of employment often is a central issue in public management debates. Third, the model may be used to explain economic growth in a region by showing how all linked sectors grow (or decline) as one sector grows (or declines).

---

[2] A small body of literature on the valuation of ecosystems exists and is growing. We do not review the literature in this paper. The interested reader is referred to Bingham *et al.* (1995) for some of the central issues.

[3] The results of some recent meta-analyses have made careless transfers of benefits from small-scale, resource-specific valuation studies to large areas of the ocean. These studies often lack credibility.

[4] For example, some activities, such as responses to an oil spill, may lead to higher dollar estimates of economic impacts--even though oil spills should be regarded as involving a net loss to society.

Input-output models are used widely in regional economic impact analyses (Loomis 1993). Importantly, the input-output model can be used to estimate the economic impacts of different management alternatives. For example, the output levels and labor or supplier requirements associated with regulatory alternatives can be used as inputs to the model to estimate economic impacts such as changes in jobs (by industry), county income, or population. The input-output approach is now being developed by the US National Marine Fisheries Service to estimate the economic impacts of fisheries regulations for federally managed fisheries. A small number of studies examine the economic impact of fisheries and marine-related activities (Steinback 1999; Radtke and Davis 1998; Storey and Allen 1993; Andrews and Rossi 1986; Briggs, Townsend and Wilson 1982; Grigalunas and Ascari 1982).

In this article, we utilize the commercial software program IMPLAN Professional® (IMPLAN) and its associated data package to estimate the economic activity of marine-related industries and sectors associated with the Northeast Shelf LME. We generate state and coastal county level economic impact results for the US coastal states from Maine to North Carolina. This exercise is motivated by two considerations. First, while realizing its limits, we believe that the input-output approach currently provides one of the most practical ways to assemble essential economic information associated with a wide range of economic activities related to an LME. This quantification of economic activities could provide information useful for estimating economic value, including the identification of sectors and problems where the application of costly valuation methodologies might be worthwhile. Second, the IMPLAN data package, which combines key data from all major surveys at the federal, state, and local levels, is so far the most comprehensive database assembled for input-output analysis.

We focus on the Northeast Shelf LME. The Northeast Shelf has been the subject of several oceanographic studies, and it is thought to be a well-defined LME (see generally Sherman, Jaworski and Smayda 1996). The Northeast Shelf extends over approximately 260,000 km^2 (Figure 7-1), supporting a coastal county population of over 40 million within a coastal state population of about 71 million. There have been few attempts to quantify economic activity specifically for this LME.[5] We analyze state and coastal county level data for the Northeast Shelf to estimate associated economic activity. Because the input-output model aggregates data from many industries, it is sometimes difficult to factor out marine-related industries from aggregate industry sectors. Thus, we believe that data from coastal counties results in a better estimate of marine-related economic activity, although this estimate still may represent an upper bound.

The article is organized as follows. In the next section, we outline the input-output methodology. In the subsequent sections, we describe the data, present the results of the model runs, and provide a brief discussion of the results. In the last section, we describe limitations of the model and future research directions.

---

[5] One example is referred to in the prologue to Sherman, Jaworski and Smayda (1996: ix). The editors state that "[t]he coastal states from Maine to North Carolina currently realize $ 1 billion of economic benefits annually from the fisheries of the ecosystem." Note that this figure is likely to be the value of commercial fisheries landings (*i.e.*, gross sales) at that time. The figure is not really an estimate of economic "benefits," but of impacts. It is comparable to the "output impacts" figure we report below. Some other, more specific, examples include economic analyses conducted for specific fishery management plans.

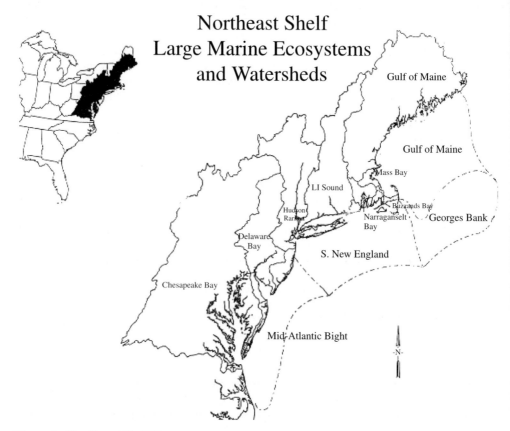

Figure 1. Northeast Shelf Ecosystem

METHODOLOGY

A static[6] Leontief input-output model is a system of linear equations:

$$(I - A)X = Y$$

where I is a $n \times n$ identity matrix; A is a $n \times n$ technical coefficient (input-coefficient) matrix; X is a $n \times 1$ column vector denoting output; and Y is a $n \times 1$ column vector denoting final demand. The idea behind the model is that the output of any industry (x_i, an element of X) is needed as an input in many other industries, or even in that industry itself. Therefore, the correct level of x_i depends on the input requirements of all the n industries as well as final demand.

---

[6] The static analysis has its limitations, because it is not able to address when and what may happen along the adjustment process, which may take a long time to complete. Also, using a static analysis, we cannot determine whether the solution is stable. When certain additional economic considerations are incorporated into the static model, it can take on a dynamic character.

The standard input-output model is based on three assumptions: (i) each industry produces only one homogeneous commodity (broadly interpreted); (ii) each industry uses a fixed input ratio (factor combination) for production of its output; and (iii) production in every industry is subject to constant returns to scale (Chiang 1974).

The elements of $\mathbf{A}$, a_{ij}, are called technical coefficients and are defined as:

$$a_{ij} = \frac{z_{ij}}{x_j}$$

where z_{ij} is the monetary value of the flow from sector i to sector j; and x_j is the total output of sector j.

The matrix $(\mathbf{I}\text{-}\mathbf{A})$ is called the technology matrix. If the technology matrix is not singular, the impact of changes in final demand $(\mathbf{Y})$ on output $(\mathbf{X})$ can be estimated as

$$\mathbf{X} = (\mathbf{I}\text{-}\mathbf{A})^{-1}\mathbf{Y}$$

where $(\mathbf{I}\text{-}\mathbf{A})^{-1}$ is called the Leontief inverse. For a comparative static analysis, we define

$$\mathbf{B} = (\mathbf{I}\text{-}\mathbf{A})^{-1}$$

and the partial derivative gives us the matrix of multipliers

$$\frac{\partial \mathbf{X}}{\partial \mathbf{Y}} = \mathbf{B}$$

For empirical analysis, an input-output table (transactions table) includes all processing sectors (industries), final demand (including consumer/household purchases, private investment, government purchases, and exports), and the payments sector (value added including labor cost, capital cost, taxes, rental payments, and profit). Total industry outlays equal the value of total industry outputs. Outlays are payments made by firms for inputs and for other purposes in the payments sector. Inputs are purchased locally (within the region) or imported from outside the region.[7] Outputs are goods or services produced by the industry. They can be consumed directly by households and others as final demand within the region, or sold to other industries as intermediate demand.

The major advantage of the input-output model is its explicit capture of all the linkages in an economy. For example, suppose a fisheries management option requires a reduction of the number of fishing vessels in a fleet. This exit of vessels leads to a decline in the local fishing industry. To capture the full effect of an industry decline on the regional economy, we need to quantify the importance of the industry to the region. Fisheries contribute to employment and household incomes. Port buildings and equipment also provide a basis for tax revenues that support local and state government programs. Further, as purchasers of inputs, the fishing industry supports a number of other industries such as boat building and repairs. When all the

---

[7] For accurate estimations of regional economic impacts, one must carefully separate the local portion from the imported portion in every purchase/payment.

linkages within the economy are considered, income and employment generated by the fishing industry have ripple effects on the overall income and employment of the region.

Specifically, an industry's contribution to the overall regional economy consists of three components: direct, indirect, and induced effects. In the case of fisheries, if a vessel is taken out of service, the associated lost jobs and income are the direct effects. Indirect effects are additional jobs and income lost in other industries, such as boat repairing, that can be indirectly credited to the lost vessel. The more inputs produced and purchased within the region, the greater the magnitude of the indirect effect. Finally, lost jobs mean lower household income or a smaller number of households in the region. Lower income leads to reduced spending on food, housing, and cars. The latter are induced effects.

Using input-output analysis, we can compute multipliers[8] for a specific industry (*e.g.*, commercial fishing). The multipliers predict changes in regional output, income, value added, and employment[9] in each industry from a given change in its final demand.[10] Because neither environmental quality nor resource stocks have been included as data in the model, the specific impact of a change in final demand on these aspects cannot be quantified. Further, the static input-output model generates only annual economic impacts rather than a discounted sum of future impacts.

Development of an input-output model from primary data is a substantial undertaking. In most cases, management agencies do not have the resources needed to develop survey-based input-output models for a local economy. Instead, they adapt existing models to their purposes. A number of ready-made regional input-output models have been developed to perform economic impact analyses (Brucker *et al.* 1990). The best known is a software package for personal computers, IMPLAN.

IMPLAN was developed at the U.S. National Forest Service (Alward, and Palmer 1983). It is a modular input-output model that works down to the individual county level for any county in the United States.[11] The IMPLAN database consists of two major parts: (1) a national-level technology matrix and (2) estimates of sectoral activity for final demand, final payments, gross output, and employment for each county. This 528-sector (based on 4-digit SIC codes), gross-domestic-based model was derived from the Commerce Department's national input-output studies. In IMPLAN, national average technology coefficients are used to develop the direct coefficients for sectors at the local level (Loomis 1993; Minnesota IMPLAN Group 1997). IMPLAN data can be used as an initial set of technological relationships among regional industry sectors. The system allows the input-output model to be modified with better information (Minnesota IMPLAN Group 1997).

---

[8] There are two types of multipliers. Type I multipliers capture direct and indirect effects; Type II capture direct, indirect and induced effects. The calculation of the Type II multipliers are realized by including households as a "processing sector."

[9] The employment multiplier is usually calculated using a direct employment coefficient that reflects the number of workers per dollar of output delivered to final demand.

[10] More specifically, we focus on the demand for output from within the region, excluding imports.

[11] The latest version of IMPLAN Pro allows input-output analysis at the national, state, county, or zip-code level.

DATA

A specific definition of the marine sector is provided by Pontecorvo *et al.* (1980) in their classic study of the contribution of the marine sector to the U.S. economy. In that study, the marine sector is defined to include those establishments in the national income accounting system that either utilize an ocean resource in a productive process, or exist because the demand for the establishment's final output is due to some attribute of the ocean. The 1980 study was updated for 1987 (Pontecorvo 1989).

We follow the Pontecorvo *et al.* approach in developing estimates of the impact of marine industries using the IMPLAN model. IMPLAN generates estimates of "value-added" effects that are directly comparable to the Pontecorvo *et al.* analysis, and we make such a comparison below. In addition to value-added, IMPLAN generates estimates of indirect and induced effects that were not estimated by Pontecorvo *et al.*

Using the IMPLAN database, we first divide the overall "marine sector" into six broad industry groups: fisheries, shipbuilding, shipping, water quality, tourism, and real estate. Several sectors in the IMPLAN database fall into these broad industry groups. These sectors are listed in Table 7-1, along with their corresponding IMPLAN sector codes.

The IMPLAN database has been constructed from several federal government databases. Each IMPLAN sector comprises data corresponding to one or more standard industrial classification (SIC) codes, which are the codes applied by the US Bureau of Economic Analysis to identify specific industrial sectors. Some of the IMPLAN sectors provide direct evidence of economic activity associated with the marine environment. These sectors include commercial fishing and processing, ship and boat building and repairing, and water transportation. We describe these sectors as the "primary tier," and we list the associated IMPLAN and SIC codes in Table 7-1.

We identify as "secondary tier" sectors all other industries that are arguably marine-related. These include IMPLAN sectors that combine marine-related SIC industries with non-marine industries, such as miscellaneous livestock (includes aquaculture), fishery services (includes fish hatcheries), and amusement and recreation services (includes public beaches, headboats, and scuba diving). In addition, we include cases in which SIC sectors combine both marine and non-marine industries. For example, SIC 7999 (amusement and recreation services, not elsewhere classified) includes a large number of small industries, only some of which are marine-related.[12] SIC sector 3812 (IMPLAN sector 400), search and navigation equipment, includes the manufacture of aeronautical, space, and defense equipment as well as marine equipment. Finally, some IMPLAN sectors are marine-related only when they are located in the coastal zone, such as water supply and sewerage systems, eating and drinking, hotels and lodging places, and real estate.

---

[12] Many of these industries may operate primarily in fresh-water environments. The relevant industries in SIC 7999 include: bath houses (independently operated), public bathing beaches, pleasure boat rentals, operation of party fishing boats, operation of fishing piers and lakes, houseboat rentals, lifeguard services, rental of beach chairs and accessories, rentals of rowboats and canoes, scuba and skin diving instruction, and swimming instruction. There is no clear way to separate out the marine component from these industries. In this report, we assume that they are marine-related.

Table 7-1: Broad Marine Industries and IMPLAN Sectors

BROAD INDUSTRY	TIER	IMPLAN SECTOR	IMPLAN CODE	SIC CODES	RELEVANT SUB-SECTOR
Fisheries	1°	Commercial Fishing	25	0912, 0913, 0919	
		Canned and Cured Seafoods	97	2091	
		Prepared Fresh or Frozen Fish and Seafoods	98	2092	
	2°	Miscellaneous Livestock	9	0273, others	Aquaculture
		Agricultural, Forestry, Fishery Services	26	0921, others	Fish Hatcheries
Shipbuilding	1°	Ship Building and Repairing	392	3731	
		Boat Building and Repairing	393	3732	
	2°	Search and Navigation Equipment	400	3812	
Shipping	1°	Water Transportation	436	4400	
Water Quality	2°	Water Supply and Sewerage Systems	445	4941, 4952	
Tourism	2°	Eating and Drinking	454	5800	
		Hotels and Lodging Places	463	7000	
		Amusement and Recreation Services	488	7999, others	Beaches, Headboats, Scuba
Real Estate	2°	Real Estate	462	6500	

We conduct the analysis at the state and coastal county levels. The state level analysis estimates economic activity associated with the Northeast Shelf LME using state data for Maine, New Hampshire, Massachusetts, Rhode Island, Connecticut, New York, New Jersey, Pennsylvania, Delaware, Maryland, the District of Columbia, Virginia, and North Carolina. The coastal county level analysis focuses on economic activity using coastal county data for these same states. Coastal counties (Table 7-2) are those counties whose populations are used by NOAA to calculate funding under section 306 of the Coastal Zone Management Act (Uravitch 1998).

We expect that the use of coastal county data for the states bordering the Northeast Shelf LME provides a better estimate of economic activity associated with the marine ecosystem. Including the entire secondary tier in a state-level analysis of the marine-related economic activity associated with the Northeast Shelf LME clearly results in an overestimate of that activity. However, it is not simple or straightforward to separate the marine-related component from the more general IMPLAN second tier industry sectors. We handle this issue by running the IMPLAN model for (i) both tiers together and (ii) only the primary tier. This gives us upper and lower bounds on our estimate of economic activity.

Table 7-2: Coastal Counties, Population and Area (mi^2)

State Name	County Name	Population	Area
MAINE	CUMBERLAND	248855	836
	HANCOCK	49386	1589
	KENNEBEC	116945	867
	KNOX	37269	366
	LINCOLN	31423	456
	PENOBSCOT	145529	3396
	SAGADAHOC	34150	254
	WALDO	35454	730
	WASHINGTON	36229	2569
	YORK	169348	991
NEW HAMPSHIRE	ROCKINGHAM	258150	695
	STRAFFORD	106368	369
MASSACHUSETTS	BARNSTABLE	199232	396
	BRISTOL	513150	556
	DUKES	12821	104
	ESSEX	682232	498
	MIDDLESEX	1405798	824
	NANTUCKET	7034	48
	NORFOLK	633992	400
	PLYMOUTH	452773	661
	SUFFOLK	647570	59
RHODE ISLAND	BRISTOL	49122	25
	KENT	162326	170
	NEWPORT	82474	104
	PROVIDENCE	580784	413
	WASHINGTON	116995	333
CONNECTICUT	FAIRFIELD	830702	626
	MIDDLESEX	147306	369
	NEW HAVEN	795485	606
	NEW LONDON	250227	666
NEW YORK	ALBANY	297980	524
	BRONX	1196046	42
	COLUMBIA	63731	636
	DUTCHESS	261512	802
	GREENE	47446	648
	KINGS	2280493	71
	NEW YORK	1525387	28
	NASSAU	1303231	287
	ORANGE	322349	816
	PUTNAM	90138	232
	QUEENS	1974383	109
	RENSSELAER	155322	654
	RICHMOND	396748	59
	ROCKLAND	277034	174
	SUFFOLK	1351843	911
	ULSTER	167223	1127
	WESTCHESTER	891044	433
NEW JERSEY	ATLANTIC	233634	561
	BERGEN	843338	234
	BURLINGTON	407931	805
	CAMDEN	507089	222
	CAPE MAY	98133	255

State Name	County Name	Population	Area
	CUMBERLAND	137748	489
	ESSEX	760615	126
	GLOUCESTER	242924	325
	HUDSON	551198	47
	MERCER	330038	226
	MIDDLESEX	698029	311
	MONMOUTH	585218	472
	OCEAN	466142	636
	PASSAIC	463558	185
	SALEM	65226	338
	SOMERSET	265158	305
	UNION	496735	103
PENNSYLVANIA	BUCKS	573130	608
	DELAWARE	548043	184
	PHILADELPHIA	1499762	135
DELAWARE	KENT	121234	591
	NEW CASTLE	467755	426
	SUSSEX	128052	938
MARYLAND	ANNE ARUNDEL	461981	416
	BALTIMORE	714495	599
	CALVERT	64521	215
	CAROLINE	28983	320
	CECIL	78317	348
	CHARLES	111626	461
	DORCHESTER	29912	558
	HARFORD	205499	440
	KENT	18816	279
	PRINCE GEORGE'S	767006	486
	QUEEN ANNE'S	36876	372
	ST. MARY'S	80984	361
	SOMERSET	24268	327
	TALBOT	32411	269
	WICOMICO	79122	377
	WORCESTER	40101	473
	BALTIMORE CITY	689432	81
WASHINGTON D.C.	WASHINGTON	554528	61
VIRGINIA	ACCOMACK	32123	455
	ARLINGTON	175035	26
	CAROLINE	21083	533
	CHARLES CITY	6786	182
	CHESTERFIELD	239371	426
	ESSEX	9250	258
	FAIRFAX	889015	396
	HANOVER	74716	473
	GLOUCESTER	33250	217
	HENRICO	232176	238
	ISLE OF WIGHT	27839	316
	JAMES CITY	40478	143
	KING GEORGE	16357	180
	KING AND QUEEN	6417	316
	KING WILLIAM	12170	275
	LANCASTER	11267	133
	MATHEWS	8824	86
	MIDDLESEX	9330	130
	NEW KENT	11673	210
	NORTHAMPTON	12979	207

State Name	County Name	Population	Area
	NORTHUMBERLAND	11151	192
	PRINCE GEORGE	28270	266
	PRINCE WILLIAM	243458	338
	RICHMOND	8475	191
	SPOTSYLVANIA	71806	401
	STAFFORD	79921	270
	SURRY	6412	279
	WESTMORELAND	16492	229
	YORK	53891	106
	ALEXANDRIA	115838	15
	CHESAPEAKE	187904	341
	COLONIAL HEIGHTS	16916	7
	FREDERICKSBURG	21899	11
	HAMPTON	138783	52
	HOPEWELL	22458	10
	NEWPORT NEWS	179163	68
	NORFOLK	236129	54
	PETERSBURG	37704	23
	POQUOSON	11680	16
	PORTSMOUTH	102100	33
	RICHMOND	197744	60
	SUFFOLK	56655	400
	VIRGINIA BEACH	429760	248
	WILLIAMSBURG	12642	9
NORTH CAROLINA	BEAUFORT	43998	828
	BERTIE	20745	699
	BRUNSWICK	60697	855
	CAMDEN	6399	241
	CARTERET	57690	531
	CHOWAN	13958	173
	CRAVEN	85163	696
	CURRITUCK	16285	262
	DARE	26074	382
	GATES	9784	341
	HERTFORD	22555	354
	HYDE	5362	613
	NEW HANOVER	139906	199
	ONSLOW	144259	767
	PAMLICO	12064	337
	PASQUOTANK	33759	227
	PENDER	35208	871
	PERQUIMANS	10737	247
	TYRRELL	3846	390
	WASHINGTON	14138	348

RESULTS

Table 7-3 presents the results from the state-level run for the Northeast Shelf LME. Both annual industrial output (1995 $ millions) and employment (thousands of employees) are shown across all industrial sectors. Industry output is the value of an industry's total production, which includes purchases by all other industries, and by consumers and government agencies for final demand

and exports. Both output and employment for these industries represent between 9 and 10 percent of their respective total economy levels in the region. Although not shown explicitly in Table 7-3, for the *primary tier* marine industries *only*, both output and employment represent less than one-half of one percent of their respective state economy totals.

Table 7-3. Northeast Shelf LME: State Level Sectoral Output, Employment, and Type II Multipliers

BROAD INDUSTRY	TIER	IMPLAN SECTOR	OUTPUT[†]	TYPE II MULT	EMPLOY-MENT[‡]	TYPE II MULT
Fisheries	1°	Commercial Fishing	880	1.87	19	1.46
		Canned and Cured Seafoods	236	1.65	2	1.76
		Prepared Fresh or Frozen Fish and Seafoods	1,294	1.63	9	2.29
	2°	Miscellaneous Livestock	638	1.64	29	1.18
		Agricultural, Forestry, Fishery Services	1,961	1.78	71	1.27
Shipbuilding	1°	Ship Building and Repairing	4,952	1.87	51	2.00
		Boat Building and Repairing	932	2.00	9	2.08
	2°	Search and Navigation Equipment	10,013	1.97	53	3.02
Shipping	1°	Water Transportation	9,394	1.95	45	3.22
Water Quality	2°	Water Supply and Sewerage Systems	1,474	1.85	7	3.22
Tourism	2°	Eating and Drinking	67,875	1.92	1,905	1.37
		Hotels and Lodging Places	26,869	1.91	427	1.73
		Amusement and Recreation Services	8,439	1.91	278	1.34
Real Estate	2°	Real Estate	200,665	1.52	927	2.36
TOTALS			335,622		3,832	
% of State Total			9.25%		9.44%	

[†] $U.S. millions (1995)
[‡] Thousands of employees

In Table 7-3, we show also the "Type II" multipliers for output and employment for each industry. Type II multipliers measure the effect of changes in final demand[13] for one industry on output in

---

[13] Final demand represents purchases by end users (consumers or firms), government agencies, and exports for consumption. Once final consumption occurs, goods and services disappear from the economy and are not available to generate further output. Final demand is *not* shown in the tables in this article.

all other linked industries and on the income of people employed in those industries. For example, for each $1.00 of final demand for boat building and repairing, $2.00 in industrial output and household income is generated in the Northeast Shelf LME "economic region." Employment multipliers are interpreted in much the same way. For example, each job in water transportation generates about three (3.22) jobs in the regional economy.

We present in Table 7-4 the same type of information as Table 7-3, focusing on the coastal county levels for the Northeast Shelf LME. Output and employment are reported in the same terms. Coastal county level output and employment are around 10 percent of the total for the coastal counties. Although not shown explicitly in Table 7-4, the primary tier output (0.7%) and employment (0.5%) percentages of the total coastal counties-level economy are fairly low.

The Type II multipliers for sectoral output are all in the same general range (between 1.48 and 2.00). There appears to be little difference between the state and coastal county level multipliers. At the coastal county level, the water transportation, search and navigation equipment, amusement and recreation, commercial fishing, and hotel and lodging place sectors have the largest output multipliers. Real estate has the lowest multiplier. At the county level, water supply and sewerage, water transportation, and search and navigation equipment have the largest employment multipliers. The miscellaneous livestock, amusement and recreation services, and eating and drinking sectors have the smallest multipliers.

Because multipliers are important for impact analyses, one must be careful not to misinterpret them. The Type II multipliers reported in Table 7-3 and Table 7-4 may be used directly to generate total impacts in a regional economy when changes in final demands are strictly limited to products made within the region. While this use is appropriate when we assess the contribution of an industry to the regional economy, an adjustment must be made to the multipliers when one wants to estimate the impact associated with changes in final demands for *local* as well as *imported* products. Generally, to assess the impact of an increase in final demand in a region, one must adjust the multiplier with a regional purchase coefficient (RPC) that reflects the percentage of demand met by local producers. The balance is provided by imports.[14] If the RPC is not applied, the total impacts will be exaggerated.

Table 7-5 presents county output and employment as a percent of state output and employment. Note that the primary tier industries exceed 80 percent of the state levels in all cases except for boat building and repairing (which is close at 79% of output; 78% of employment). This result suggests that, at the *state level*, the primary tier industries provide a good estimate of economic activity, but only for the primary tier industries. Among the secondary tier industries, coastal county outputs for all other sectors, except for miscellaneous livestock (39%), represent more than 50 percent of the state outputs. This result implies that, while these sectors are important contributors to economic activity associated with the Northeast Shelf LME, including the second tier industries in a state level analysis results in an overestimate of the influence of the Gulf of Maine ecosystem on the region.

---

[14] For example, if the change in demand is $1 million, the Type II multiplier is 2.00, and the RPC is 0.75, then the adjusted multiplier is 1.50 (0.75 times 2.00), and the impact on the local economy is $1.5 million (not $2 million).

Table 7-4: Northeast Shelf LME: Coastal County Level Sectoral Output, Employment, and Type II Multipliers

BROAD INDUSTRY	TIER	IMPLAN SECTOR	OUTPUT[†]	TYPE II MULT	EMPLOYMENT [‡]	TYPE II MULT
Fisheries	1°	Commercial Fishing	855	1.83	18	1.42
		Canned and Cured Seafoods	234	1.65	2	1.73
		Prepared Fresh or Frozen Fish and Seafoods	1,187	1.61	8	2.18
	2°	Miscellaneous Livestock	250	1.53	11	1.14
		Agricultural, Forestry, Fishery Services	1,457	1.68	45	1.27
Shipbuilding	1°	Ship Building and Repairing	4,872	1.77	50	1.86
		Boat Building and Repairing	735	1.71	7	1.79
	2°	Search and Navigation Equipment	6,554	1.85	35	2.71
Shipping	1°	Water Transportation	7,694	1.96	37	3.12
Water Quality	2°	Water Supply and Sewerage Systems	878	1.78	4	3.23
Tourism	2°	Eating and Drinking	38,993	1.80	1,042	1.33
		Hotels and Lodging Places	18,563	1.83	265	1.71
		Amusement and Recreation Services	5,413	1.84	171	1.31
Real Estate	2°	Real Estate	146,250	1.48	619	2.30
TOTALS			233,935		2,314	
% of Coastal County Total			10.65%		9.81%	

[†] $U.S. millions (1995)
[‡] Thousands of employees

Table 7-5: Northeast Shelf LME: Coastal County Level Sectoral Output and Employment as a % of State Level

BROAD INDUSTRY	TIER	IMPLAN SECTOR	OUTPUT[†]	EMPLOY-MENT[‡]
Fisheries	1°	Commercial Fishing	97	95
		Canned and Cured Seafoods	99	100
		Prepared Fresh or Frozen Fish and Seafoods	92	89
	2°	Miscellaneous Livestock	39	38
		Agricultural, Forestry, Fishery Services	74	63
Shipbuilding	1°	Ship Building and Repairing	98	98
		Boat Building and Repairing	79	78
	2°	Search and Navigation Equipment	65	66
Shipping	1°	Water Transportation	82	82
Water Quality	2°	Water Supply and Sewerage Systems	60	57
Tourism	2°	Eating and Drinking	57	55
		Hotels and Lodging Places	69	62
		Amusement and Recreation Services	64	62
Real Estate	2°	Real Estate	73	67

[†] $U.S. millions (1995)
[‡] Thousands of employees

We present in Table 7-6 the state-level output and employment impacts for the Northeast Shelf LME. These estimates are broken down into direct, indirect, and induced impacts.[15] Our upper bound estimate (primary and secondary tiers) of the marine-related economic activity due to the LME is $507 billion, employing 5.9 million persons. Our lower bound estimate (primary tier only) is $31 billion, employing 296 thousand. We examine too the impacts that result from broad industry groupings (including both tiers, where applicable). For example, the fisheries industry group has an $8 billion impact, employing 170,000 persons.

---

[15] Since industry sectors (*e.g.*, fishing and seafood processing) are linked in an input-output model, there would be a problem of double counting which leads to over-estimating the indirect and induced impacts, when we calculate multi-sector cumulative impact. To avoid double counting, we cut the linkages among sectors in a specific industry group by setting RPC = 0 for these sectors using IMPLAN editing functions. We construct separate models for each of the five sector groups (*e.g.*, Fisheries) shown in tables 6, 7, 9, and 10.

Table 7-6: Northeast Shelf LME: State Level Output and Employment Impacts Ascribed to Aggregated Marine Sectors

	OUTPUT IMPACT[†]				EMPLOYMENT IMPACT[‡]			
	Direct	Indirect	Induced	Total	Direct	Indirect	Induced	Total
Tiers 1&2	335,621	94,510	77,263	507,394	3,834	1,112	984	5,929
Tier 1	17,687	5,863	7,380	30,931	135	64	96	296
Fisheries*	5,009	1,150	2,249	8,407	130	11	29	170
Ship-building* & Shipping	25,290	9,695	11,585	46,570	159	97	151	406
Water Quality, Tourism & Real Estate	305,322	85,086	66,728	457,136	3,545	1,018	849	5,413

[†] $U.S. millions (1995)
[‡] Thousands of employees
*Includes tiers 1 and 2

In Table 7-7, we present the same kind of information as in Table 7-6, focusing on coastal county impacts. Our upper bound estimate (primary and secondary tiers) of the marine-related economic activity in coastal counties due to the LME is $339 billion, employing 3.6 million persons. Our lower bound estimate (primary tier only) is $26 billion, employing 245,000 persons. The fisheries sector represents about $6 billion in output impacts at the coastal county level, with 112,000 employees.

Table 7-8 examines the coastal county output and employment impacts as a percent of the state level impacts. Again, the tier 1 coastal county impacts are a large proportion of the state level impacts. The combined water quality, tourism, and real estate sectors have the smallest percentages.

In Table 7-9 and Table 7-10, we present the state and coastal county level *value-added* estimates. These tables are organized exactly like those preceding, showing direct, indirect, induced, and total value-added impacts. In the IMPLAN model, total value-added is defined as industry output less the sum of inter-industry sales and imports, [16] and it is equivalent to the measure used to estimate gross state product (GSP). Thus the last columns in Table 7-9 and Table 7-10 can be used to compare to the estimate of GSP for the region (about $2.2 trillion in 1995). The total for tiers 1 and 2 can be thought of as an upper bound on marine-related value-added, and the total for tier 1 only can be thought of as a lower bound. Table 7-11 shows coastal county value-added impacts as a percentage of state level impacts.

---

[16] Value-added includes employment compensation, proprietary income, other property type income, and business taxes.

Table 7-7: Northeast Shelf LME: Coastal County Level Output and Employment Impacts Ascribed to Aggregated Marine Sectors

	OUTPUT IMPACT[†]				EMPLOYMENT IMPACT[‡]			
	Direct	Indirect	Induced	Total	Direct	Indirect	Induced	Total
Tiers 1&2	233,935	58,662	46,599	339,196	2,313	679	569	3,561
Tier 1	15,577	4,656	5,878	26,110	123	49	74	245
Fisheries*	3,983	886	1,566	6,435	84	8	20	112
Shipbuilding* & Shipping	19,855	6,639	8,093	34,587	129	65	101	295
Water Quality, Tourism & Real Estate	210,096	52,172	39,192	301,461	2,100	615	478	3,193

[†] $U.S. millions (1995)
[‡] Thousands of employees
*Includes tiers 1 and 2

Table 7-8: Northeast Shelf LME: Coastal County Level Output and Employment Impacts as a % of State Level Impacts

	OUTPUT IMPACT (%)				EMPLOYMENT IMPACT (%)			
	Direct	Indirect	Induced	Total	Direct	Indirect	Induced	Total
Tiers 1&2	70	62	60	67	60	61	58	60
Tier 1	88	79	80	84	91	77	77	83
Fisheries*	80	77	70	77	65	73	69	66
Shipbuilding* & Shipping	79	68	70	74	81	67	67	73
Water Quality, Tourism & Real Estate	69	61	59	66	59	60	56	59

*Includes tiers 1 and 2

Table 7- 9: Northeast Shelf LME: State Level Value-Added Impacts Ascribed to Aggregated Marine Sectors

	VALUE-ADDED IMPACT[†]			
	Direct	Indirect	Induced	Total
Tiers 1&2	204,802	53,541	48,848	307,192
Tier 1	6,548	3,375	4,609	14,532
Fisheries*	2,712	614	1,403	4,729
Shipbuilding* & Shipping	9,480	5,445	7,234	22,159
Water Quality, Tourism & Real Estate	192,610	48,285	42,127	283,022

[†] $U.S. millions (1995)
*Includes tiers 1 and 2

Table 7-10: Northeast Shelf LME: Coastal County Level Value-Added Impacts Ascribed to Aggregated Marine Sectors

	VALUE-ADDED IMPACT†			
	Direct	Indirect	Induced	Total
Tiers 1&2	144,536	34,611	30,005	209,151
Tier 1	5,966	2,755	3,741	12,462
Fisheries*	2,115	480	995	3,590
Shipbuilding* & Shipping	7,630	3,878	5,150	16,658
Water Quality, Tourism & Real Estate	134,791	30,852	25,188	190,831

† $U.S. millions (1995)
*Includes tiers 1 and 2

Table 7-11: Northeast Shelf LME: Coastal County Level Value-Added Impacts Ascribed to Aggregated Marine Sectors as a % of State Level Impacts

	VALUE-ADDED IMPACT†			
	Direct	Indirect	Induced	Total
Tiers 1&2	71	65	61	68
Tier 1	91	82	81	86
Fisheries*	78	78	71	76
Shipbuilding* & Shipping	80	71	71	75
Water Quality, Tourism & Real Estate	70	64	60	67

† $U.S. millions (1995)
*Includes tiers 1 and 2

DISCUSSION

The industries that are directly related to the use of the Northeast Shelf LME and its resources have a substantial influence on the economies of the states of New England and the mid-Atlantic. Coastal county level data are more useful than state level data to estimate the regional economic influence of the marine environment. Our best upper bound estimates for economic activities associated with the Northeast Shelf LME are $339 billion in total output impacts, 3.6 million persons, and $209 billion in value-added impacts. This latter estimate is 9.5% of the total GSP for the coastal states ($2.2 trillion in 1995).[17] A lower bound estimate, using value-added impacts from the first tier only, is less than one percent of the total GSP for the region.

---

[17] Gross *county* product (GCP) for the counties bordering the Northeast Shelf LME is $1.4 trillion. Our upper bound estimate is 14.9% of GCP.

Our upper bound estimate is considerably larger than the Pontecorvo (1987) estimate (2.6%) of the ocean sector as a percentage of total industry contribution to GNP at the national level. There are at least three reasons for this discrepancy. First, the IMPLAN sectors include some SIC sectors that arguably are non-marine. Inclusion of these sectors would inflate our estimate relative to that obtained by Pontecorvo *et al.*[18] However, examination of coastal county level data should ameliorate this problem, because second tier industries are more likely to have a marine connection when located in coastal counties. Second, Pontecorvo *et al.* estimate marine production as a direct value-added impact. Our upper bound direct value-added impact (Table 7-10) is $145 billion, accounting for roughly 6.5 percent of the regional GSP. Third, Pontecorvo *et al.* examine marine-related production as a proportion of national value-added. National value-added necessarily includes production from industries in areas of the country that have no marine connections, thereby reducing the relative marine contribution. We would expect to find a higher proportion of marine-related value-added in coastal states. The true level of economic activity probably lies somewhere in between our estimate and that of Pontecorvo *et al.* Further research, involving the disaggregation of both IMPLAN and SIC sectors, is required to resolve this issue.

Fisheries are often thought to be the most important use of the marine environment. However, our estimates show that, including the secondary tier industries, fisheries account for only 2 percent of total output impacts, 3 percent of employment impacts, and 2 percent of value-added impacts. Note also that seafood processing, defined here to be a tier 1 industry sector, may involve significant amounts of imports when local fish stocks are overexploited. Although the processing activity itself generates important economic impacts, if the fish being processed do not derive from the Northeast Shelf LME, then it is inappropriate to attribute those impacts to the LME. More work needs to be done to discover what proportion of the New England and mid-Atlantic processing sector depends specifically upon the Northeast Shelf.

Shipbuilding and shipping represent about 10 percent of total output impacts, 8 percent of employment impacts, and 8 percent of value-added impacts. This grouping represents important industries that are almost completely reliant[19] upon the existence of the ocean as an economical transportation medium. It is important to note, however, that the reliance of these industries upon an ecosystem, per se, is more tenuous. In fact, the growth of the shipping and shipbuilding industries may be limited because of interactions with the ecosystem. For example, the occurrence of oil and hazardous waste spills, waste disposal, transport of non-indigenous species, and ship strikes of marine mammals, among others have all contributed to the development of a more stringent regulatory environment. While many of these regulations are worthwhile and act to protect important ecosystem features, they may limit the potential economic impacts from this industry sector. In future work, it will be important to identify those industry sectors that contribute to marine environmental protection vis-à-vis the shipbuilding and shipping sectors,

---

[18] Note that Pontecorvo *et al.* (1980) include additional sectors, such as offshore oil and gas and naval expenditures, that are not included in our analysis. Removal of these sectors from the Pontecorvo *et al.* model would lower their estimate of the contribution of the marine sector to the national economy.

[19] Note that only the marine component of the search and navigation equipment sector depends upon the adjacent LME. However, even if we are able to disaggregate the marine business from this sector, it would represent an overestimate of economic activity associated with the Northeast Shelf LME because some of the manufactures are sold to firms or consumers operating in regions other than the Northeast Shelf.

such as oil spill prevention or electronic charting.[20] Of course, the recreational boating industry is dependent upon a healthy ecosystem.

Most impacts occur in the water quality, tourism and real estate sectors: 89 percent of output impacts, 90 percent of employment impacts, and 91 percent of value added impacts. These industry groupings are all second tier. The tourism industries tend to aggregate many different kinds of activities, and it is difficult to separate those activities that are distinctly marine-related from those that are not. However, a case can be made that the marine environment and its associated coastal zone represent important attributes of a multifaceted tourism "experience," whose output and employment impacts would be much diminished in their absence. Nevertheless, more work needs to be done to identify the components of these industries that are directly related to the marine environment. Water quality and real estate are much easier to categorize as primary tier industries when located in the coastal zone.

In the future, critical interactions between industrial sectors and the ecological health of the Northeast Shelf will affect associated economic activity in opposing ways. A better managed and healthier large marine ecosystem leads to higher levels of output, value added, and employment impacts in industries such as fisheries, tourism, boat building, water quality, and real estate. On the other hand, actions taken to improve the health of the ecosystem may limit the growth of the shipbuilding and shipping sectors. A possible restriction on the disposal of dredged materials from New York Harbor is an excellent case in point. Too, restrictions imposed on the commercial fishing industry may limit or reduce output and employment from that sector—at least until stocks recover. These interactions and the range of possible effects on economic activity in the Northeast Shelf region need to be examined more closely in future efforts.

LIMITATION OF THE MODEL AND FUTURE RESEARCH

In this paper, we have presented a positive description of the economic activities associated with the Northeast Shelf LME. Although the information generated from the IMPLAN model can be useful for policymakers in understanding the economic impacts of marine-related sectors on the coastal economy, due to a number of limitations, the model is not directly useful for making ecosystem management decisions. Instead, we believe the model could be an important building block for the development of an integrated ecological-economic analytical framework.

As noted, a major limitation of a conventional input-output model (*e.g.*, IMPLAN) is its exclusion of the effects of environmental degradation and resource depletion. In order to address this issue, economists have taken some initial steps toward expanding the input-output model by including environmental sectors explicitly (Leontief 1970). Progress in developing resource and environmental accounting and some important issues have been summarized in recent work by Nordhaus and Kokkenlenberg (1999).

Other limitations of the input-output approach relate to its underlying assumptions. For example, naive impact analyses assume that labor and resources have no alternative uses. An increase in factor demand will be met by local supplies or imports at fixed costs (*e.g.*, wage rate is constant).

---

[20] We expect that many of these sectors are already embedded in the shipbuilding and shipping industries.

However, with full employment, an increase in wage payments and an increase in output in one sector comes largely at the cost of relative reductions in wages and output in other sectors.

Finally, conventional input-output models are static and deterministic. However, ecosystem management involves decision-making under uncertainty and possibly irreversibility (Chavas 2000) and in a dynamic context (*e.g.*, inter-temporal resource allocation). For example, the economic benefits of resource conservation efforts may not be realized all at once. The selection of a discount rate may be critical in comparing benefits and costs in different periods (Starrett 2000).

For LME management, there exists a need for development of an integrated economic-ecological framework (Arrow *et al.* 1995). Such a framework would extend the traditional bioeconomic approach (Clark 1990). It would consist of two major components that model, respectively, the economic system and the ecosystem. It should capture two general types of linkages between the two systems. The first linkage represents the supply of ecosystem resources, goods, and services to a coastal economy (*e.g.*, fish stocks as inputs to the fish harvesting industry), and the second describes the impacts of economic activities on the ecosystem (*e.g.*, marine pollution, bycatch, and destruction of fish habitat).

The integrated model could be designed to be used to describe existing economic and ecological conditions and to demonstrate the potential wealth to society that may be derived from the consumption of marine resources, goods, and services associated with a well-managed marine ecosystem (*cf.*, Edwards and Murawski 1993). The model could be useful for assessing the change in wealth associated with changes in the quality and quantity of natural and environmental resources in the ecosystem. Further, the integrated model would be useful for exploring a variety of policy-relevant research questions. For example, a change in final demand for the output of a particular industry could be traced back to determine its impact on the structure of the ecosystem. On the other hand, a change in the structure of the ecosystem could be followed through to determine its economic impacts. Because there may be more than one feasible ecosystem state, the economic impacts of alternative states might be compared.

Although the concept of a dynamic general equilibrium model is clear, it is difficult to construct such a model to capture the many interactions between ecological and economic systems. Most classical bioeconomic models involve the dynamic control of nonlinear biosystems (see Clark 1990). Because of complexity, these models include a small number of variables (*e.g.*, biomass and either fishery yield or fishing effort). Starrett (2000) has argued that we are still far away from the capability of constructing a dynamic general equilibrium model, much less analyzing it.

Realizing the tradeoff between the number of variables and nonlinear dynamics in modeling, we believe an interesting area for future research is to explore the possibility of merging a regional input-output model of a coastal economy with a model of a marine food web (*viz.* Jin *et al.* 2003). This type of analysis reprises the seminal work conducted by Walter Isard and his colleagues more than three decades ago (Isard *et al.* 1968). It makes sense to revisit this approach now because of the improved input-output framework (*e.g.*, the IMPLAN model) and the development of marine ecosystem models for New England. Given these developments, creating a linear version of the integrated economic-ecological model enhances the potential for making sound public policy decisions, and could serve as a foundation for the development of dynamic analysis.

ACKNOWLEDGEMENTS

An early version of this chapter was presented at the Workshop on the Human Dimensions of LMEs at the W. Alton Jones Campus, University of Rhode Island, 12-15 February 2000. The authors thank K. Sherman, A.R. Solow, and J. Uravitch for helpful comments and assistance on earlier versions of this chapter. Prepared with sponsorship from the US Department of Commerce, NOAA, Narragansett Laboratory (Ref. Ord. No. 40ENNF800239), the Social Sciences Branch of the Northeast Fisheries Science Center, and the Marine Policy Center, WHOI. WHOI Contribution No. 9936.

REFERENCES

Alward, G. and C. Palmer. 1983. IMPLAN: An input-output analysis system for Forest Service planning, In U.S. Forest Service, IMPLAN Training Notebook. Fort Collins, Colorado: Land Management Planning, Rocky Mountain Forest and Range Experiment, U.S. Forest Service.

Andrews, M. and D. Rossi. 1986. The economic impact of commercial fisheries and marine-related activities: A critical review of northeastern input-output studies. *Coastal Zone Management Journal* 13(3/4): 335-367.

Arrow, K., B. Bolin, R. Costanza, P. Dasgupta, C. Folke, C. Holling, B. Jasson, S. Levin, K-G. Maler, C. Perrings, and D. Pimentel. 1995. Economic growth, carrying capacity, and the environment. *Science* 268:520-521.

Bingham, G., R. Bishop, M. Brody, D. Bromley, E. Clark, W. Cooper, R. Costanza, T. Hale, G. Hayden, S. Kellert, R. Norgaard, B. Norton, J. Payne, C. Russell and G. Suter. 1995. Issues in Ecosystem Valuation: Improving Information for Decisionmaking. *Ecological Economics* 14: 73-90.

Briggs, H., R. Townsend and J. Wilson. 1982. An input-output analysis of Maine's fisheries. *Marine Fisheries Review* 44(1):1-7.

Brucker, S.M., S.E. Hastings and W.R. Latham. 1990. The variation of estimated impacts from five regional input-output models. *International Regional Science Review* 13:119-39.

Chavas, J. 2000. Ecosystem valuation under uncertainty and irreversibility. *Ecosystems* 3:11-15.

Chiang, A.C. 1974. *Fundamental Methods of Mathematical Economics.* New York: McGraw-Hill.

Clark, C.W. 1990. *Mathematical Bioeconomics: the Optimal Management of Renewable Resources.* 2nd edition. New York: John Wiley & Sons, Inc.

Edwards, S.F. and S.A. Murawski. 1993. Potential economic benefits from efficient harvest of New England groundfish. *North American Journal of Fisheries Management* 13: 437-449.

Grigalunas, T. and C. Ascari. 1982. Estimation of income and employment multipliers for marine-related activity in the Southern New England marine region. *New England Journal of Agricultural and Resource Economics* 11(1):25-34.

Isard, W., K.E. Bassett, C.L. Choguill, J.G. Furtado, R.M. Izumita, J. Kissin, R.H. Seyfarth and R. Tatlock. 1968. Ecologic-economic analysis for regional development. Mimeo. Cambridge, Mass.: Regional Science and Landscape Analysis Project, Department of Landscape Architecture, Harvard University (December).

Jin, D., P. Hoagland, and T.M. Dalton. 2003. Linking economic and ecological models for a marine ecosystem. *Ecological Economics* 46(3):367-385.

Leontief, W. 1970. Environmental repercussions and the economic structure: an input-output approach. *American Economic Review* 52 (August): 263-271.

Loomis, J.B. 1993. *Integrated Public Lands Management.* New York: Columbia University Press, pp. 171-191.

Minnesota IMPLAN Group. 1997. IMPLAN Professional®: User's Guide, Analysis Guide, and Data Guide. Stillwater, Minnesota.

Nordhaus, W. and E. Kokkenlenberg, eds. 1999. *Nature's Numbers: Expanding the National Economic Accounts to Include the Environment.* Washington, DC: National Academy Press.

Pontecorvo, G., M. Wilkinson, R. Anderson and M. Holdowsky. 1980. Contribution of the Ocean Sector to the United States Economy. *Science* 208: 1000-1006.

Pontecorvo, G. 1989. Contribution of the Ocean Sector to the United States Economy: Estimated Values for 1987: A Technical Note. *Marine Technology Society Journal* 23(2): 7-14.

Probst, D.B. and D.G. Gavrilis. 1987. Role of Economic Impact Assessment Procedures in Recreational Fisheries Management. *Transactions of the American Fisheries Society* 116: 450-460.

Radtke, H.D. and S.W. Davis. 1998. Description of Oregon's Commercial Fishing Industry in 1996 and 1997. Report 1 in a series of 5. Corvallis, Oregon: The Research Group.

Sherman, K., N.A. Jaworski and T.J. Smayda, eds. 1996. *The Northeast Shelf Ecosystem: Assessment, Sustainability, and Management.* Cambridge, Massachusetts: Blackwell Science, p. ix.

Starrett, D.A. 2000. Shadow pricing in economics. *Ecosystems* 3:16-20.

Steinback, S.R. 1999. Regional economic impact assessment of recreational fisheries: An application of the IMPLAN modeling system to marine recreational party and charter boat fishing in Maine. *North American Journal of Fisheries Management* 19:724-736.

Storey, D.A. and P.G. Allen. 1993. Economic Impact of Marine Recreational Fishing in Massachusetts. *North American Journal of Fisheries Management* 13:698-708.

Uravitch, J. 1998. Personal communication. Washington: Coastal Programs Division, Office of Ocean and Coastal Resource Management, NOAA (2 November).

Large Marine Ecosystems, Vol. 13
T.M. Hennessey and J.G. Sutinen (Editors)

8

Portfolio Management of Fish Communities in Large Marine Ecosystems[1]

Steven F. Edwards, Jason S. Link and Barbara P. Rountree

ABSTRACT

The single-species approach to fisheries management is under fire as valuable fish stocks around the world continue to collapse despite 25 years of extended federal jurisdiction and intensive government regulations. Consensus is mounting for ecosystem-based approaches, but few tangible alternatives have been offered. We propose a two-part portfolio approach to multi-species management which integrates the fish and fisheries, socioeconomic, and governance modules of the LME model. The portfolio framework is a technical methodology which explicitly integrates fish stocks that are joined by ecology (*e.g.*, predation) and fishing technology (*e.g.*, mixed-species fisheries), and selects combinations of species biomass and other stock attributes (*e.g.*, fish size, recruitment patterns, growth rates) that balance expected aggregate returns for society against the risks associated with ecological, market, and institutional uncertainties. The technical framework is complemented by property rights institutions that remove the wedge between harvest and stock rights now in place in government regulatory regimes. New property rights institutions are needed which create incentives for harvesters to regard fish stocks as capital assets that potentially produce benefits for society indefinitely, internalize spillovers among fisheries caused by ecological and technological interactions, and reduce uncertainty through investments in information.

INTRODUCTION

The predominant approach that governments and several international organizations use to manage fishery resources treats species in isolation from each other. In the United States, the Magnuson-Stevens Fishery Conservation and Management Act (Magnuson-Stevens Act) hints at multi-species management of stock complexes and the prey of managed species, but, in practice, management plans generally define overfishing, bycatch, and optimum yield on the basis maximum sustainable yield (MSY) from individual stocks of species.[2]

---

[1] Cf. an article similar to this one was published by Edwards, S.F., J.S. Link, and B.P. Rountree. Portfolio management of wild fish stocks. *Ecological Economics* 49 (2004):317-329.

[2] The National Standard Guidelines define MSY as "the largest long-term average catch or yield that can be taken from a stock or stock complex under prevailing ecological and environmental conditions" (Federal Register 63(84), 1998, p. 24229). Also, the Essential Fish Habitat Final Rule allows for "limits on the take of species that provide structural habitat for other species assemblages or communities and limits on the take of prey species" (Federal Register 67(12), 2002, p. 2378). Although these perspectives speak to ecosystem-based

The Large Marine Ecosystem (LME) model of fisheries management (Sherman and Duda 1999) contrasts with the long-standing and often criticized (Larkin 1996) single-species approach. Thermodynamics alone suggests that there is not enough primary production in a system to support all populations at MSY levels. Further, simple theoretical models of ecological and technological interactions demonstrate that there can be an infinite number of MSYs [and of the economics counterpart, maximum economic yields (MEY)] for each stock, contingent on the populations of other species and environmental states (Anderson 1975; Hannesson 1983; Huppert 1979; Larkin 1966; May *et al.* 1979; Mitchell 1982; Pontecorvo 1986). These theoretical points find support in empirical work. For example, estimates of Georges Bank haddock biomass which corresponded to MSY during 1931-1993 varied greatly depending on the time-period chosen (Applegate et al. 1998). Also, Brown *et al.* (1976) reported that the sum of MSY estimates from individual stock assessments of several demersal finfish species in the Northwest Atlantic by far exceeded the MSY estimate for the complex.

Research on multi-species VPAs (*i.e.*, virtual population analysis; Pope 1991), ecosystem simulation models such as Ecosim (Walters *et al.* 1997), and LMEs (Sherman and Alexander 1986) expose difficult questions about what the mix of species and population sizes in an ecosystem should be, and what sorts of institutional arrangements can operationalize multi-species management. There is no reason to expect biological yield goals to coincide with society's preferences. Indeed, simple extrapolation of MSY-thinking to fish communities could extinguish slow-growing species and top predators. Accordingly, Beddington *et al.* (1984: 236), state that "... efforts to partition and reduce the 'fishery problem' to its biological essentials are misguided. And, by extension, the 'solutions' thereby determined and presented by biologists are too often suboptimal, if not irrelevant, to the actual needs of those trying to make the best use of the resource." The notion of optimum yield as defined, for example, in the Magnuson-Stevens Act, is not helpful here because it is conditional on MSY, as well as too vague to be operational and evaluated: "[u]nfortunately, this concept is often referred to without any clear understanding as to what objective is to be 'optimized'" (Clark 1976: 3).[3] Further, rigid notions of sustainability can be questioned on the basis of social welfare (Solow 1991) and dramatic regime shifts that have not altered conventional measures of biodiversity (Steele 1998).

We agree with Hanna's (1998) suggestion that fisheries management should be reoriented from single species to portfolios of species. Our two-part portfolio approach to multi-species management views LME resources as risk-bearing capital assets that can provide society with benefits indefinitely. One part is a technical framework which evaluates tradeoffs in fishery benefits (protein, community stability, employment, or income) due to ecological interactions (*e.g.*, predation) and unspecialized harvest technologies (*e.g.*, mixed-species fisheries), and which further balances the expected aggregate benefits from manipulating species biomass and other valuable stock attributes (*e.g.*, fish size, recruitment patterns, growth rates, gene

---

management, the standard notion of MSY permeates these rules.

[3] The Magnuson-Stevens Act defines optimum yield as "the amount of fish which - (A) will provide the greatest benefit to the Nation, particularly with respect to food production and recreational opportunities, and taking into account the protection of marine ecosystems; (B) is prescribed as such on the basis of the maximum sustainable yield from the fishery, as reduced by any relevant economic, social, or ecological factor; and (C) in the case of an overfished fishery, provides for rebuilding to a level consistent with producing the maximum sustainable yield in such fishery."

pool, food habits, habitat requirements) against risk. The portfolio framework requires society's objectives and ecological constraints to be clearly defined and capable of being evaluated.

The other part of the portfolio approach puts the technical framework into action with institutional arrangements that treat fish stocks as economic assets, that create incentives for parties to maximize benefits of the fish community as a whole, and that protect investments in information which reduce uncertainty, including through adaptive management.

The next section presents the ecological context of the portfolio approach to multi-species fisheries management. This is followed with a general description of the technical framework and then the requirements of compatible institutional arrangements. We end with a summary and conclusions.

ECOLOGICAL FOUNDATIONS OF THE PORTFOLIO APPROACH

Wild populations of finfish and invertebrates normally undergo natural fluctuations in biomass and other attributes (Beddington 1986; Beddington *et al.* 1984). Much of this variability is thought to be due to abiotic (Myers 1998; Rothschild 2000) and biotic (Bailey and Houde 1989; Rice *et al.* 1993; Sissenwine et al. 1984) factors operating during larval and early juvenile stages prior to recruitment to a fishery. Bax (1991, 1998) concluded from his reviews of several marine ecosystems that mortality from fishing is commonly only a small fraction of the total mortality to which fish are subjected across their entire life history.

Much of larval and juvenile fish mortality is due to predation. This is illustrated by energy budgets for the Georges Bank region of the Northwest Atlantic Ocean (*e.g.* Cohen et al. 1982; Overholtz *et al.* 1991; Sissenwine 1986; Sissenwine *et al.* 1984) where approximately three-quarters of the fish production was consumed by fish followed by marine mammals (10%), fishermen (10%), and sea birds (5%). An implication of this information is that there likely are many possible natural stable states of multi-species communities (*e.g.*, Beddington 1986; Steele 1984, 1998; van de Koppel *et al.* 2001).

Fishing can have major impacts on the species composition and biomass structure of an adult fish community, however, thereby multiplying the number of possible stable states for a society to choose from. Over 70% of the world's target species are fished excessively, even to the point of collapse (NRC 1999). Of particular concern are biomass flips to low-value species (*e.g.*, Fogarty *et al.* 1991; Sherman and Alexander 1990) and what Pauly *et al.* (1998) call "fishing down the food web"—*i.e.*, the gradual transition from stocks of long-lived, high trophic level, piscivivorous bottom fish toward short-lived, low trophic level invertebrates and planktivorous pelagic fish. Other species in heavily fished communities are affected as well due to compensatory mechanisms that co-evolved over time. For example, competitive release may have allowed low-value elasmobranchs (dogfish sharks and skates) to replace stocks of gadid and flounder species on Georges Bank that were harvested in excess during the 1960s-80s (Fogarty and Murawski 1998; Link *et al.* 2002).

Recognizing the occurrence and importance of biotic interactions suggests opportunities to increase the benefits that a society derives from fisheries through an integrated strategy that selectively harvests a fish community. Yet, there are some ecological constraints and limitations on society's objectives. First, only a small fraction of the energy at one trophic level (10% is generally used) is converted into biomass at the subsequent trophic level. Thus, it is energetically expensive to produce top predators (and game fish), such as swordfish, tunas, striped bass, and even cod, for consumers (and recreationists). Second, some systems are dominated by keystone predators, such as walleye pollock in the East Bering Sea, whose harvesting should be carefully considered in terms of possible cascade effects (Payne *et al.* 1990) and competitive release of other species (Fogarty and Murawski 1998; Link *et al.* 2002). Third are ecosystem constraints that protect ecosystem productive capacity (*e.g.*, primary productivity) and functions (*e.g.*, nutrient cycling, energy flow) (Arrow *et al.* 1995; Holling 1993). And fourth, trophic relationships can be complex and difficult to model or understand quantitatively (but see Hollowed *et al.* 2000 and Whipple *et al.* 2000).

THE PORTFOLIO FRAMEWORK

Dramatic shifts in community biomass and species composition are typical of open access fisheries whereby harvesters sequentially discover, develop, deplete, and abandon the most valuable fish stocks (Hanna 1997). This behavior is responsible for the collapse of valuable stocks, such as herring and cod (*e.g.*, see Hilborn *et al.* 2001), for biomass flips in species composition (*e.g.*, see Sherman 1986), and for driving other species, such as Atlantic halibut, beyond the point of economic extinction. Opportunistic and short-sighted approaches to fishing are counterproductive and annual returns too variable and risky when the most valuable assets are exhausted.

The work of some biologists (*e.g.*, Christensen 1996; Sainsbury 1991) and economists (*e.g.*, Anderson 1975; Hannesson 1983; Mitchell 1982) suggests important economic gains for society from manipulating the stock sizes of predator, prey, and competitor species. Put simply, "... there may be too many of some fish and too few of another" in fisheries management regimes where fish stocks are not the property of harvesters (*i.e.*, are non-exclusive to harvesters) or enforcement of government regulations is too costly (Cheung 1970: 53).

For example, Gulland (1982) roughly calculated that total dockside revenues in North Sea fisheries could be doubled by drawing down stocks of piscivores (cod, dogfish, saithe) and the group of industrial species (sand eel, Norway pout, sprat) which compete with those that feed mainly on invertebrates (haddock, plaice, whiting, sole), and by maintaining stocks of herring and mackerel (which feed on fish larvae but are themselves important prey) at low-to-moderate levels. This strategy was designed to maintain the stocks of high-priced, invertebrate-feeding species at high levels of biomass which could be harvested at moderate-to-high rates. He further concluded that integrated exploitation of the North Sea fish community could provide gains that are "much larger than would be likely to be obtained by treating each species as if it lived in isolation" (Gulland 1982: 15). This conclusion is somewhat bolstered by Sumaila's (1997) empirical simulations of industry profit from cooperative harvest of cod and capelin stocks in the Barents Sea.

The technical portfolio framework is presented here as a general, conceptual model for multi-species management (Table 8-1). We work with income when discussing portfolio returns and risk under separate sub-headings, but any of a society's objectives could be substituted, as noted above.

Table 8-1. The portfolio approach to multi-species management of wild fish stocks

Portfolio framework	• fish stocks are real assets capable of generating a flow of returns indefinitely • society's objectives and constraints are clearly defined and capable of being evaluated • combinations of stock sizes and other attributes for each species are evaluated for their effects on aggregate returns • tradeoff between expected aggregate returns and risk (i.e., variation in returns)
Institutional arrangements	• assign stock rights (management and exclusion) to harvesters in order to create the long time-horizons necessary to treat fish stocks as assets • internalize spillovers caused by ecological (e.g., your fish ate my fish) and technological (e.g., your gear caught my fish) interactions by centralizing production decisions • respond to uncertainty adaptively • invest in information that reduces uncertainty

Portfolio returns

In general, portfolios are used to find the most desirable combination of assets given their individual value and risk properties (Elton and Gruber 1995). Whether financial securities or real capital, portfolios can be designed to manage risk by balancing the mix of assets. However, in contrast to selecting financial securities from independent companies or governments, fish stocks which have co-evolved over many years and are harvested, in many cases, by unspecialized technologies in mixed species fisheries should be managed jointly due to spillover interactions between fisheries in order to optimize value. That is, harvesting one species generally has important implications for other species and fisheries that are integrated through ecology and technology.

Populations of wild finfish and shellfish are real assets to society— *i.e.*, they are capital stocks that can potentially yield benefits indefinitely to the owners of stock rights (governments, commons, or private individuals) and the public. From an economics standpoint, the value of a renewable asset, such as a fish stock, is the present value of income (revenues minus costs) from *future* harvests, appropriately discounted and balanced against the income from current harvest (a dividend) (Clark 1976; Fisher 1981). Stock value is not tantamount to current dockside revenues as commonly thought. Dockside revenue can be quite high while income is negative and future harvesting opportunities are being foreclosed by stock depletion.

The economic value of a fish stock is a function of current and future prices and extraction costs, and is subject to production technology and the resource growth rate (Fisher 1981). Prices are derived from people's demand (preferences) for seafood and other commodities and services. Further, prices are influenced by product attributes, such as species and fish size.

Extraction costs are determined by market prices for inputs, such as fuel and insurance, and by extraction technology and stock attributes. Other costs associated with administration, management, and enforcement (*i.e.*, transaction costs) certainly are important when considering alternative institutional arrangements in the next section, but they are not essential at this point.

Although generally taken for granted, value is also conditional on the property rights arrangements that govern resource use. Wild fish stocks have zero economic value in rule-of-capture regimes (open access, regulated open access) because it is prohibitively expensive for individuals to exclude others and conserve the asset for future use (Lueck 1998). Other property rights arrangements can be differentiated by the extent that they allow control of the suite of resource attributes (Cheung 1970; Edwards, this volume), such as biomass, age structure, and others mentioned above. Further, harvest technology reflects property rights. For example, the predominance of draggers and other mobile gear in fisheries is partly due to the premium on fishing power which accompanies incentives to race for fish in rule-of-capture fisheries.

Ecological and technological interactions are sufficient reasons to manage wild fish stocks for their aggregate benefits because stocks affect each others' dynamics (Anderson 1975; Hannesson 1983; Huppert 1979). As mentioned above, fish populations are mingled either directly (*e.g.*, predation, competition) or indirectly (one or more levels removed from the direct interaction) by trophic structure. Thus, harvesting one species can affect the stock sizes and capitalized values of many other species in the community. Similarly, many types of fishing gear are unspecialized or only moderately specialized and thereby catch a number of species simultaneously, some incidentally. Finally, many attributes other than biomass and size structure potentially contribute to stock value as already mentioned.

We next explore the influence of ecological and technological interactions on income from a fish community with a simple hypothetical 3-species model with two fisheries. The equations for fishery stock dynamics are specified as Gordon-Schaefer models with Type I predation as found in May *et al.* (1979):

$$dX_1/dt = r_1X_1[1 - X_1/K_1] - q_1E_1X_1 - aX_1X_2$$

$$dX_2/dt = r_2X_2[1 - X_2/K_2] - q_2E_{2+3}X_2 + bX_1X_2 - cX_2X_3$$

$$dX_3/dt = r_3X_3[1 - X_3/K_3] - q_3E_{2+3}X_3 + dX_2X_3$$

where X is stock biomass, r is the instantaneous growth rate, K is carrying capacity, q is catchability, E is fishing effort, and a, b, c, and d are predator-prey coefficients. The food web is linear (*i.e.*, species 3 eats species 2 which eats species 1). Species 2 and 3 are caught by the same fishery. Prices are a function of the landings of each species (market substitution, although not perfect), and increase up the food chain. Harvesting costs are a constant per unit of fishing effort. Parameter values were taken from Larkin's (1966) two-species model and otherwise chosen so that isoclines intersect (assures equilibria) and the three equations converge when disturbed from equilibria (structural stability). Being a hypothetical model (with realistic properties), values of income and effort are meaningless except in comparisons.[4]

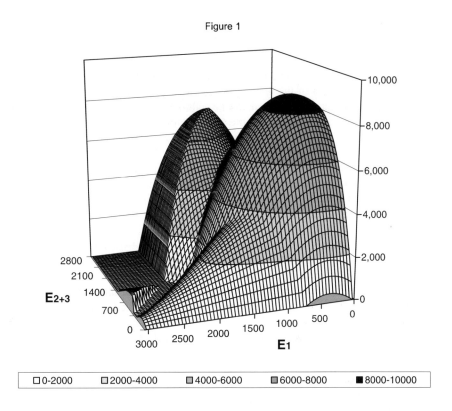

Figure 1

Legend: □ 0-2000   □ 2000-4000   ▨ 4000-6000   ▨ 6000-8000   ■ 8000-10000

Figure 8-1. Equilibrium portfolio income in a model of a hypothetical 3-species fish community with a linear trophic web and two fisheries. Shading is in intervals of 2000. Portfolio income is maximized where effort in the prey fishery is 350 and effort in the mixed-species fishery that targets the two predators is 1400. All values are internally consistent, but arbitrary. See the text for more details.

Equilibrium incomes in both fisheries were simulated from combinations of fishing effort that ranged from zero to levels that extinguished the species (Figure 8-1). Aggregate income is maximized (8658) at a level of effort in the fishery for prey species 1 (350) which is less than half of the effort (760) estimated to maximize income in the prey fishery from a single-species model fit to the simulated data.[5] Similarly, fishing effort in the mixed-species fishery for species 2 and 3 (1400) is greater than either single-species result (1043 for species 2, and 671 for species 3). These comparisons reveal several problems with the single-species approach.

---

[4] Contact the first author for a full description of the model.
[5] Notice that there is a local maximum at very high levels of effort in the mixed-species fishery. This surface has sharp edges because the mixed fishery becomes uneconomical as the prey stock which is targeted by the first fishery is fished down.

First, the independent single-species estimates of effort in the mixed-species fishery are far apart. Second, the total of combined incomes under the single-species approach (6220 when effort in the mixed fishery is 671) is considerably less than in the portfolio approach. Further, although not shown here, the single-species biomass targets can not be achieved simultaneously. Finally, optimizing aggregate income requires the fishery for prey species i to cut back from its self-interested levels. We will return to some of these results later in the paper.

Returning to the portfolio framework, the economic return to a real asset such as a fish stock is measured by the sum of the change in the value of the asset over, say, a year, plus income received from harvests, all divided by the value of the asset at the beginning of the period. The return concept is relatively simple to grasp for financial securities and privately-owned real assets, such as farmland, cattle, and oil pools, because market data on stock prices and dividends are readily available. It is harder to operationalize for fish stocks because government management regimes that emphasize regulations are costly to enforce, leading to the dissipation of resource value (Cheung 1970; Gordon 1954). Nevertheless, the concept does relate to resource rent which is the overriding concern of fisheries economics. Weaknesses in governance and data should not preclude developing the portfolio approach for fisheries.

The return to a portfolio of fish stocks is the sum of the individual asset returns. Portfolio managers would attempt to optimize the aggregate return from exploitation of a fish community by comparing tradeoffs in the attribute levels of individual stocks similar to what Gulland (1982) described for the North Sea. That is, fishing effort would be used to manipulate natural populations of predators and prey to desired levels of species biomass, fish size structure, and other stock attributes. However, managers would not necessarily select the highest return for the portfolio without also considering return variability.

Portfolio risk

The portfolio construct is useful partly because it formalizes the deliberate planning for the aggregate benefits from uses of integrated fishery assets. However, the principal result from portfolio theory is the understanding that the risk of a combination of assets is very different from a simple average of the risks of individual assets (Elton and Gruber 1995). There is a risk reduction from holding a portfolio of assets that do not move in perfect unison. Risk concerns future outcomes that take place with known or uncertain probability distributions.

There are three distinct classes of risk that affect fishery returns: availability of fish, market price, and institutions (Pontecorvo 1986). Variability in the availability of fish results because wild biological assets are not manufactured by humans in controlled environments, such as in catfish or tilapia farms. The recruitment process is highly uncertain due to a randomly varying environment—*e.g.*, storms, currents, warm core rings, El Nino and other oscillations, productivity of phytoplankton and zooplankton—and unobservable trophic dynamics which impact the survival of larval and juvenile fish (Sissenwine 1984). These factors also affect the survival of adult fish in ways that are poorly understood from a quantitative standpoint. Stock attributes are hard to predict because of these complex relationships.

Being economic assets, uncertainty about future prices, input costs, and technology also make returns risky. Dockside prices are influenced strongly by global supplies of seafood and fish meal. Further, dockside prices and demand elasticities differ among species because they are not perfect substitutes. The price of fuel, a major input, is also determined by global markets. Most of the significant advances in fishing technology, including diesel power, refrigeration, and electronics, have been adapted from other industries.

Finally, government policies and future management rules and entitlements for fishermen probably are as unpredictable as the other sources of risk (Brewer 1984). Property rights and governance institutions determine time horizons and the costs of integrating harvest rates across species, as discussed later.

Elton and Gruber (1995) show how to estimate the expected value and variance of returns for a portfolio of risky individual assets. Expected value combines possible outcomes for each stock (in economic or other social value, not biomass or other biological characteristics) with their probability of occurring, weighted by the stock's contribution to the portfolio. The variance on a portfolio's return is the combination of each stock's variance and covariances with other stocks, also weighted by each stock's share in the portfolio. Covariance is also the product of pairwise standard deviations and correlation.

Covariance is critical to opportunities to manage risk in fishery (and other) portfolios. Assets with returns that move together (*i.e.*, are positively correlated) will have a positive covariance. Also, assets that are independent will have zero covariance. In both of these cases, combining assets that are mingled by ecology or technology may improve expected returns, but it does not reduce risk. Opportunities to reduce portfolio risk result when assets' returns move in opposite directions. In this case, the covariance is negative, meaning that there is a tendency for assets to somewhat compensate for each other. That is, when the returns for some assets are down, you can expect others' returns to be up, thereby dampening variation.

In theory, the so-called opportunity set of all possible combinations of fish stocks that comprise stable states in the presence of fishing would plot as a cloud of points in a two-dimensional space of expected portfolio return and variance. Portfolios that offer a smaller return for the same risk would be dropped from consideration, as would those with the same expected return but larger risks (see Elton and Gruber 1995), depending on ecological constraints placed on the objective function. The result would be a line called the efficient frontier between the minimum variance portfolio and the maximum expected return portfolio.

Each point on the frontier is associated with a unique set of fish stocks and attributes, and is a feasible choice depending on risk preferences. The efficient frontier is illustrated in Figure 8-2 (see 2a and 2b) for our hypothetical trophic chain model with random log-normal recruitment for each species. Gains in the expected value of aggregate returns in the portfolio are achieved at the cost of disproportionate increases in variance.

It may be advantageous to deplete the stocks of species that contribute most to portfolio variance in order to achieve a lower, but steadier return, as constrained by ecological considerations. For example, Beddington *et al.* (1984) observed that species high in the food web generally live longer which tends to make their population densities less variable. This characteristic would complement the higher value that consumers generally place on upper trophic level fish.

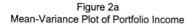

Figure 2a
Mean-Variance Plot of Portfolio Income

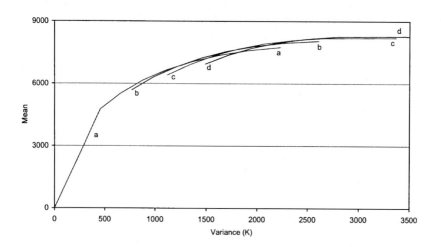

Figure 2b
Mean-Variance Plot of Portfolio Income

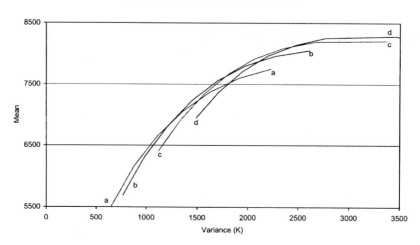

Figure 8-2. Mean-variance plot of portfolio income in a model of a hypothetical 3-species fish community with a linear trophic web and two fisheries. Variance results from random log-normal recruitment and price models with a random normal residual process. Effort (arbitrary units) in the prey fishery is constant along each curve: (a) 0; (b) 100; (c) 200; and (d) 300. Effort in the mixed-species fishery for predators increases along each curve towards the maximum mean. The efficient frontier would be an outer envelope touching each surface. Figure 2b provides a close-up view.

Another consideration for risk in a portfolio of fish stocks concerns two ways that ecological relationships affect covariance. First, the structural linkages in a food web will cause some populations to move in unison and others to move in opposite directions. The overall influence of such pairwise covariances on total portfolio variance will be complicated by ripple effects and feedbacks throughout the food web. Second, stochastic processes, such as spawning success and recruitment, which affect one stock will also affect the variance and covariances of related species because stock dynamics are simultaneous. That is, the biomass of each population is a function of the stochastic processes of the other species.

Likewise, harvesting is a technological production (or supply) function with its own stochastic process that affects the populations of all species in a mixed species fishery. Steele (1984: 259) commented on these structural and stochastic phenomena as follows: "We should replace the single equilibrium assumption by the recognition of possible multiple states . . . with the changes between them occurring rapidly, and with the frequency of change increasing with increased predatory fishing pressure. The expected variability is then very different from noise about some single steady state. The external and environmental fluctuations are responsible not only for variations about a particular equilibrium state but contribute significantly to the changes in state."

Regulations and other management rules derived from single-species approaches can also affect variability in portfolio returns. Input and output regulations, such as allowable fishing time and catch limits, which are derived from single-species models can not achieve single-species targets simultaneously due to unaccounted for ecological interactions, even if implemented and enforced perfectly. Further, the levels of fishing effort that correspond to MSYs of species caught in the same gear will most likely be quite different. This was illustrated above where some results of the hypothetical 3-species/2-fishery model were presented. As a result, regulations affecting stock sizes and harvest rates will constantly be revised as managers chase implausible goals. Also, other rules, such as area closures and hard catch limits, which either safeguard selected species, require species to be discarded, or close a mixed species fishery, will cause aggregate returns to fall short of their potential and exacerbate return variability. For example, most of the Atlantic sea scallop biomass is located inside areas off the U.S. New England coast which were closed in 1994 to rebuild depleted stocks of Atlantic cod, haddock, and yellowtail flounder. Access by the sea scallop fleet has been sporadic.

Regarding market risk, there is growing evidence that dockside prices are co-integrated which would contribute a tendency for expected returns to move together and increase risk. This effect would be augmented by unspecialized fishing technologies with joint production because costs would be spread across all species that are landed. However, technology and fishing practices vary as a function of property rights and governance arrangements. It is conceivable that institutions which are capable of implementing portfolio management would induce technological innovations that specialize in fewer species. For example, line trawling for Atlantic cod may gain a comparative advantage over dragging for mixed species.

Finally, the rules and decision-making process prescribed by a management institution can worsen risk when they are not compatible with the ecosystem. For example, McGlade (1989) found that the regulatory process for haddock management in Nova Scotia exacerbated oscillations in recruitment by introducing time delays into the system. She recommended a change in management regimes to one that integrates the behaviors of fishermen, processors, and managers.

COMPATIBLE PROPERTY RIGHTS REGIMES

Technical frameworks for multi-species management, such as the portfolio framework, require institutional arrangements that value assets in the ecosystem and coordinate conflicting uses on an ecosystem scale (Table 8-1; Hanna 1998). Property rights regimes that match society's objectives with the capacity constraints of the environment are needed (Arrow *et al.* 1995).

By way of definition, property rights are entitlements to specified uses of resources that are accepted by society and protected against encroachment by others (Bromley 1992). Property rights regimes are a type of institution, or what Ciriacy-Wantrup and Bishop (1975: 714) define as a "social decision system that provide[s] rules for adjusting and accommodating, over time, conflicting demands ... from different interest groups in a society." Property rights regimes are integrated with resource exploitation because they form people's expectations about the outcomes of their behavior and thereby create behavioral incentives. In our case, property rights regimes define the relationship between harvesters and stocks of wild fish.

In much of the world, including in the United States by way of the Magnuson-Stevens Act, governments reserve all rights to marine fishery resources on behalf of (as agents for) the public.[6] A fisherman's de facto ownership of harvest rights to his target species typically is not complemented with the combined stock rights (see Schlager and Ostrom 1992; Scott 1986) of management and the authority to exclude others from the fishery in most cases, except where aquaculture is practiced. This wedge between harvest rights and stock rights creates the very short time-horizons characterized by Gordon (1954:135) almost a half century ago: "... the fish in the sea are valueless to the fisherman, because there is no assurance that they will be there for him tomorrow if they are left behind today." Stocks are undervalued, fishermen over-invest in harvesting capital well beyond the productive capacity of fish stocks, and society invests too little in constructive institutional capital.

Ecological and technological interactions are two dimensions of the more general common pool problem which plagues sustainable uses of fishery and many other resources (*e.g.*, oil pools, rivers, groundwater, atmosphere) whenever too many people with only access or use rights make exclusion too difficult (*i.e.*, too costly, illegal, or socially unacceptable) and their individual uses subtract from overall potential gains (see Libecap 1998). In the absence of a more complete system of property rights, fishermen do not have the authority or incentive to invest in fish stocks (including through restraint) because it is too costly to negotiate, monitor, and enforce contracts that are not recognized by society or law. For example, a New England dragger who increases fishing pressure on silver hake and spiny dogfish in order to promote the growth of haddock and flounders, or who switches to trawl lines in order to reduce impacts on habitat, will not reap any benefits from these investments because others will take the prey species left behind or pull trawls and dredges over the same area. Similarly, a seiner in a herring fishery with IFQ (individual fisherman quota) harvest rights would not be

---

[6] Property rights and governance arrangements are changing in some countries and fisheries, however. For example, shares of total catch quotas are allocated as legal harvest rights in New Zealand and Iceland. There are a few examples of similar individual fisherman quotas systems and, increasingly, harvest cooperatives in the United States, but the federal fisheries laws clearly state that these rights are permissions.

compensated for not harvesting his share which would support the growth of predator species harvested by others. These relational margins that could increase benefits from the fish community as a whole are dissipated in the common pool.

Nor do regulatory regimes create personal accountability among managers (Hanna 2002). For the most part, managers do not incur the costs of their short-sighted decisions, and they are not entitled to the gains that would result from integrated management. The different fisheries (defined by target species and gear) and other stakeholders (*e.g.*, environmental organizations) compete over harvest opportunities in a negative-sum conflict that does not increase the overall production of fishery products in an economy. Management decisions are rigid and unresponsive to changes in the environment, stock attributes, or markets because any change that deviates from the status quo creates uncompensated losers. A further consequence of poor accountability is that management plans frequently contain many broad objectives that are often incompatible and difficult to evaluate.

Regulatory regimes that follow the single-species approach have not managed ecological and technological interactions in ways that promote aggregate benefits in fisheries. Managing tradeoffs from ecological interactions is impractical because there is no process to compensate fishermen who would be required to reduce mortality on prey species, to reduce stocks of certain predators, or to change fishing gear and practices in order to improve habitats. Technological interactions are addressed with restrictions such as single-species catch limits (trip limits, fishery total-allowable-catch limits) and so-called bycatch controls that close mixed-species fisheries, prevent access by most fisheries to certain areas, or result in wasteful discarding. For example, the three large areas on Georges Bank and in Southern New England waters that were closed to rebuild stocks of Atlantic cod, haddock, and yellowtail flounder to MSY target levels significantly restricts access by scallop fishermen even though the abundance of scallops is very high in these areas.

Short of the politically infeasible take over of harvesting by government enterprises, governments will need to devolve stock rights to the fishing industry in order to change harvesters' incentives and lower the costs of contracting common pool problems. Harvesters need the assurance secured by suitable property rights arrangements that they are entitled to the future returns from current investments in stock management, and that their entitlements and contracts will be enforced by law.

Elsewhere, one of us has examined the implications of ecological relationships and other resource attributes for beneficial property rights regimes in marine fisheries (Edwards, this volume, 2004). Property rights theory recognizes that all resources have multiple attributes that are potentially valuable if managed properly (*e.g.*, fish size structure, growth rate, gene pool, diet, habitat requirements, etc.), but it is prohibitively costly to define and enforce ownership of property rights to all attributes, especially those that are uncertain and involved with spillovers. One function of government, however, is to facilitate property rights arrangements that increase economic productivity by reducing the so-called transaction costs of collective management or contracts (Coase 1960).

Situations marked by frequent spillovers, such as predation by species in one fishery on the species harvested in another fishery, and considerable uncertainty, including recruitment variability and gaps in the quantitative knowledge of ecological and technological interactions, require institutions that centralize production decisions. Centralized production internalizes spillovers in common pool situations because gains from tradeoffs are captured. Centralized production also economizes on the resources used to balance asset levels because the producer incurs full costs. This contrasts markedly with the current management process and rules which artificially divide resource uses and underachieve society's objectives because it is politically impractical to adjust harvest rights that benefit one fishery at the expense of another.

The prevalence of environmental and stock uncertainties in fisheries also requires a flexible decision-making process that can adapt to unexpected conditions (Pontecorvo 1986). Furthermore, experimental manipulations of fish stock levels and adaptation to the new information is an important response to uncertainty (Walters 1986). Making harvesters (or other claimants) the so-called residual claimants to the benefits (and costs) of their decisions in a highly uncertain environment would create incentives to invest in information to reduce portfolio risk and improve expected benefits even though the outcomes and payoff period are unsure.

SUMMARY

The portfolio approach is recommended as a model to manage stocks of wild fish species that interact ecologically, are caught jointly, and are characterized by stock-attribute uncertainty. The biomass and other attributes of species would be manipulated to balance the expected aggregate returns of a fishery portfolio against return variability given clearly defined and measurable objectives and constraints, including ecological safeguards (Arrow *et al.* 1995; Holling 1993). This technical framework requires compatible institutions to be implemented, however. Property rights institutions should be designed to create the necessary incentives for harvesters to value fish stocks as assets, to optimize returns from the system as a whole, and to manage environmental and ecological uncertainties productively. The dearth of examples of adaptive multi-species management worldwide, such as removal of large carnivorous fish off Spain to create a shrimp fishery (May 1984) and experimental management of an Australian multispecies fishery (Sainsbury *et al.* 1997), suggests that most extant management institutions are not compatible with ecosystem-based management of fishery resources.

Government regulatory regimes may not be able to implement the portfolio approach due to the high transaction costs (negotiation, monitoring, enforcement) of redefining implicit harvest rights arrangements among the many competitive and politically active stakeholders. The wedge between harvest and stock rights in most government regulatory regimes makes fish stocks non-exclusive from the harvesters' perspective (see Cheung 1970). Under these circumstances, fish stocks are not economic assets because future benefits are dissipated by overuse, including the potential benefits of managing ecological and technological interactions.

The portfolio approach requires production and management to be centralized in a commons or corporate structure (depending on social circumstances) where decisions are being made by harvesters (Edwards, this volume, 2004). Commons and corporate production and governance arrangements are better prepared to internalize spillovers caused by ecological and technological interactions because the incidence of benefits and costs are personal. Further, as residual claimants to gains and losses, communal and corporate harvesters can weigh the value of reducing natural as well as economic uncertainties, including through adaptive management experiments.

We recommend the portfolio approach for management of LME fishery (and other) resources not only because of its distinct advantages over the single-species approach: it clarifies objectives, it assimilates interactions that naturally integrate fish and fisheries, and it manages uncertainty in beneficial ways. Another advantage of this general, integrative approach to resource management is that it can be extended to incorporate other, non-renewable resources in the environment, such as oil and gas and hard minerals. As above, the focus would be on the aggregate benefits of the resources in an LME for society, and it would include other technological interactions and spillovers, including pollution. As with the single-species approach to fisheries management, the interactions between fisheries and other extractive sectors of the ocean economy are currently fractured by incompatible property rights and governance arrangements.

REFERENCES

Anderson, L.G. 1975. Analysis of open access commercial exploitation and maximum economic yield in biologically and technologically interdependent fisheries. *Journal of the Fisheries Research Board of Canada* 32: 1825-1832.

Applegate, A., S. Cadrin, J. Hoenig, C. Moore, S. Murawski, and E. Pikitch. 1998. Evaluation of existing overfishing definitions and recommendations for new overfishing definitions to comply with the Sustainable Fisheries Act. Northeast Fisheries Science Center, Woods Hole, Massachusetts. 179 p.

Arrow, K., B. Bolin, R. Costanza, P. Dasgupta, C. Folke, C.S. Holling, B-O. Jansson, S. Levin, K-G. Maler, C. Perrings, and D. Pimentel. 1995. Economic growth, carrying capacity, and the environment. *Science* 268: 520-521.

Bailey, K.M. and E.D. Houde. 1989. Predation of eggs and larvae of marine fishes and the recruitment problem. *Advances in Marine Biology*. 25: 1-83.

Bax, N.J. 1991. A comparison of the fish biomass flow to fish, fisheries, and mammals on six marine ecosystems. *ICES Marine Science Symposia* 193: 217-224.

Bax, N.J. 1998. The significance and prediction of predation in marine fisheries. *ICES Journal of Marine Science*. 55: 997-1030.

Beddington, J.R. 1986. Shifts in resource populations in large marine ecosystems. In: K. Sherman and L.M. Alexander. *Variability and Management of Large Marine Ecosystems*. AAAS Selected Symposium 99. Westview Press, Boulder, Colorado. 9-18.

Beddington, J.R., W.E. Arntz, R.S. Bailey, G.D. Brewer, M.H. Glantz, A.J.Y. Laurec, R.M. May, W.P. Nellen, V.C. Smetacek, F.R.M. Thurow, J.-P. Troadec, and C.J. Walters. 1984. Management under uncertainty. In: R.M. May. Exploitation of Marine Communities. Life Sciences Research Report 32. Springer-Verlag, New York. 227-244.

Brewer, G.D. 1984. The wider dimensions of management uncertainty in world fisheries. In: R.M. May. Exploitation of marine communities. Life Sciences Research Report 32. Springer-Verlag, New York. 275-286.

Bromley, D.W. 1992. The commons, common property, and environmental policy.

Environmental and Resource Economics 2:1-17.

Brown, B.E., J.A. Brennan, M.D. Grosslein, E.G. Heyerdahl, and R.C. Hennemuth. 1976. The effect of fishing on the marine finfish biomass in the Northwest Atlantic from the Gulf of Maine to Cape Hatteras. *ICNAF Research Bulletin* 12: 49-68.

Cheung, S.N.S. 1970. The structure of a contract and the theory of a non-exclusive resource. *Journal of Law and Economics 13: 49-70.*

Christensen, V. *1996.* Managing fisheries involving predator and prey. *Reviews in Fish Biology and Fisheries* 6: 417-442.

Ciriacy-Wantrup, S.V. and R.C. Bishop. 1975. Common property as a concept in natural resource policy. *Natural Resources Journal* 15: 713-727.

Clark, C.W. 1976. *Mathematical Bioeconomics: The Optimal Management of Renewable Resources.* John Wiley & Sons, New York.

Coase, R.H., 1960. The problem of social cost. *The Journal of Law and Economics* 3:1-44.

Cohen, E., M. Grosslein, M. Sissenwine, F. Steimle, and W. Wright. 1982. Energy budget of Georges Bank. *Canadian Special Publications in Fisheries and Aquatic Sciences.* 59, 169p.

Edwards, S.F. 2003. Property rights to multi-attribute fishery resources. *Ecological Economics* 44: 309-323.

Elton, E.J. and M.J. Gruber. 1995. *Modern Portfolio Theory and Investment Analysis.* John Wiley and Sons, Inc., New York. 5th edition.

Fisher, A.C. 1981. *Resource and Environmental Economics.* Cambridge University Press, NY.

Fogarty, M.J. and S.A. Murawski. 1998. Large scale disturbance and the structure of marine systems: fisheries impacts on Georges Bank. *Ecological Applications* 8: S6-S22.

Fogarty, M., E.B. Cohen, W.L. Michaels, and W.W. Morse. 1991. Predation and the regulation of sand lance populations: an exploratory analysis. *ICES Marine Science Symposia.* 193:120-124.

Gordon, H.S. 1954. The economic theory of a common-property resource: the fishery. *Journal of Political Economy* 62:124-142.

Gulland, J.A. 1982. Long-term potential effects from management of the fish resources of the North Atlantic. *Journal du Conseil International pour l'Exploration de la Mer* 40: 6-16.

Hanna, S. 1997. The new frontier of American fisheries governance. *Ecological Economics* 20:221-233.

Hanna, S. 1998. Institutions for marine ecosystems: economic incentives and fisheries management. *Ecological Applications* 8:S170-S174.

Hanna, S. 2002. The economics of fisheries management: behavioral incentives and management costs. Proceedings of the Pew Ocean Commission Workshop on Marine Fisheries Management. Pew Foundation, Washington, D.C. 40-44.

Hannesson, R. 1983. Optimal harvesting of ecologically interdependent fish species. *Journal of Environmental Economics and Management* 10:329-345.

Hilborn, R., J.-J. Maguire, A.M. Parma, and A.A. Rosenberg. 2001. The Precautionary Approach and risk management: can they increase the probability of success in fisheries management? *Canadian Journal of Fisheries and Aquatic Sciences* 58: 99-107.

Holling, C.S. 1993. Investing in research for sustainability. *Ecological Applications.* 3: 552-555.

Hollowed, A.B., N. Bax, R. Beamish, J. Collie, M. Fogarty, P. Livingston, J. Pope, and J.C.

Rice. 2000. Are multispecies models an improvement on single_species models for measuring fishing impacts on marine ecosystems? *ICES Journal of Marine Science.* 57: 707-719.

Huppert, D.D. 1979. Implications of multipurpose fleets and mixed stocks for control policies. *Journal of the Fisheries Research Board of Canada* 36:845-854.

Larkin, P.A. 1966. Exploitation in a type of predator-prey relationship. *Journal of the Fisheries Research Board of Canada* 23:349-356.

Larkin, P.A. 1996. Concepts and issues in marine ecosystem management. *Reviews in Fish Biology and Fisheries* 6:139-164.

Libecap, G.D., 1998. Common property. In: P. Newman, ed. *The New Palgrave Dictionary of Economics and the Law.* Stockton Press, New York. Volume **1**:317-324.

Link, J.S., L.P. Garrison and F.P. Almeida. 2002. Interactions between elasmobranchs and groundfish species (*Gadidae* and *Pleuronectidae*) on the Northeast U.S. Shelf. I. evaluating predation. *North American Journal of Fisheries Management* 22:550-562.

Lueck, D.L. 1998. First possession. In: P. Newman, ed. *The New Palgrave Dictionary of Economics and the Law.* Stockton Press, New York. Volume 2:32-144.

May, R.M. 1984. Introduction. In May, R.M. Exploitation of marine communities. Life Sciences Research Report 32. Springer-Verlag, New York. 1-12.

May, R.M., J.R. Beddington, J.W. Horwood, and J.G. Shepherd. 1978. Exploiting natural populations in an uncertain world. *Mathematical Biosciences* 42:219-252.

May, R.M., J.R. Beddington, C.C. Clark, S.J. Holt, and R.M. Lewis. 1979. Management of multispecies fisheries. *Science* 205:267-277.

McGlade, J.M. 1989. Integrated fisheries management models: understanding the limits to marine resource exploitation. *American Fisheries Science Symposium* 6:139-165.

Mitchell, C.L. 1982. Bioeconomics of multispecies exploitation in fisheries: management implications. Canadian Special Publication of Fisheries and Aquatic *Sciences* 59: 57-162.

Myers, R.A. 1998. When do environment-recruitment correlations work? *Reviews in Fish Biology and Fisheries* 8:285-305.

NRC (National Research Council). 1999. Sustaining Marine Fisheries. National Academy Press, Washington, D.C.

Overholtz, W.J., S.A. Murawski, and K.L. Foster. 1991. Impact of predatory fish, marine mammals, and sea birds on the pelagic fish ecosystem of the northeastern USA. *ICES Marine Science Symposium* 193:198-208.

Pauly, D., V. Christensen, J. Dalsgaard, R. Froese, and F. Torres Jr.. 1998. Fishing down marine food webs. *Science* 279:860-863.

Payne, P.M., D.N. Wiley, S.B. Young, S. Pittman, P.J. Clapham, and J.W. Jossi. 1990. Recent fluctuations in the abundance of baleen whales in the southern Gulf of Maine in relations to selected prey. *Fishery Bulletin* 88:687-696.

Pontecorvo, G. 1986. Cost benefit of measuring resource variability in Large Marine Ecosystems. In Sherman K. and L.M. Alexander. *Variability and Management of Large Marine Ecosystems* 269-279. AAAS Selected Symposium 99. Westview Press, Boulder, Colorado.

Pope, J.G. 1991. The ICES multispecies assessment working group: evolution, insights, and future problems. *ICES Marine Science Symposia* 193:22-33.

Repetto, R. 2002. Creating asset accounts for a commercial fishery out of equilibrium: a case study of the Atlantic sea scallop fishery. *Review of Income and Wealth* 48:245-259.

Rice, J.A., T.J. Miller, K.A. Rose, L.B. Crowder, E.A. Marschall,, A.S. Trebitz, and D.L.

DeAngelis. 1993. Growth rate variation and larval survival: inferences from an individual based size-dependent predation model. *Canadian Journal of Fisheries and Aquatic Sciences* 50:133-142.

Rothschild, B.J. 2000. Fish stocks and recruitment: the past thirty years. *ICES Journal of Marine Science* 57:191-201.

Sainsbury, K.J. 1991.Application of an experimental approach to management of a tropical multispecies fishery with highly uncertain dynamics. *ICES Marine Science Symposia* 193: 301-320.

Sainsbury, K.J., R.A. Campbell, R. Lindholm, and A.W. Whitelaw. 1997. Experimental management of an Australian multispecies fishery: examining the possibility of trawl-induced habitat modification. In: Pikitch, E.K., D.D. Huppert, M.P. Sissenwine, and M. Duke. *Global Trends: Fisheries Management.* American Fisheries Society Symposium 20. AFS, Bethesda, MD. 107-112.

Schlager, E. and E. Ostrom. 1992. Property rights regimes and natural resources: a conceptual analysis. *Land Economics* 68:249-262.

Scott, A. 1986. Catch quotas and shares in the fish stock as property rights. In: Miles, E., R. Pealy and R. Stokes (Editors). *Natural Resource Economics and Policy Applications: Essays in Honor of James A. Crutchfield.* University of Seattle Press, Seattle, Washington. 61-96.

Sherman, K. 1986. Biomass flips in Large Marine Ecosystems. In: Sherman, K. and L.M. Alexander. *Biomass Yields and Geography of Large Marine Ecosystems.* Westview Press, Boulder, Colorado. 327-333.

Sherman, K. and A.M. Duda. 1999. Large marine ecosystems: An emerging paradigm for fishery sustainability. *Fisheries* 24:15-26.

Sherman, K. and L.M. Alexander (editors). 1986. *Variability and Management of Large Marine Ecosystems.* AAAS Selected Symposium 99. Westview Press, Inc., Boulder, Colorado. 319p.

Sherman, K. and L.M. Alexander (editors). 1990. *Biomass Yields and Geography of Large Marine Ecosystems.* AAAS Selected Symposium 111. Westview Press, Inc., Boulder, Colorado. 493p.

Sissenwine, M.P. 1984. The uncertain environment of fishery scientists and managers. *Marine Resource Economics* 1:1-30.

Sissenwine, M.P. 1986. Perturbations of predator-controlled continental shelf ecosystems. In: Sherman K. and L.M. Alexander. *Variability and Management of Large Marine Ecosystems.* AAAS Selected Symposium 99. Westview Press, Boulder, Colorado. 55-85.

Sissenwine, M.P., E.B. Cohen, and M.D. Grosslein. 1984. Structure of the Georges Bank ecosystem. *Rapports et Procès_Verbaux des Réunions Conseil International pour l'Exploration de la Mer* 183:243-254.

Solow, R.M. 1991. Sustainability: an economist's perspective. Eighteenth J. Seward Johnson Lecture in Marine Policy. Woods Hole Oceanographic Institution, Woods Hole, Massachusetts.

Steele, J.H. 1984. Kinds of variability and uncertainty affecting fisheries. In: May, R.M. Exploitation of Marine Communities. Life Sciences Research Report 32. Springer-Verlag, New York. 245-262.

Steele, J.H. 1998. Regime shifts in marine ecosystems. *Ecological Applications.* 8(S):S33-S36.

Sumalia, U.R. 1997. Strategic dynamic interaction: the case of Barents Sea fisheries. *Marine*

Resource Economics 12:77-94.

van de Koppel, H., P.M.J. Herman, P. Thoolen, and C.H.R. Heip. 2001. Do alternate stable states occur in natural ecosystems? Evidence from a tidal flat. *Ecology* 82: 3449-3461.

Walters, C.J. 1986. *Adaptive Management of Renewable Resources*. Macmillan Press, New York.

Walters, C.J., V. Christensen, and D. Pauly. 1997. Structuring dynamic models of exploited ecosystems from trophic mass-balance assessments. *Reviews in Fish Biology and Fisheries* 7:139-172.

Whipple, S., J.S. Link, L.P. Garrison, and M.J. Fogarty. 2000. Review of models of predator-prey interactions and fishing mortality. *Fish and Fisheries* 1:24-40.

Large Marine Ecosystems, Vol. 13
T.M. Hennessey and J.G. Sutinen (Editors)

9

Fish Habitat: A Valuable Ecosystem Asset

Harold F. Upton and Jon G. Sutinen

INTRODUCTION

Fisheries of the Northeast Shelf ecosystem are among the most productive and valuable in the world. Although overfishing has been identified as the primary reason for fish population declines of the last four decades, the maintenance of productive fish populations also depends on the quality and quantity of fish habitat. Fish habitat can be viewed as an ecological asset that is an essential input for the generation of valuable goods and services, such as fish products and recreational opportunities that are utilized by humans. The values of these goods and services are reflected in the marketplace by the choices people make when purchasing seafood or taking angling trips. Although the values of end products are readily identified by users, the values of ecosystem assets such as fish habitat are less easily identified because habitat is being used indirectly. How valuable is fish habitat? What are the benefits of protecting and restoring fish habitat? To help answer these questions, this chapter presents a method for estimating the value of groundfish habitat in the Northeast Shelf large marine ecosystem.

This chapter is organized as follows. The next section provides the biological and economic context of the problem. The second section presents a conceptual framework to explore linkages between ecological and economic systems, and provides estimates of the potential benefits (costs) of New England groundfish habitat protection (degradation). The final section discusses future research directions.

Biological Context

Many economic activities physically alter or pollute estuaries, salt marshes and benthic communities—the areas often associated with nurseries, shelter, and food production of many fish species. Eventually the products that humans utilize directly, such as fish, will be impacted because coastal ecosystems are composed of webs of interrelated components and processes. For example, the loss or degradation of inshore winter flounder spawning habitat such as bays and estuaries will likely result in lower offshore production of adult flounder. The indirect nature of these impacts and uncertainties, related to linkages between ecosystem components, often makes it difficult to assess the values of ecosystem assets such as fish habitat.

Habitat, the place where a species lives, is essential for the survival and maintenance of any wildlife population. Habitat is "the physical and biological environment used by an individual, population, a species or a group" (Hunter 1996). For example, areas may be defined as blue whale habitat or in a more general sense waterfowl habitat (Hunter 1996). However, if a broader term like wildlife habitat is used, the term loses meaning because nearly all habitats support some type of wildlife. A natural community, such as a redwood forest, is an assemblage of interacting organisms and forms a "distinctive living system with its own composition, structure, environmental relations, development and functions" (Whittaker 1975). An ecosystem is composed of a group of interacting organisms, (a community) and the physical environment that they inhabit at a given time (Hunter 1996, Whittaker 1975). The ecosystem is "a functional system of complementary relationships that transfer and circulate energy and matter" (Whittaker 1975). Adjacent or different ecosystems are divided according to the assemblage of organisms inhabiting them. Divisions between different ecosystems are comprised of transition zones across which there are relatively few interactions between species.

The distinctions between habitat, community and ecosystem are somewhat subtle but important. Habitat refers to the kind of environment in which a species occurs. A species may inhabit different habitats within its range, especially during different phases of its life history. An ecosystem may or may not correspond to the habitat of a species. In many cases, a species will inhabit the specific patches of an ecosystem that are most favorable to the maintenance of its basic function. For example, the habitat of a species of beetle may consist of the bark of a specific kind of tree within a forest ecosystem. A frog might inhabit two distinct ecosystems such as a pond and a forest with its occurrence in space largely dependent on its stage in development.

Terrestrial ecologists have moved from a focus on individual species to greater emphasis on habitats and ecosystems in recognition of the need to maintain ecosystem integrity, structure and function. The web-like linkages of ecosystems illustrate that conservation of a single component is related to the conservation of other system components and processes. For marine populations, especially in the case of fisheries, management has often emphasized individual species or species groups. This is likely due to our more limited knowledge of marine systems and the inherent bias associated with our terrestrial origins. Yet, as in terrestrial cases, these systems are characterized by the relationships between species and their abiotic and biotic environment. Therefore, an understanding of ecosystem interactions and processes is necessary to sustain ecosystem composition, structure and function. The research on, and conservation of, marine fish habitat are potential major steps in this direction.

Economic Context

The habitats of different organisms are ecosystem assets, which are inputs or requirements for system processes that provide valuable goods and services (Freeman 1993). Although a given asset may not have a direct market value, assets such as fish habitat are essential for the eventual utilization and associated utility provided by the end product. The animal's habitat provides it shelter, food and a place for reproduction. For many marine species, the inputs to the natural production process that habitat provides are difficult to duplicate. For example, currently, cod culture is not an economically viable alternative due to the physical and biological requirements during different stages of its life history. These requirements cannot be duplicated without great cost. Even when aquaculture is successful, as in the case of salmon, intensive culture requires inputs such as net pen enclosures, high quality feeds and a supply of fingerlings. Potential environmental degradation related to aquaculture poses an additional set of social concerns.

Habitat loss, or loss of associated ecosystem function, is a type of market failure that occurs for several reasons. Institutional elements such as property rights structures are often a primary reason for market failure. Private property rights in parts of the coastal zone, river systems and most areas seaward of mean high tide are poorly defined. The public use doctrine has often led to management that is based on either political expediency or command control solutions. A related concern is the public good character of this system, characterized by widely dispersed public benefits from which it is difficult to exclude individuals. In addition, imperfect information exists because linkages between ecosystem and community components are difficult to quantify due to their complexity. For example, factors such as environmental variability and direct fishing mortality may be difficult to identify and separate from each other. The lack of definitive answers often leads to paralysis in the public policy arena, especially when specific users will be impacted directly by the policy.

The quality and quantity of fish habitat are threatened by a diverse array of human activities such as agriculture, residential and coastal development, fishing operations, industrial pollution and shipping. Yet, damages to habitat are not easily linked to impacts on specific ecosystem resources that are valued directly, such as fish, or to the costs that these damages impose on society. Markets fail in instances where prices do not communicate the value of ecological assets. Decisions to undertake a given economic activity that impacts the asset without consideration of the asset's value to fisheries production will generate an inefficient allocation of resources. The unintended effects of an economic activity, for example on fishermen who are not directly involved with the production or consumption of the commodity produced, are termed externalities. Pollution from a factory may harm fish habitat, resulting in costs to the fishing industry. Externalities may impose costs on several different sectors, such as costs to tourism from the loss of aesthetic value of shoreline areas, as well as costs to fisheries from the loss of productivity.

At the same time, high human population densities and associated economic activities are more concentrated in coastal regions and increase the probability of anthropogenic impacts. Habitat degradation may be divided into two major categories, direct physical alteration and pollution. Physical alteration includes changes to water flow, to the physical structure of land or the removal of biological organisms. Dredging for navigation of shipping, fishing vessels, and pleasure craft results in direct physical alteration of the bottom and in the need for disposal of dredge spoils. Filling of coastal wetlands for residential or other development results in the degradation or loss of inshore areas such as marshes and estuaries.

Pollution may take many forms including nutrients, noxious chemicals and heat discharges, and includes point-source discharges from sewage treatment plants, factories or power plants and non-point sources such as agriculture, parking lots and residential areas. Nutrient pollution from agriculture, municipal sewage, and residential development (storm water runoff) results in water quality changes in inshore areas. Increased algal growth from greater nutrient levels changes light penetration and alters community composition. This may result in the loss of "bioengineers," important structural species such as submerged aquatic vegetation. Contamination due to pollution from agriculture, sewage and storm water runoff may also render shellfish such as clams and oysters unfit for human consumption. Wastes from industrial processes may cause direct mortality or inhibit growth and reproduction of marine organisms. Oil or chemical pollution from spills or storm water runoff may result in immediate impacts from a given event or long-term sub-lethal impacts on marine organisms from chronic exposure.

Recent Policy Developments

Fishery management agencies, such as the National Marine Fisheries Service, have not always viewed the conservation of marine fish habitat as a priority. Activities that threatened marine fish habitat continued in part because NMFS was not usually equipped with either the resources or the mandate to deal with these problems. Perceptions have been changing, however, during last two to three decades as our understanding of ecological relationships has improved and agency priorities have changed. The NMFS Office of Habitat Conservation works to protect and conserve fish habitat by reviewing licensing, permitting, legislative activities and administrative activities (Waste 1996). Further progress was made with the 1996 amendments to the Sustainable Fisheries Act, which mandates the National Marine Fisheries Service and Regional Fishery Management Councils to identify and protect Essential Fish Habitat (EFH). The law requires NMFS and the councils, to the extent practicable, to take action that minimizes fishing-related adverse effects on EFH and to identify other actions to encourage the conservation and enhancement of EFH. Although a large amount of work has been completed since 1996, few specific management measures have been passed solely for the purpose of EFH protection.

One of the most controversial issues associated with EFH implementation involves fishing-related habitat impacts. As reflected in the EFH provisions, recent symposia, and publications during the last decade, there has been increasing concern among fishermen, environmentalists and managers that mobile fishing gear may produce negative impacts on benthic communities (Dorsey and Pederson 1998). Bottom-tending gear such as bottom trawls and scallop dredges are towed along the bottom. These gears have become a concern because of the large areas of sea floor that are swept by them. Given NMFS estimates of fishing effort, in recent years bottom trawls and scallop dredges in the Gulf of Maine swept an area equivalent to the entire Gulf area ($65,000$ km^2) annually (Dorsey and Pederson 1998). These gears swept an area estimated at more than three times the area of Georges Bank ($41,000$ km^2) annually during the same period (Dorsey and Pederson 1998). Generally, it has been concluded that bottom gear reduces habitat complexity by smoothing out structures on the bottom and by removing bottom fauna that contributes to sea floor complexity (Dorsey and Pederson 1998). Although more scientific attention has been attracted to this problem, there is a great deal of uncertainty surrounding its magnitude and severity.

ESTIMATING THE VALUE OF FISH HABITAT

We extend and apply a bioeconomic model developed by Edwards and Murawski (1993) to estimate the potential losses or gains to the New England groundfish fishery that are associated with changes in habitat quantity and quality. This is a preliminary attempt to establish the possible magnitude of the economic value of changes in EFH.

The model below integrates biological and economic systems by relating economic net benefits to the dynamics of the fish resource. Fish harvest connects the biological and economic components of the model. The economic net benefit of seafood harvest is the difference between the total benefits of seafood consumption and the total opportunity costs of producing seafood. Economic net benefit or value includes consumer surplus, producer surplus and economic rent. Consumer surplus is a measure of consumer benefits that considers the difference between the maximum willingness to pay and the actual price at which the good is purchased. Producer surplus is the amount earned by producers that exceeds the opportunity costs of the inputs (e.g., labor and capital) that are used to harvest fish. Resource rent is the economic value of the fishery resource.

We use the bioeconomic model to calculate the value of changes in habitat for two cases: (1) when the fishery is operated under open access; and (2) when the fishery is managed to achieve maximum economic yield. Open access (OA) and maximum economic yield (MEY) are extremes on opposite ends of a continuum of fishery management institutions. Open access is defined as unregulated harvest by perfectly competitive fishing firms with no barriers to entry or exit. The potential economic value of the resource (rent) is lost due to the lack of institutions that promote cooperative actions or property rights structures. The problem can be recast in the context of a large number of individual firms that are only concerned with private costs and benefits. The firms do not consider the cost that they impose on other users of the resource because there is no guarantee that the individual can benefit from constraining their exploitation rate (Hanley *et al.* 1997).

Fishery management entails recognition by the firms that there is an opportunity cost associated with the removal of a portion of the stock. In the open access case, the opportunity cost of the stock – its value to future production – is not considered by users when making decisions regarding harvest. Most fisheries are highly regulated by gear, areas and size limits while access may be limited by permit. Yet, often the value of the stock to future productivity is not explicitly considered. Therefore, entry or investment of current participants tends to move toward an equilibrium where resource rent is dissipated, even when overfishing is avoided.

Maximum net economic yield (MEY) is attained when the economic net benefits of the resource is maximized over time. Maximum economic yield as used in this paper refers to the maximization of net economic value of a fishery (NEV) and does not necessarily consider all social factors associated with the optimal level of exploitation. In many real world cases such as the Northeast Shelf, fisheries have been managed as regulated open access, which have outcomes that lie between the OA and MEY cases.

Changes in the net economic value (NEV) of a fishery resource due to changes in habitat quality or quantity may be assessed by linking habitat to the bioeconomic model. For example, changes in habitat may be linked to changes in growth, recruitment or carrying capacity. These changes in natural productivity can then be translated into changes in harvest, which are then related to economic considerations such as consumer demand and production of fish products. The specific details of the model are presented in the following section.

A Case Study of New England Groundfish

Edwards and Murawski (1993) employed a bioeconomic model to quantify potential economic benefits from the efficient harvest of New England groundfish. The conditions that determine MEY from the fishery are derived by solving the following dynamic optimization problem:

Maximize $\displaystyle\int_0^\infty e^{-\delta t}[B(Q,t)-C(Q,X,t)]dt$
{Q}

S.T. $\dfrac{dX}{dt}=F(X)-Q$

$X\geq 0, Q\geq 0$

In these equations X represents the harvestable biomass, Q represents landings, δ represents the social discount rate, $B(Q, t)$ represents total social benefits of seafood consumption, $C(X,Q,t)$ represents total social opportunity costs of seafood production and $F(X)$ represents the natural rate of growth of the fish biomass. The net economic value is the difference between total social benefits $B(Q,t)$ and total social opportunity costs of seafood production $C(X,Q,t)$. Both are functions of time due to growing demand because of increasing numbers of consumers and changing tastes and preferences, and technological changes which lower the cost of seafood production.

The rule or condition for generating the MEY from the fishery is given by the following equation (Clark and Munro 1975):

$$\delta = \frac{F_X(X)-C_X(Q,X,t)+C_{QX}(Q,X,t)\cdot[F(X)-Q]-[B_{Qt}(Q,t)-C_{Qt}(Q,X,t)}{B_Q(Q,t)-C_Q(Q,X,t)} \quad (1)$$

The subscripts in equation (1) denote the partial differentiation (for example, C_{QX} is the cost function $C(X,Q,t)$ differentiated with respect to landings Q and biomass X). Conditions that satisfy equation (1) would produce maximum net economic value in the fishery.

Edwards and Murawski (1993) substituted empirical models into equation (1) to find the MEY for the groundfish fishery. They used a Pella and Tomlinson stock production model to simulate growth of the groundfish resource:

$$F(X) = H \cdot X^m - K \cdot X \tag{2}$$

where H, m and K are parameters in the growth function. The production function is assumed to be linear and given by:

$$Q = q(t) \cdot EX \tag{3}$$

where E is a composite index of fishing effort and $q(t)$ is the catchability coefficient, the percentage of the resource that is harvested by a unit of fishing effort. Advances in fishing technology were modeled as a function of time by the following equation:

$$q(t) = q \cdot [1.02]^t \tag{4}$$

where t is the year, starting at zero in 1976, and q is a constant.

The cost function was obtained by solving (3) for E and substituting the result into the cost function for effort, which yields

$$C = c \cdot [Q/q(t) \cdot X] = C(X,t) \cdot Q \tag{5}$$

where the lower-case c is the unit cost of fishing effort.

Since microeconomic cost data were unavailable it was assumed that fishing effort expands until the average cost of harvesting fish is equal to landings price in an open access fishery with identical firms. Therefore, the unit cost of fishing effort was determined by:

$$c = P q(t) \cdot X \tag{6}$$

This approach approximates the average cost of marginal firms, but it overestimates costs for inframarginal firms such as highliners. Therefore, the producers' surplus attributable to these firms was not included in either the open access or optimal cases.

Finally landings price was modeled with the following linear form:

$$P = f + g \cdot Q + h \cdot t \tag{7}$$

where f, g and h reflect the behaviors of consumers and seafood producers beyond the harvesting sector; and t represents the year, as above.

Parameters for these empirical equations were estimated for groundfish species combined, and Atlantic cod, yellowtail flounder and haddock individually. Edwards and Murawski then compared the value of current landings to the potential maximum economic value when the resource is harvested to maximize economic net benefits. When compared for 1989, the actual net economic value was estimated at $10 million while potential maximum net economic value was estimated at $149 million, of which $128 million was economic rent. The actual harvest rate of 112,000 metric tons in 1989 was not sustainable. The sustainable harvest level at the 1989 resource size was less than half the amount that was actually taken, with an economic value of only $2 million of consumer surplus. The optimal case was then considered of over a thirty-year time period, assuming no benefits for the first ten years that are needed for stock recovery. Using a 2% discount rate, the present value of the stream of economic net benefits for this time period was estimated to be approximately $2.0 billion.

Estimates of Habitat Values

How does the degradation of fish habitat affect the net economic value of the New England groundfish fishery? To answer this question, we assumed that carrying capacity is directly proportional to habitat quality and/or quantity. The carrying capacity was varied as a percentage of its current level, as determined by the parameters used by Edwards and Murawski (1993). This was done for both the open access case and the maximum economic yield case.

The open access equilibrium was calculated by equating total revenues with total costs:

$$P \cdot Q = c \cdot E \tag{8}$$

Since Edwards and Murawski (1993) assume that the fishing fleet is made up of identical perfectly competitive firms, consumer surplus was the only component of net economic value that it was possible to consider in the open access case.

We varied the level of carrying capacity from 15% to 130% of its estimated current level in 5% intervals. Then, the open assess equilibrium and associated net economic value were calculated at those 5% intervals over range (15% - 130%). The same range of carrying capacities was also used for the maximum economic yield case. Stock size, harvest and associated net economic value were calculated for both cases. In the MEY case the net social benefits were composed of two components, consumer surplus and economic rent but, again, no producer surplus was considered due to assumptions associated with costs. Resulting net economic value for these ranges were reported for a single year, 1989, and for the next 30 years with discount rates of 2 and 5 %. For the optimal case, it was assumed that no harvest would occur during the first 10 years to allow for stock recovery from overfishing.

For the OA (open access) case at the current carrying capacity, net economic value for 1989 was estimated at $6.7 million. A habitat improvement to 110% of the current level yields an annual net economic value estimate of $7 million. Estimated net economic value with a 10% reduction in groundfish habitat yields a net economic value of approximately $6.4 million. The arc elasticity in this range was between .44 and .52 (a one-percent change in habitat quality corresponds to a .5 % change in net economic value). If habitat were degraded by 10%, estimated annual losses would be $300,000 assuming open access equilibrium is reached for each stock level. When discounted at 2 and 5 % over the next 30 years, losses were $6.7 and $4.7 million, respectively (Figures 9-1 and 9-2).

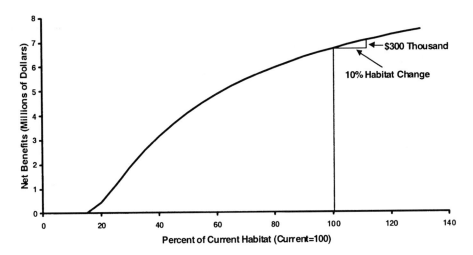

Figure 9-1. Annual net benefits as a function of habitat condition under open access.

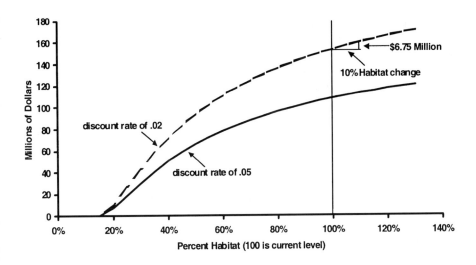

Figure 9-2. Net present value over a 30 year period as a function of habitat under open access.

For the maximum economic yield case at the current carrying capacity, net economic value was estimated at $169 million of which $20 million was consumer surplus and $149 million was economic rent. An improvement to 110% the current level yields an annual net economic value estimate of $188 million. Estimated net economic value with a 10% reduction in groundfish habitat yields a net economic value of approximately $149 million. The arc elasticity in this range was approximately 1.2 (a one percent change in habitat corresponds to a 1.2 % change in net economic value). If habitat has been degraded by a factor of 10%, annual losses would be approximately $20 million. When discounted at 2 and 5 % over the next 30 years with no harvesting in the first 10 years, losses ranged from $266 to $156 million, respectively (Figures 9-3 and 9-4).

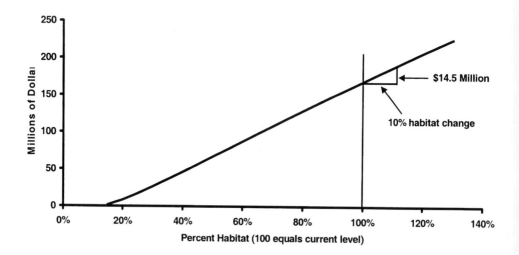

Figure 9-3. Annual net benefits as a function of habitat condition under MEY

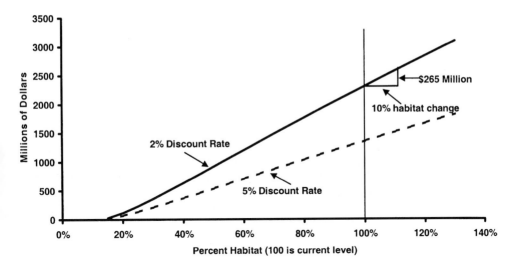

Figure 9-4. Net present value over a thirty-year period as a function of habitat under MEY

The principal estimates of the added benefits (costs) of habitat restoration (degradation) are summarized in Table 1.

Table 1. Benefits (Costs) of Habitat Restoration (Degradation)

A 10% change in habitat:	Open Access	Maximum Economic Yield
Change in NEV/year	$300,000	$20 million
Change PV of NEV	$4.7 - $6.7 million	$156 - $266 million

DISCUSSION AND FUTURE DIRECTIONS

Perhaps the most striking results are the differences between the open access and the maximum economic yield estimates of net economic value. Annual gains from a 10 % improvement in habitat are $300,000 for the open access case and $20 million in the maximum economic yield case. If producer surplus could be calculated and added in both cases, the relative proportions would change but the difference would remain at the same order of magnitude. The principal conclusion that we draw from this is that as long as New England groundfish fishery management tends toward open access, efforts to conserve fish habitat will result in relatively small levels of benefits within likely levels of habitat degradation. This is an important consideration, especially if extensive and potentially costly management measures are proposed without consideration of fundamental changes to the management of the fishery. This example illustrates that management of the fishery and its associated habitat should be integrated.

This conclusion may be even more meaningful if fishing effort causes habitat degradation through interactions with benthic communities. Our example varies habitat quality with the assumption that the cause is not related to the fishery that is harvesting the resource. However, habitat degradation could be modeled as a function of fishing effort. Sea scallop dredges and bottom trawls are the main mobile gear types on Georges Bank and the Gulf of Maine. Potential habitat impacts of these gear types are the concerns that are the most closely related to Essential Fish Habitat provisions. The annual open access equilibrium groundfish effort level was 38,000 standard fishing days in 1989, and the actual effort level was 49,000 fishing days. The maximum economic yield effort level was estimated at 18,600 fishing days. It is likely that similar results in effort reduction would be associated with the social optimum in the sea scallop fishery. If the level of fishing effort is associated with changes in habitat productivity, then management of the fishery and its values are directly related to concerns associated with fish habitat productivity and conservation. Renewable resource exploitation is usually modeled as a homogenous resource where entry and exit to the system are dependent on profits obtained from the exploitation of this one uniform pool. Ecological systems are actually made up of heterogeneous patches which may exhibit unique population dynamics within each patch and movement of individuals among patches. The habitat type of specific patches may be especially important for recruitment or spawning. A more realistic model would consider these biological factors for each habitat type, the entry and exit that occurs due to the opportunity cost of alternatives outside the fishery, and entry and exit for a given patch due to the opportunity cost of exploitation in other patches (Sanchirico and Wilen 1999).

In addition, fishing effort is not evenly distributed over fishing grounds and habitat quality is not homogeneous. As one would expect, fishing is concentrated in those areas that produce the greatest catch, and that lack impediments to towing. Impacts also are not uniform because bottom types and associated communities vary in fish productivity and susceptibility to damage. For example, mobile gear towed over soft sandy bottom is not as likely to generate impacts on finfish recruitment as hard bottom habitat (DeAlteris *et al.* 1999). This study might be extended to consider changes in habitat quality and quantity within specific patches and the resulting changes in the distribution of fishing effort. Further progress will depend on the degree of spatial resolution that is available with respect to historical fishing effort, and the establishment of linkages between habitat degradation and fish productivity. Finally, further modeling work could consider habitat quality and quantity as a function of fishing effort.

REFERENCES

Clark, C.W., and G. Munro. 1975. The economics of fishing and modern capital theory: a simplified approach, *J. Environmental Economics and Management 2: 92-106.*
DeAlteris, J., L. Skobe and C. Lipsky. 1999. The significance of seabed disturbance by mobile fishing gear relative to natural processes: A case study in Narragansett Bay, Rhode

Island. Pages 224-237 *in* L. Benaka, editor. Fish habitat: essential fish habitat and rehabilitation. American Fisheries Society, Symposium 22, Bethesda, Maryland.

Dorsey, Eleanor M. and Judith Pederson. 1998. In Dorsey, E.M. and J. Peterson, editors. *Effects of Fishing Gear on the Sea Floor.* Conservation Law Foundation, Boston, MA. Introduction 1-6.

Edwards, S.F. and S.A. Murawski. 1993. Potential economic benefits from efficient harvest of New England groundfish. *North American Journal of Fisheries Management.* 13:437-449.

Freeman, A.M. 1993. *The Measurement of Environmental and Resource Values Theory and Methods.* Resources for the Future, Washington, D.C.

Hanley, N., J.F. Shogren and B.White. 1997. *Environmental economics In Theory and Practice.* Oxford University Press, New York, NY.

Hunter, M. L. 1996. *Fundamentals of Conservation Biology.* Blackwell Science, Cambridge, Massachusetts. 150.

Kahn, J. 1987. "Measuring the Economic Damages Associated with Terrestial Pollution of Marine Ecosystems," MRE 4(3):193-210.

Sanchirico, J. N. and J.E. Wilen. 1999. Bioeconomics of spatial exploitation in a patchy environment. *Journal of Environmental Economics and Management* 37:129-150.

Waste, S.M. 1996. The NMFS Office of Habitat Conservation: Protecting the habitats of living marine resources. *Fisheries* 21:24-29.

Whittaker, R.H. 1975. *Communities and Ecosystems.* MacMillan Publishing Co., Inc.:New York. 2.

Large Marine Ecosystems, Vol. 13
T.M. Hennessey and J.G. Sutinen (Editors)

10

The Economic Values of Atlantic Herring in the Northeast Shelf Large Marine Ecosystem

Jung Hee Cho, John M. Gates, Phil Logan, Andrew Kitts, and Mark Soboil[1]

ABSTRACT

In this chapter we develop a method to estimate the value of an ecosystem reserve for Atlantic herring in the U.S. Northeast Shelf Large Marine Ecosystem. Herring and other small pelagic species are important forage for other fish and marine mammals in the ecosystem. A reduction in the commercial total allowable catch (TAC) would create an 'Ecosystem Reserve' (ER) for herring, which may increase the residual stock of herring that would be available to other species in the ecosystem. An increase in ER (*i.e.* a reduction in TAC) may imply reduced benefits of commercial harvest; and, as an offset, we may realize added benefits for the ecosystem from the ER. Conversely, an increase in TAC may result in increased commercial benefits at the expense of reduced ecosystem benefits. The approximate magnitudes of these two changes—added and reduced benefits—are the focus of this chapter.

INTRODUCTION

The Atlantic herring, *Clupea harengus*, is a pelagic species that is widely distributed in the continental shelf waters along the Atlantic coast, from the Gulf of Maine to Cape Hatteras. Schools of adult herring undertake extensive seasonal migrations. They spend the summer in the north and the winter in the south. Their migration route covers much of the Northeast Large Marine Ecosystem (NELME). The natural mortality of herring due to predation during the first year of life is believed to be a major factor affecting recruitment to the fishery at age 2 years.

While herring are hardly a staple food of North Americans, the herring fishery illustrates certain issues common in the environmental economics of fisheries, viz. what, if any, values do prey stocks have in *situ* as "feed stocks" in the ecosystem? This question could be asked also of menhaden (the most massive of US fisheries until the Alaskan pollock fishery developed in the 1980s), or the Atlantic mackerel fishery whose stocks, like those of herring, are at historic high levels. All three are pelagic species and have relatively low market values

[1] Cho is Senior Researcher, Fisheries Economics Research Division, Korea Maritime Institute in Seoul. Soboil

is a Ph.D candidate and Gates is Professor, in the Department of Environmental & Natural Resource Economics,

URI. The research received support from the NOAA/URI Cooperative Marine Education and Research program

and the RI Agricultural Experiment Station.

for human consumption. In the case of mackerel, they are also major predators on other fish species in the ecosystem. Figure 10-1 shows that the herring portion of diet for cod increased since 1982. Figure 10-2 shows the importance of herring as forage species in the ecosystem.

Other quantitative information exists which indicates the importance of herring as a food source. First, stomach content data collected from the National Marine Fisheries Service (NMFS) bottom trawl surveys in late 1969 to 1972 show herring to be an important prey item for cod, pollock, haddock, silver hake and white hake (Langton and Bowman 1981). Second, Overholtz *et al.* (1999) estimated that silver hake, cod, and dogfish consumed an average of about 1500, 200 and 4300 metric tons (mt) of herring each year from 1988-1992 on the Northeast U.S. Continental Shelf, respectively. Furthermore, five species of whales, three species of dolphins, harbor porpoises and harbor seals consumed on average 19,300 mt of herring a year from 1988-1992. Thus, about 30,000 mt of herring are consumed by other fishes, marine mammals, and marine birds.[2] This figure was equal to about 50 % of the annual commercial harvest in 1997.

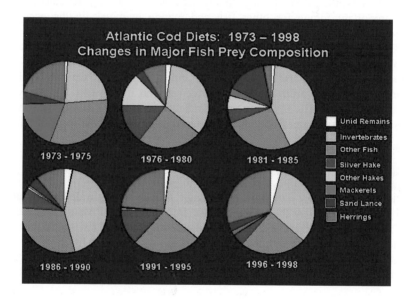

Figure 10-1. Projection of the change in cod prey, 1973-1998[3]

---

[2] This estimate is probably low due to the low stock of herring on Georges Bank at the time of the survey.

[3] Source, National Marine Fisheries Service URL : www.nefsc.nmfs.gov/pbio/fwdp/FWDP.htm

Consumption of Prey Species 1977-1997

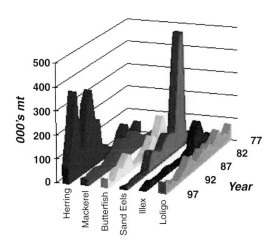

Figure 10-2. Consumption of prey species, 1977-1997[4]

The abundance and biomass of Gulf of Maine/Georges Bank herring have undergone significant changes over the past several decades. During their exploitation by foreign distant-water fleets between 1967 and 1976, their biomass on Georges Bank was reduced by 75% (Sissenwine *et al.* 1984). While the herring stocks declined, sandlance (*Ammodytes dubius*) abundances, particularly in the southern regions from the Gulf of Maine to Cape Hatteras, began to increase (Meyer *et al.* 1979). Due to the competition between herring and sandlance for food and space, the subsequent explosion of sandlance is thought to be an ecosystem response to the disappearance of the herring (Kenney *et al.* 1999). In recent years with better fishery management and the delimitation of the Economic Exclusive Zone (EEZ), herring stocks have been allowed to grow, with current biomass levels at all time highs.

The NELME also contains a rich cetacean fauna such as (but not limited to) four species of large baleen whales: the humpback whale (*Megaptera novaeangliae*), the fin whale (*Balaenoptera physalus*), the sei whale (*B. borealis*), the pilot whale (*Globicephala melas)*; the right whale (*Eubalaina glacialis*), and three species of smaller odontocete (toothed) cetaceans: the white-beaked dolphin (*Lagenorhynchus albirostris*), the Atlantic white-sided dolphin (*L. acutus*), and the harbor porpoise (*Phocoena phocoena)*. Their aggregation into particular regions within this NELME has been shown to be determined by the abundance of pelagic fish stocks such as herring and sandlance[5]. They are known to be important prey for several species of cetaceans inhabiting this area.

---

[4] Source, National Marine Fisheries Service URL : www.nefsc.nmfs.gov/pbio/fwdp/FWDP.htm

[5] Although there may be some species feeding on fish and plankton, it is important to distinguish ER affects from the perspective of toothed whales in contrast to plankton-feeding baleen whales (Kenneth Sherman, Personal communication).

An increasing number of studies support the idea that toothed cetacean distributions are controlled by the distribution and abundance of prey (Lynch and Whitehead 1984; Whitehead and Carscadden 1985). Payne *et al.* (1986) compared humpback whale sightings in the northern Gulf of Maine to those in the southwestern region from 1972-1982. They observed a sharp decline from 1977 onwards in the most northern regions while concurrently observing that the southwestern regions experienced a significant increase in sightings. They concluded that this shift in humpback distribution was correlated with the decline in herring stocks in the north and the more abundant sand lance populations in the southwest. Another example involving humpback whales is that after the collapse of heavily exploited offshore capelin stocks, the distribution of humpback whales shifted from offshore feeding grounds on the Grand Banks to near shore Newfoundland waters. With the decrease in use of the southwestern region by humpback whales, a return to the northern regions was documented as the herring stocks recovered (Payne *et al.* 1990). Therefore, this reduction in stock biomass supported the hypothesis that changes in the abundance of fish stocks (= the availability of prey) can cause significant shifts in cetacean distribution. Kenney *et al.* (1999) have suggested that human induced factors such as overfishing have thus contributed to their shifts in distribution. Over the last couple of decades there has also been a species shift from the once predominant white-beaked dolphin inhabiting the Gulf of Maine region in the mid 1970s, to the white-sided dolphin (Katona *et al.* 1983). White-beaked dolphins prefer schooling fish composed primarily of herring, and their subsequent decline was concurrent with the reduction of herring stocks. One hypothesis attributes this species shift to the changes in prey species.

Atlantic white-sided dolphins are now one of the most abundant cetaceans in the NELME. Their distribution is principally determined by seasonal inshore/offshore migrations that parallel that of their prey, which during the winter months move offshore. Thus, this pattern suggests that the white-sided dolphin distribution could also be largely controlled by the sandlance's seasonal concentrations (Selzer and Payne 1988).

The ecological interactions between herring and its predators have not been explicitly considered in fishery management plans.[6] Flaaten (1998) introduced bioeconomic analysis using a simple two-species predator-prey model. Here, both the biological and economic factors were used as determinants for an efficient management plan. Since then there has been increasing pressure to adapt an ecosystem approach that would look beyond the value of species in human consumption. We should note also that approximately three-quarters of Atlantic herring landings are used as lobster bait (NEFMC 1999). This further complicates matters since some fisheries scientists believe that excessive fishing effort on lobsters that involves baiting of traps is actually increasing the growth of lobster biomass.[7]

An economically efficient management plan for herring would consider the dependence of bottom creatures, marine mammals, and marine birds on the herring stock. Therefore, this chapter presents only a framework for estimating the economic value of reserved herring in

---

[6] However, herring consumption by other species is implicitly imbedded in the parameters of the surplus production model for herring.

[7] Personal communication, Dr. Saul Saila.

the NELME using economic concepts such as *willingness–to-pay* (WTP). Such a framework may be helpful to fishery managers who are trying to adopt an ecosystem approach, as when the value of herring reserved for other species is estimated as explicitly as possible, enabling better societal choices about market versus non-market economic trade-offs.

CONCEPTS AND METHODS

The value of herring stock held as ecosystem reserves is not readily available because such reserves are not sold in a market like private goods and services, such as newspapers, gasoline, and food. Consequently, market prices cannot be used to infer directly the value of herring in the ocean. In order to estimate the economic value of reserves in the ocean, non-market valuation methods must be employed. The non-consumptive use value of marine mammals is so labeled because observers are able to enjoy whales and other mammals by viewing them but without consuming them to the exclusion of others. From studies of whale-watching, we do know that non-consumptive uses of marine mammals are important to many people. In actuality, this activity delivers a bundle of experiences which people value including all the biota observed on a given trip. Although whales are no longer commercially hunted in American waters, they are still of considerable non-use value to society.[8] Emphasis has shifted from commercial exploitation (consumptive use) to public fascination (non-consumptive use).[9] Determination of this non-consumptive use value through an estimate of the total net benefits associated with the recreation will provide a measure of the benefit for management efforts which seek to weigh the costs of whale preservation against the benefits of direct commercial exploitation of whales and indirect exploitation associated with conflicting types of resource development (Samples 1986).

In 1985, close to one million people went whale-watching in New England, spending almost eighteen million dollars to purchase tickets alone. The large scale to which this industry has grown since its inception in 1975 implies that Americans have a growing interest in and place a significant value on this non-consumptive use of whales. Several different methodologies can be used to measure non-consumptive use values.

Methods include *revealed preference* (indirect or actual behavior-based) methods including the Travel Cost Method (TCM) and *stated preference* (direct) methods. Each of these approaches has its advantages and disadvantages. Stated preference methods, such as Contingent Valuation (CV), are useful for estimating the total value of a resource. CV methods of valuation can be important, particularly when non-use value is thought to be significant. CV is a popular hypothetical market technique. Through a survey using a questionnaire,[10] people are asked to reveal their willingness-to-pay for the provision of a good such as wildlife and recreation areas. On the other hand, revealed preference methods, such as the travel cost method, are based on actual behavior and avoid some of the challenges

---

[8] Non use value refers to values that obtained with no use the resource. For example some people may obtain the value from knowing whales swimming in the ocean.

[9] For a review of the history of whaling (for Right whales, *Balena glacialis*), see Reeves *et al.* 1999.

[10] Individuals are asked how much they would be willing to pay to use the services of the resource in question if faced with a hypothetical market situation.

posed by popular stated preference methods (see, *e.g.*, Mitchell and Carson 1989). The basic premise of the TCM is that other things being equal, the number of trips an individual takes to a recreational site depends upon the costs of the activity and its quality. We expect that people living closer to a site will travel to it more frequently than those living further away because their travel cost per trip is relatively low. Trips would also increase for greater levels of quality, other things being the same. A virtue of the TCM is that it can improve our understanding of recreationists' decisions and their behavioral responses to quality and price changes.

We can suppose that aggregate herring predation is a function of herring quantity,[11] proposing the following functional relationship between the stock of herring as prey and the stock of predators:

$$P_S = f(H_S, E)$$

where P_S is the population of predators, H_S is quantity of herring for marine mammals, not for commercial fishing, and E represents variables that affect the stock of whales, such as water temperature, pollution, disease, and poaching. The reason for this assumption is that it is possible that when the herring stock is low for other species in the ocean, the predator stock may be low due to a deficit in prey. Even though there are other factors affecting whale stocks, and the magnitude of their effect on whales is not clear, we can simplify the bioeconomic model by making this assumption. If our assumption is reasonable, then we would expect the correlation between P_S and H_S to be high. Actually this is undoubtedly too simplistic a formulation. In an ecosystem, species will have large differences in life spans. Herring fisheries tend to fluctuate widely over a period of a decade or less and herring are not the sole source of food for predators. For herring predators, a decade is only a fraction of their life spans. We should expect then that the influence of ER on herring predators is a marginal effect, a delayed one. The value of whale-watching can be estimated using the travel cost or contingent valuation methods. However, this estimated value depends on the possibility of seeing whales while at sea. If the possibility of seeing whales is very low due to the decreasing stock of herring as a food source for the whale, the value of whale-watching decreases. Conversely, if the possibility of seeing whales is high, the value of the recreational experience increases. Additionally, more travelers may go whale-watching from ports like Gloucester in Massachusetts. In short, we hypothesize that the per trip value of the recreational experience increases and the participation rate may increase when the possibility of seeing whales increases. By hypothesizing a relationship between the herring stock and the aggregate value of whale-watching, we can state the demand function for whale-watching trips as follows:

$$V = f(p, q, s)$$

where V is trips (*i.e.*, services) provided by whales to users, p is the *price* (*i.e.*, the cost of participation) of a trip, q is the quality of the whale-watching experience as perceived by the tripper, and s represents a set of socioeconomic and other variables (including income) that affect the demand for whale-watching. Examples of the quality of whale-watching include the number of whales sighted, weather, or the distance from port to the whale-watching site.

---

[11] Herring quantity is stock size minus commercial harvest.

Given an estimate of V(p, q, s), several measures of user benefits can be calculated, depending upon the issue of interest. The Law of Demand in economics asserts that the quantity demanded of a good is inversely related to its price. The greater the price, the less will be the quantity demanded. In Figure 10-3 an individual tripper's demand curve for whale-watching is the curve $P^* V_1$. If the tripper's average cost per trip is P_0, the tripper will take V_0 trips per season. Therefore, the total annual expenditure for V_0 trips is measured by the rectangle $(OP_0 AV_0)$. Total Consumer Surplus is the area under the estimated demand curve from the equilibrium number of trips at a given p_0 to the choke price p^* where demand is zero, corresponding to the number of trips, V_0. Total consumer surplus is $P_0 P^* A$. The CS is a measure of the non-use value of the activity to the individual and of the net economic value to society. This is the value of a whale-watching experience. For example, the value of whale-watching (= CS) in the California coast[12] is $43.50 per person per day, estimated by TCM (Loomis *et al.* 2000).

Expenditures by whale watchers on food, fuel, and equipment are often of interest to policy makers and others. Although expenditures are a cost and not a benefit to users, expenditures are a descriptive indicator of the scale of whale-watching-related economic activities in the local or national economy and often are of interest for local or regional policy purposes. Expenditures to engage in whale-watching (including the value of a recreationist's time) are given by V_o*P_o. The relationship between expenditures, total CS, and the incremental CS due to a quality change is shown in Figure 10-3.

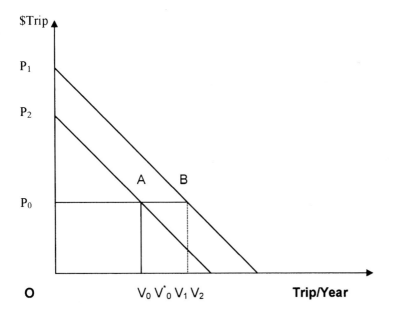

Figure 10-3. Expenditure, consumer surplus, and demand curve shift due to quality change

---

[12] Survey areas were San Diego (Point Loma National Seashore), Monterey, Half Moon Bay (both south of San Francisco), and Point Reyes National Seashore (north of San Francisco).

We could use the TCM to estimate the non-market value of the reserved herring stock. To estimate how WTP changes with a quality improvement, such as the number of marine mammals and sea birds due to an increase in herring stock as prey, we have to compute the change in consumers' surplus. Increased quality or quantity of ecosystem services shifts the demand curve outward to $P_1 V_2$. As the quality of a good improves, more trips (V^*_0) will be made at a given price, P_o. The area, (P^*P_1BA), between the original demand curve and the new demand curve represents the amount that visitors would pay for that increase in quality due to the reserved herring stock for other species. This increase in WTP is a measure of the marginal value of ER.

We would expect an estimate of willingness to pay (WTP) for whale-watching to change with increases in abundance of whales due to increased ecosystem reserves. We might expect diminishing returns to ecosystem reserves of herring for two reasons: (1) the productivity of herring reserves in sustaining increases in marine mammals is probably subject to diminishing returns and, (2) the WTP of whale watchers probably increases at a decreasing rate with whale abundance. Thus, we assume that the WTP increases with ecosystem reserves but the marginal or added WTP decreases (Figure 10-4).

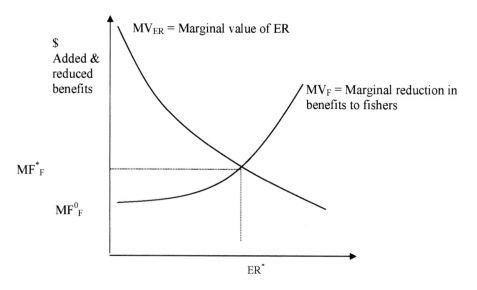

Figure 10-4. Marginal value of ER and harvest

The opportunity cost[13] of ER measures the cost of Ecosystem Reserves when the reserved herring would have been caught by commercial fisherman for human consumption and lobster bait. In other words, the opportunity cost of marginal reductions in TAC to "reserve"

---

[13] "Opportunity cost" is a very flexible, widely used concept in economics. It measures the value or cost of something by what is given up--the lost opportunity.

amounts for the ecosystem is measured by reduced profits to herring fishermen that are made possible by a larger herring stock. There are well established methods for estimating these benefits. They are based on the "resource rent' of incremental change in discounted profit which would follow a change in Ecosystem Reserves (See Clark 1990). These values are the opportunity cost of ecosystem reserves in terms of reduced net profit for fishermen. The concept of opportunity cost is a reciprocal one; the opportunity cost of increased TAC includes the reduction (if any) in ecosystem benefits.

In general, we can regard increases in ecosystem reserves as reductions in TAC allowances for the herring fishery. Because of this inverse relationship, increases in ecosystem reserves are associated with declines in marginal value of ER, and increases in marginal value of the TAC (Figure 10-4). Suppose, in the absence of ecosystem reserves, the economic optimum is at ER = 0 where $MV_F = MV°_F > 0$, and MV_F is less than the marginal value of ER. A decrease in ecosystem reserves would have greater marginal value of ER than the TAC increase of the herring harvesters. Thus, at ER = 0, an increase in ecosystem reserves is economically efficient in that ecosystem uses could bid reserves away from herring harvesters. They could do so until reserves reach ER^*. At this point MV^*_F is equal to MV_{ER} and marginal values of herring stocks are equalized in both uses. In general, we can presume that ER^* involves herring stocks larger than at ER = 0. So, using estimates of marginal value of ER and MV_F, and connecting the marginal values of herring stocks in the fishery and the ecosystem, we can find, in principle, the optimal level of ecosystem reserves of herring, ER^*. It is not necessary to actually quantify the concepts we have outlined unless it is deemed useful as a case study to illustrate methods. An alternative would be to encourage devolution to rights-based approaches for firms in both the herring harvest and whale-watching sectors, but we will attempt a crude approximation.

ANALYSIS AND RESULTS

Two steps are required for estimating the marginal value of herring reserves. The first step is to measure the opportunity cost of ER if left as part of the TAC. Second, we must estimate the reduced value of ER in generating ecosystem benefits. We estimated values of added benefits and reduced benefits for a 10% decrease in the cetacean, sea birds and commercial fish species population such as cod, silver hake and dogfish.[14] Following that, we will repeat the exercise at alternative percentages of 5, 10, 15 and 20 under an assumption of linearity. This is done not because we believe a linear response is realistic but to give an order of magnitude indication of the numbers. Since we have already established a direct relationship between the herring stock size and the cetacean, birds and fishes population, we can assume that a reduction in the herring population will affect the cetacean, birds, and fishes population sizes. If we assume that approximately 29,000 mt of herring is consumed per year by marine mammals, birds and other fishes, a reduction of 10% in the cetacean/birds/other fishes populations could be equivalent to a savings of 2,900 mt (= about 3 kiloton) in herring. Therefore, an increase (= decrease TAC) of about 3 kiloton for ER for marine mammals, birds and other fishes means that there is the opportunity cost of ER to society.

---

[14] Even though there are more species that are consuming herring as a prey, such as pollock, haddock, white hake, etc., we have excluded these from this exercise.

To illustrate the economic effects of reduced Ecosystem Reserves and their opportunity cost, we consider a hypothetical 3 kiloton decrease in Ecosystem Reserves (= 3 kiloton increase in commercial harvest TAC). Given the dependency of species on herring (see Applied Science Associates *et al.* 1994), this quantity corresponds to a 10% decrease in commercial species such as cod, silver hake and in marine mammals and sea birds. On this basis we can use the Applied Science Associates *et al.* (1994) reference to calculate the reduced non market benefits. The value to society for a 10% decrease of marine mammals, birds and commercial fish populations are -$24,580, -$298,380 and, -$36,596 respectively, per year[15] in the New England states and Atlantic provinces. For the commercially harvested species we assume a 10% profit margin on gross ex-vessel value. This is a rather conservative allowance; 10-15% profit on capital investment would be typical of returns in the US economy, but return on sales would typically be lower. For the otter trawl fleet, profits as a percent of sales have been estimated to be 17% (Dr. P. Logan, personal communication). Our results are summarized in the following Table 10-1. For the added benefit, the derived demand for Atlantic herring harvest has been less than the TAC (or MSY) so the added benefit of an increase in TAC is essentially zero but the reduced benefits of ecosystem reserves are also zero in this case; an increase in an ineffective TAC has no effect on either side of the benefit-cost ledger. The results indicate a net loss of $319,100 from an increased TAC when the herring TAC is ineffective (as seen in Table 10-2).

Thus, we can say that a decrease in ER due to an increase in the TAC implies that reduced benefits for the ecosystem are greater, in some circumstances, than added benefit for commercial herring fishing. However, it must be admitted that these values are not precise, the natural environment is not static, and the dynamics of the herring-cetacean/birds interaction have been ignored in this exercise. We have not discounted because we do not know the dynamics well enough. Clearly, whales have a much longer response time than do fish. However, if the linkage is primarily one of attraction to coastal areas, then response time may be quite rapid. Secondly, since we know that there is a marginal effect that has not been introduced, our estimates are quite crude. Also, because we did not take into account all commercial species value, the reduced value of ER to society may be larger than $319,100 dollars per year. These results suggest that a more refined analysis may be appropriate. We will next rescale these numbers for a larger percentage of re-allocations from ER. Since we do so under an assumed linearity, the rescaling (Table 10-3) is a trivial numerical exercise. However, the larger the extrapolation, the more important become the distinctions between marginal and average effects if the true response is non-linear.

---

[15] See The CERCLA Type A Natural Resource Damage Assessment Model for Coastal and Marine Environments. However, this value was in 1991 dollars. For commercial fish species, we assume that a profit margin is 10 percent of landing value in 1998.

Table 10-1. Consumptive and Non-consumptive use value of Atlantic herring in the ecosystem

Commercial species	Consumption of herring (mt/year)	Value from commercial fishing[5] ($/mt)	Non-consumptive use value[1] ($)	Added benefit[4] ($)
Cod	200	2,291	.	4,582
Silver hake	1,500	893	.	13,395
Dog fish	4,300	433	.	18,619
Marine mammals			24,580[3]	24,580
Finback whale	5,000	.	.	.
Humpback whale	2,600	.	.	.
Pilot whale	2,800	.	.	.
Sea birds	3,250	.	298,380	298,380
Others[2]	19,650	.	.	.
Total	29,000	361.7	322,960	359,556

Note:

1. The unit is $ for 10 % of population change.

2. Some commercial species such as pollock, haddock, and white hake, and other marine mammals have no quantitative information on consumption of herring as a prey. Also there is no value estimated.

3. There is only total non-consumptive use value for marine mammals.

4. The benefit from commercial fishing was estimated by assuming profit is 10 percent of total revenue i.e. total revenue is equal to ex-vessel price multiplied by landing.

5. The commercial landings for cod, silver hake, and dogfish are 11,122; 14,959; and 936 respectively. The landing values are $ 2.5 million, $13.4 million, and $ 0.4 million for cod, silver hake, and dog fish.

Table 10-2: Partial Budget for 10 percent increase in TAC for Atlantic herring.

B. Reduced Benefits (kilo-dollars)		A. Added Benefits (kilo-dollars)	
Marine mammals	-24.6	Herring Harvests	
Sea Birds	-298.4	TAC effective	40.5[16]
Commercial Fishes	-36.6	TAC ineffective	0
Subtotal A:	-359.6	Subtotal B:	TAC effective→ 40.5 TAC ineffective → 0
Net Change = A + B			
TAC effective	-319.1		
TAC ineffective	-359.6		

Table 10-3: Sensitivity Analysis of the Level of Ecosystem Reserves

		Percent Increase in TAC *			
		5	10 **	15	20
			kilo-dollars		
B. Reduced Benefits		-180	-360	-539	-719
A. Added Benefits		20	41	61	81
C. Net Change = B + A		-160	-319	-479	-638

*The table assumes a linear response. While linearity is plausible for sufficiently small changes, it is increasingly questionable as we increase the percentage change.
** Column based on Table 2

CONCLUSIONS

There are extra-market values for non-consumptive enjoyment of sea birds and marine mammals. For example, people are willing to sacrifice market goods to maintain or to conserve whale populations for themselves (existence value) and future generations (bequest value). However, these existence and bequest values are excluded in this chapter. We presented a framework for estimating the value of changes in Ecosystem Reserves of herring based (partly) on non-market valuation methods. This framework could be useful to fishery managers when they decide on herring fishery management plans that take into account ecosystem reserves of herring for other species. However, to quantify non-market values more precisely, future research on the relationship between herring stock and its predators and a refinement of the whale-watching/whale-herring interactions is required. What we have shown by manipulating our simple model is that consideration of the marginal value of ER implies that optimal herring stocks could be higher than when such extra market values are ignored. At this time, we do not know how great a difference these considerations would make. Currently, the stock of herring is at historic highs. At the same time, the world markets for herring have relatively low prices for reasons thought to be due to changing tastes among consumers and healthy stocks off Scandinavia and Japan. Under these circumstances, the opportunity cost of Atlantic herring ecosystem reserves is low if not zero. An increase in export demand for herring would be associated with changes in potential benefits to certain groups. Herring fishers would benefit from increased profits on resource rents. If increased harvests of herring reduce near-shore biota which feed on herring or which are enjoyed by humans for their non-consumptive values, there would also be benefit decreases.

In a market economy, such benefit trade-offs are handled via voluntary transactions between interested parties. Such transactions entail minimal transaction costs. Where markets are imperfect or non-existent *(e.g.* the extra-market values of herring), participants can be expected to pursue their interest through compulsion and collective action.

In principle, the introduction of market mechanisms would make it relatively easy for a whale-watching industry and environmental organizations to bid herring ecosystem reserves away from the herring harvest sector—if the marginal WTP in the whale-watching sector exceeds the marginal profits in the commercial herring harvest sector. Attempts to achieve such ends through political processes could result in substantial rent-seeking costs, and a shifting of much of these costs to the general taxpayer rather than to the primary beneficiaries. In market processes rent-seeking will also occur, but the magnitude of dead weight losses is upper-bounded by potential benefits. In political markets, bounds are limited only by the ability of activists to shift program costs to the general taxpayer.

REFERENCES

Applied Science Associates, Inc., A.T. Kearney, Inc. and HBRS, Inc. 1994. The CERCLA type A natural resource damage assessment model for coastal and marine environments. Technical Document Submitted to U.S. Department of the Interior

Clark, C.W. 1990. Mathematical Bioeconomics: The Optimal Management of Renewable
    Resources. Second edition, Pure and Applied Mathematics series. New York;
    Chichester, U.K., Brisbane and Toronto: Wiley, Wiley-Interscience, pp. xiii, 386.

Flaaten, O. 1998. On the bioeconomics of predator and prey fishing. *Fisheries Research*
    37(1-3): 179-191(Amsterdam).

Katona, S.K., V.A. Rough, and D.T. Richardson. 1983. A Field Guide to the Whales,
    Porpoises and Seals of the Gulf of Maine and Eastern Canada: Cape Cod to
    Newfoundland (3rd ed.). New York: Scribner's.

Kenney, R.D, G.P. Scott, T.J. Thompson and H.E. Winn. 1997. "Estimates of prey
    consumption and trophic impacts of cetaceans in the USA northeast continental shelf
    ecosystem," *Journal of Northwest Atlantic fishery science* 22: 155-171.

Langton, R.W. and R.E. Bowman. 1981. Food of eight northwest Atlantic pleuronectiform
    fishes," NOAA Technical Report No. NMFS SSRF-749, NOAA/NMFS, SEATTLE,
    WA (USA). pp. 19.

Logan, P., Northeast Fisheries Science Center, National Marine Fisheries Service, Woods
    Hole, MA. Personal communication.

Lynch, K., and H. Whitehead. 1984. Changes in the abundance of large whales off
    Newfoundland and Labrador, 1976-1983, with special reference to the finback whale.
    Doc. SC/36/02. Cambridge, UK: International Whaling Commission.

Meyer,T.L., R.A. Cooper and R.W. Langton. 1979. Relative abundance, behavior, and food
    habits of the American sand lance, *Ammodytes americanus*, from the Gulf of Maine.
    Fish. Bull. 77(1): 243-253.

Mitchell, R.C.and R.T. Carson. 1989. *Using Ssurveys to Value Public Goods: The Contingent
    Valuation Method.* Washington, D.C.: Resources for the Future. pp. xix, 463

New England Fishery Management Council (NEFMC). 1999.

Overholtz, W.J. 1999. Precision and uses of biological reference points calculated from stock
    recruitment data. *North American Journal of Fisheries* 19(3): 643-657.

Payne, P.M, J.R. Nicolas, L. O'Brien and K.D. Powers. 1986. "The distribution of the
    humpback whale, *Megaptera novaeangliae*, on Georges Bank and in the Gulf of
    Maine in relation to densities of the sand eel, *Ammodytes americanus*," Fishery
    Bulletin. 84(2): 271-277.

Payne, P.M, D.N Wiley, S.B Young, S. Pittman, P.J Clapham and J.W Jossi. 1990. Recent
    fluctuations in the abundance of baleen whales in the southern Gulf of Maine in
    relation to changes in selected prey. *Fishery Bulletin* 88(4): 687-696.

Selzer, L.A. and P.M. Payne. 1988. The distribution of white-sided (*Lagenorhynchus acutus*)
    and common dolphins (*Delphinus delphis*) vs. environmental features of the
    continental shelf of the northeastern United States. *Marine Mammal Science.* 4(2):
    141-153.

Sissenwine, M.P, B.E Brown, M.D. Grosslein and R.C. Hennemuth. 1984. The multispecies
    fisheries problem: A case study of Georges Bank, Mathematical Ecology. Proc.,
    Trieste, 1982. Lecture Notes Biomath. 54: 286-309.

Whitehead, H. and J.E. Carscadden. 1985. Predicting inshore whale abundance - whales and
    capelin off the Newfoundland coast. *Canadian Journal of Fisheries and Aquatic
    Sciences* 42(5): 976-981.

Large Marine Ecosystems, Vol. 13
T.M. Hennessey and J.G. Sutinen (Editors)
© 2005 Elsevier B.V. All rights reserved.

11

Eutrophication in the Northeast Shelf Large Marine Ecosystem: Linking Hydrodynamic and Economic Models for Benefit Estimation

Thomas A. Grigalunas, James J. Opaluch, Jerry Diamantides and Dong-Sik Woo

INTRODUCTION

Pollution of coastal waters by nutrient (phosphorus and nitrogen) over-enrichment is a major cause of coastal water degradation in the Northeast Shelf Large Marine Ecosystem and throughout the United States (National Research Council 2000; Boesch *et al.* 2001, Sherman *et al.* 1996) and is an important issue in coastal waters virtually worldwide (*e.g.* Gren *et al.* 2000; Korea Ocean Research and Development Institute 1996). Nutrient over-enrichment causes several problems, among the most common of which is eutrophication, which has been defined as "...the process of increasing organic enrichment of an ecosystem where the increased supply of organic matter causes changes to that system (Nixon 1995, cited in National Research Council 2000).

Eutrophicaton can lead to, or contribute to, algal blooms, depletion of dissolved oxygen[1], reduced water clarity, loss of seagrasses, and other injuries to water and sediment quality. These injuries, in turn, can cause shellfish area closures due to pathogens and lost biological productivity with reduced commercial and recreational fish catches (Dept. of Health 1992; Epstein *et al.* 1998); losses to mariculture operators due to perceptions of contamination (Wessells 1995); adverse amenity effects from diminished water clarity (Opaluch *et al.* 1999; Diamantides 2001; Poor *et al.* 2000), and in the case of harmful algal blooms, human health and the health of marine mammals are at risk (Shumway 1990; Epstein *et al.* 1998; National Research Council 2000).

The general consequences of eutrophication are widely recognized (*e.g.* National Research Council 2000; Boesch *et al.* 2001), as are measures to control nutrient pollution sources through reduced use of fertilizer, changes in tilling practices, use of buffer zones, residential zoning, and improved management of individual septic systems and sewage treatment plants (Besedin 1996). However, many of these control measures can be expensive (*e.g.* Peconic Estuary Program).

---

[1] Depletion of dissolved oxygen--hypoxia--occurs when algae die, fall to the bottom, and decompose, a process that consumes oxygen, a key attribute of a biologically healthy ecosystem. In the extreme, oxygen may be totally depleted—anoxia—which can lead to major fish kills.

Analyses of the costs and benefits of nutrient reduction could contribute to policy in this area, by helping to target areas and control measures to suit the problems faced, but it is extremely difficult in practice to incorporate consideration of benefits. This is because of (1) the general lack of adequate data on the many non-point and point sources of nutrient pollution that affect water quality in a particular locale; (2) the inherent difficulty in linking policies to control different pollution sources to site-specific changes in ambient water quality, and (3) the fact that some important sources of nitrogen, such as faulty septic systems, can be associated with fecal coliform, so that joint "bads" result: for example, over-enrichment of coastal waters, loss of dissolved oxygen, and closures of shell fishing grounds due to threats by pathogens to human health. Given the multitude of issues, many empirical difficulties arise in estimating the site-specific change in benefits from the implementation of control measures.

Additional complications arise from the complementary or conflicting nature of some coastal and water resource policies. For example, policies to preserve agricultural lands provide important public good benefits to the community at large (Beasley *et al.* 1986; Halstead 1984; Kline and Wichelns 1994; Ready *et al.* 1997; Opaluch, Mazzotta, and Grigalunas 2001). However, farmland preservation can impose costs on nearby property owners concerned with surface and ground water quality, odors, or other farm-related dis-amenities (Johnston *et al.* 2001).[2]

A rich literature examines the effects of water quality on recreational benefits using indirect or revealed preference methods (*e.g.* Braden and Koldstadt 1991; Freeman 1993; Diamantides 2001; Poor *et al.* 2000). However, few studies have been able to link multi-source pollution control to changes in ambient water quality attributes at different sites, and to the resulting change in human behavior and benefits at these sites. In large part, this gap is due to lack of information on sources of pollution and the effects of pollution source control policies on ambient conditions in coastal waters, as noted. This unfortunate gap in the literature makes it difficult to assess the benefits from alternative control polices and hence difficult (if not impossible) to weight the relative benefits and costs of proposed policy measures for different areas.

In sum, for all of the reasons mentioned above, economists have been hard pressed to contribute information on the benefits of proposed nutrient source control measures since benefits assessment in such cases requires considerable (and generally unavailable) information from science studies on the sources, pathways, and fates of pollution. These constraints hinder the effective use of some incentive-based approaches to control this type of pollution (Randall and Taylor 2000; Morgan, Coggins and Eidmen 2000).

This chapter provides some results on estimating benefits from controlling coastal eutrophication. The work links the results of a hydrodynamic model with an economic model for the Peconic Estuary System (PES). Our overall goal in this chapter is to estimate recreational benefits from alternative water quality policies to reduce a major nutrient--nitrogen. First, we describe the PES study area and provide some background on nutrient pollution in the PES, focusing on total nitrogen, which has been identified as the major source of water quality concerns (TETRA TECH 2000). Then, we outline the economic framework.

---

[2] Thus this requires the use of two policy instruments, one to encourage preservation of agricultural lands and a second to manage pollution at the site.

The model incorporates each recreational participant's subjective perceptions of water quality at recreation sites in the PES. The model results show that objective measures of water quality from sampling can be used to explain subjective perceptions of water quality and recreational behavior by swimmers. Exploiting this important link, we analyze several water quality policies being considered for the PES. A preliminary assessment of the impact of these policies on water quality parameters was simulated by TETRA TECH (2000) in a project done for the PES National Estuary Program. Given estimates of changes in water quality, the economic model predicts changes in users' subjective perceptions, in their recreational behavior, and in the associated benefits.

Our analysis is restricted to benefits to swimmers from controlling nitrogen pollution; other potential benefits, such as preserving or reopening shellfish grounds are not considered. Further, we focus on the issue of water clarity, which earlier research has found to be a key water quality attribute for amenity users (Diamantides 2001; Poor *et al.* 2000; Freeman 1993). As a result, we ignore benefits from improvement in other quality attributes and in other services (including possible non-use value) provided by coastal waters that could be affected by nitrogen reduction. Also, only a representative sub-set of source-control policies assessed by TETRA TECH is considered. Consequently, we do not analyze all possible nutrient control policy options being considered for the PES, and the results presented likely underestimate, perhaps significantly, the benefit of nitrogen reduction in the study area.

THE PECONIC ESTUARY

Located on the east end of Long Island, in Suffolk County, the PES watershed is made up of five towns (Riverhead, Southhold, East Hampton, South Hampton, and Shelter Island) and part of a sixth, Brookhaven. Some 130,000 year-around residents live in the PES study areas, with year-round residential development clustered in the west end (Riverhead) and in scattered communities (Greenport, Sag Harbor, Southold, Noyack, Montauk, Riverhead, North Sea, and East Hampton etc.) along the North and South Forks of the estuary. The estuary itself is comprised of five bays (from west to east: Flanders Bay, Great Peconic Bay, Little Peconic Bay, Shelter Island, and Gardeners Bay) and has more than 100 embayments and creeks. In total these water bodies cover some 100,000 acres (SCDH 1992).

Marine-related economic activity, especially tourism and recreation, is an important part of the local economy (Grigalunas and Diamantides 1996). Reflecting the importance of tourism, the population is highly seasonal, and more than doubles during the peak tourism season. At the same time, agriculture, including vineyards, nurseries and vegetable farms, remain an important part of the area economy (Grigalunas and Diamantides 1996)

Water quality in the PES has been generally good. However, from the mid-1980s to the early 1990s, serious Brown Tide incidents[3] are believed to have caused a collapse in the then-important scallop industry, and substantial areas (over 3,300 acres) of shell fishing grounds have been closed due to pollution (Suffolk County 1992). Brown Tide also has been a substantial concern to residents and recreational users of the PES waters (Opaluch *et al.* 1999;

---

[3] Brown Tide is caused by explosive blooms of the microalgae organism *Aureococcus anophageffrens.* The bloom derives its name from the stark, dark brown color of the affected water body.

Diamantides 2001). Pollution from agriculture remains an important concern due to intensive use of fertilizer. Enormous development pressures have raised concerns about the effects of loss of open space from this additional development and the consequences of these changing land-use patterns on water quality throughout the PES. Further, atmospheric deposition of nitrogen from power plants has been recognized as an additional important source of pollution of PES waters.

Sources of Nutrient Pollution in the PES

Studies conducted as part of the National Estuary Program have examined the sources, pathways, and fates of nitrogen, as well as ambient water and sediment quality throughout the PES (Suffolk County Dept. of Health Services 1999; TETRA TECH 2000). Nitrogen is transported in surface water and groundwater, but the principal source of nitrogen is groundwater pollution, estimated to be some 6,500 pounds per day. Of the 6,500 pounds, 32% is in the western PES and 68% is in the eastern PES, due primarily to the concentration of agriculture in this area (Suffolk County Dept. of Health Services 1999, p.1).

Agriculture is the single most nitrogen-intensive pollution source per acre, although the quantity of pollution varies between crops. In the aggregate, agriculture (41%) and residential development (40%) account for most of total nitrogen loadings into the PES, with lesser contributions from commercial, industrial and institutional land uses (<10%). Nutrient pollution from residential properties reflects the limited availability of sewage treatment plants in the PES. More modest contributions in the aggregate are from open space (including golf courses) and other sources, including storm water runoff and sewage treatment plants. Other important sources of nitrogen pollution are sediment flux loads in the western PES (primarily due to wastes from now defunct duck farms) and atmospheric nitrogen deposition in the eastern PES.

Reflecting strong development pressures, groundwater loadings of total nitrogen may have risen by 200% since the 1950s in the eastern PES due to dramatic population growth and pervasive use of inorganic nitrogen fertilizer in agriculture (Suffolk County Dept. of Health Services 1999, p.2). In the western part of the estuary, decreases in duck farm TN loadings appear to be offset by increases in non-point source total nitrogen loadings (Suffolk County Dept. of Health Services 1999, p.45).

Management policies to address nitrogen pollution must consider the relative contribution of different sources per acre, the current pattern of land use, development trends and projected land use patterns, and constraints on developable lands and their uses. All of this must be done in an appropriate framework that considers potential recreational and other benefits in a geographical setting.

Economic Concepts and Model

We illustrate the basic problem using a hypothetical (Hicksian income-compensated) demand function for an individual for recreation trips to the PES. The number of trips in a period depends upon its own price, the price of visiting substitute sites, site quality, and initial utility. Price is assumed to reflect the cost of participation and encompasses all incremental costs, including the opportunity cost of the user's time.

$$D_i = D(Pi, Ps, Q, Uo) \qquad (1)$$

Where

D_{iB} = Demand for recreation trips to swimming site i, before a quality change
Pi = the price of visiting site i,
Psj = the price of visiting substitute sites j = 1,…J, j ≠ i
Q = quality of the site before a quality change
U_o = the individual's initial level of utility

Given an initial choke price, P*, consumer surplus (CS) before the quality improvement is measured as the area, A, under the individual's demand curve and above the price of participation (Figure 11-1). A change in quality from Q_B to Q_A at site i shifts the demand for trips to the site from D_{iB} out to D_{iA}, causing an increase in trips, r, from r_B to r_A. The new choke price is P**, and the quality change increases CS. The increase in CS reflects the additional benefits, B, accruing to the user at the original number of trips r_B, plus benefit in area C due to the additional trips induced by the improved quality, (r_B - r_A).

Key empirical issues in the recreational-demand model implemented for the work reported on in this chapter concern (1) perceptions of water quality by recreational users, (2) how behavior is affected by quality perceptions, (3) the link between these *subjective* perceptions of quality and *objective* measures of quality, and (4) the estimated recreational behavioral changes due to the cause-and-effect link between source controls and objective water quality measures. We discuss these next.

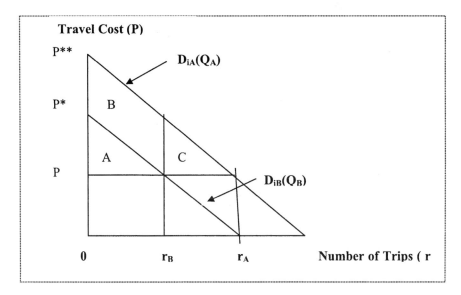

Figure 11-1. Effect of Quality Change at Site

Implementing the Recreation Model[4]

A survey was developed during the winter and spring of 1994-1995 using standard procedures: focus groups, pre-testing of draft questions and of the survey instrument as a whole, and use of a pilot study (for details (see Opaluch *et al.* 1999; Diamantides 2001). The final survey was administered during the week of 22-29 August 1995, using convenience intercept sampling at pre-selected locations. In total 1,345 useable responses were obtained from residents, second homeowners, and day-trippers.

The survey asked respondents about their participation in 16 Outdoor recreational activities over the past year. Then, respondents were asked detailed information concerning major recreational activities. Included among the questions asked in this part of the survey was the point of origin and destination for the *most recent* outdoor recreational trip in each category (swimming, boating, etc.) as well as the costs associated with that trip. Respondents also were asked standard socio-demographic questions about their age, sex, and income level, for example.

The data is grouped into ten sites: Flanders Bay north and south, Great Peconic Bay north and south, Little Peconic Bay north and south, Shelter Island Sound north and south, and Gardiners Bay north and south. Swimming locations were allocated to the ten sites according to locations indicated by the survey respondent. Only observations that indicate an identifiable swimming location for trips taken during August 1995 were included in the data set.

Water quality sampling stations were allocated to the ten location groups according to SCDHS sampling station maps, as described in Table 11-1:

Table 11-1. PES Sampling Station Locations

Location Group	Sampling Station ID#
Flanders Bay North	170 (Central), 220, 240
Flanders Bay South	170 (Central)
Great Peconic Bay North	101, 130 (Central)
Great Peconic Bay South	130 (Central)
Little Peconic Bay North	102, 103, 105, 113 (Central)
Little Peconic Bay South	113 (Central)
Shelter Island Sound North	106,107,108,109,111,112,114, 115,119,122
Shelter Island Sound South	118,121,126,127,131
Gardiners Bay North	116
Gardiners Bay South	132,133,134

---

[4] This section draws heavily on Diamantides (2000).

The number of samples recorded from each sampling station varies. An average of the data recorded at each station during August 1995 (the period during which the survey was administered) was taken. Sampling stations with a central position, as opposed to north or south, were included in the averages of *both* north and south location groups.

Data for Brown Tide, total Kjeldahl nitrogen, and total coliform bacteria were used to construct indices of water quality relative to threshold values for each parameter to make the measures more comparable. The average Brown Tide cell count for each of the ten location groups was divided by 250,000 cells/ml, the visibility threshold for Brown Tide. The resulting value was then multiplied by 1, if the respondent indicated that his or her swimming was affected by Brown Tide, or by 0 if not. In this way, the Brown Tide cell count is modeled to affect the water quality ranking of only those respondents who indicated they had knowledge of and were affected by Brown Tide.

The average total Kjeldahl nitrogen reading for each of the ten location groups was divided by .5mg/l, the nitrogen guideline established by the SCDHS. The average total coliform bacteria reading for each of the ten location groups was divided by 400MPN/100 ml, a threshold for bathing water quality. The total Kjeldahl nitrogen index and the total coliform bacteria index were added together to form a joint nitrogen-bacteria index. This joint index was used in the ordered logit model instead of the individual indices because high co-linearity in field measurements of nitrogen and coliform bacteria precluded separate estimation of the effects of these two water quality variables. All indices were multiplied by 100 to preclude the use of fractions. Average input values for each location group are presented in Table 11-2.

Table 11- 2. Average Input Value for the Water Quality Model (Ordered Logit)

Location Group	Brown Tide Index	Nitrogen – Bacteria Index	Secchi Disk Depth (Feet)
Flanders Bay North	5.57	818.09	3.2
Great Peconic Bay North	18.33	169.85	3.13
Great Peconic Bay South	14.81	117.75	3.5
Little Peconic Bay North	28.04	126.39	3.5
Little Peconic Bay South	16.21	126.76	3.5
Shelter Is. Sd. North	39.51	136.44	4.35
Shelter Is. Sd. South	47.5	256.26	4.37
Gardiners Bay North	32.41	94.21	5.75
Gardiners Bay South	33.37	124.7	5.08

The literature has long recommended that perceptions of quality, *i.e.*, the subjective measure, is the appropriate input into models of economic behavior (Smith *et al.* 1983; Smith 1989; Freeman 1995). The research reported on herein extends the literature by modeling the objective–subjective relationship and using predicted, subjective measures as the water quality input into the travel cost model. The advantage of this approach is that changes in objective water quality measurements due to management actions are converted into subjective measures. The predicted subjective measures are then appropriately applied to the travel cost model.

The ordered logit model provides coefficient estimates and estimates of the threshold parameters μ_i's. These threshold parameters are combined with the coefficient estimates to identify transition points between the subjective water quality assessments (poor, fair, good, excellent). The area to the left of μ_i, the first transition point, is proportional to the probability of selecting poor; the area between the first and second transitions points represents the probability of selecting fair; etc.

The results of the ordered logit model are presented in Table 11-3. All estimation was performed with the discrete choice model in LIMDEP™ (Greene 1995). The coefficients carry the expected signs and are significant at the 10% level with the exception of Secchi Disk measurements which are significant at the 14% level. The Chi-squared statistic was calculated as –2*[restricted log-likelihood – unrestricted log-likelihood]. The likelihood ratio index (ρ^2) was calculated as 1 – [unrestricted log-likelihood / restricted log-likelihood].

Table 11- 3. Results of the Water Quality Model (Ordered Logit)

Log Likelihood	-225.26			Chi-Squared	27.50
Res. L. Likelihood	-239.00			D. Frdm.	3
Likelihood Ratio (ρ^2)	.058			Sig. Level	.000005
Variable	Coefficient	Std. Err.	T Stat.	Sig. Level	Mean of X
Constant	2.0679	1.1136	1.857	0.06332	
Brown Tide Index	-0.02297	0.00499	-4.605	0.00000	33.83
Nitro. - Bact. Index	-0.00147	0.00089	-1.652	0.09860	173.4
Secchi Disk	0.34515	0.23232	1.486	0.13737	4.397
μ 1	2.2812	0.25915	---	---	
μ 2	4.6968	0.3522	---	---	

The marginal effects are calculated for each water quality ranking and for each explanatory variable. Since the probabilities sum to one, the marginal effects of increasing the probability in any rank are offset by decreased probabilities in other ranks. The marginal effects resulting from the water quality model are presented in Table 11-4.

Table 11- 4. Marginal effects of the water Quality Model (Ordered logit)

Variable	Poor	Fair	Good	Excellent
Constant	-.1386	-.3690	.3133	.1942
Brown Tide Index	.0015	.0041	-.0035	-.0022
Nitrogen – Bacteria Index	.0001	.0003	-.0002	-.0001
Secchi Disk	-.0231	-.0616	.0523	.0324

The probabilities for the various water quality rankings under existing conditions are presented in Table 11-5. Table 11-6 presents the probabilities for each of the PES water bodies. The rank order probabilities are given for nine location groups only, as no survey respondents reported swimming at Flanders Bay-south.

Table 11-5. Baseline Probabilities for Nine Location Groups

Location Group	Prob. Poor	Prob. Fair	Prob.Good	Prob. Excellent
Flanders Bay North	0.14	0.47	0.34	0.05
Great Peconic Bay North	0.08	0.37	0.45	0.10
Great Peconic Bay South	0.06	0.32	0.49	0.13
Little Peconic Bay North	0.08	0.38	0.45	0.10
Little Peconic Bay South	0.06	0.33	0.49	0.12
Shelter Is. Sd. North	0.08	0.38	0.45	0.10
Shelter Is. Sd. South	0.11	0.43	0.39	0.07
Gardiners Bay North	0.04	0.25	0.53	0.18
Gardiners Bay South	0.05	0.30	0.50	0.14

*Probabilities may not sum to one due to rounding

Table 11- 6. Baseline Probabilities for PES Water Bodies

PES Water Body	Prob. Poor	Prob. Fair	Prob. Good	Prob. Excellent
Flanders Bay	0.14	0.47	0.34	0.05
Great Peconic Bay	0.08	0.37	0.45	0.10
Little Peconic Bay	0.08	0.37	0.45	0.10
Shelter Is. Sd.	0.08	0.39	0.44	0.09
Gardiners Bay	0.05	0.28	0.52	0.16

Rank order probabilities for each of the five PES water bodies are calculated as the weighted average of the probabilities for the corresponding location groups. For example, the probability that water quality at Gardiners Bay will be ranked good is based on the weighted average of the probability for good at Gardiners Bay-North and Gardiners Bay-South. The weights are based on the total number of observations for each location group. The probabilities for the five PES water bodies are presented in Table11-8.

Count Data Model

A count data model based on the Poisson distribution is used to estimate the demand for recreational activities in the PES. Data for the travel cost model comes from the Recreational Use Survey and from the results of the water quality model. The survey questionnaire asks respondents to indicate the number of times they went swimming in 1995 at a variety of water bodies on the East End. These water bodies include Flanders Bay, Great Peconic Bay, Little Peconic Bay, Shelter Island Sound, Gardiners Bay, Long Island Sound, Block Island Sound, and the Atlantic Ocean.

It is important to note that a large proportion of the participants in the sample visited more than one site. Following the procedure outlined in Bockstael *et al.* (1987), different site choices by the same individual were included in the data as if made by different individuals. In this way, the data set is expanded and variation within the data set is increased. Expanding the data set this way also increases the incidence of zero trips to some locations. In total, 199 respondents provided sufficient information to be included in the travel cost model. Given that there are five PES swimming locations, 995 observations were used in the travel cost model estimation.

The travel cost variable was created in the standard manner (Freeman 1993). Travel costs to swimming locations are based on distance traveled, travel time, and the household income according to the formula:

*Travel Cost = Round Trip Distance * $0.32 + {Round Trip Distance/40mph * 40% hourly wage}.*

The opportunity cost of time was calculated as 40% of the wage for employed participants and as zero for retirees. The opportunity cost of time was applied to travel time but not to time on site. Travel expenses were calculated at $0.32 per mile, which is the Federal automobile reimbursement rate for 1995. The round trip distance traveled is double the one-way highway distance between the respondent's point of origin and each swimming location. This distance was determined on a case-by-case basis using a road atlas and assessing the most direct travel route. Verification of distance traveled was made possible when respondents answered the question concerning distance traveled on their most recent swimming occasion. Travel time was estimated by dividing the round trip distance by 40 miles per hour. This speed was used as a reasonable compromise between highway speeds of 55 mph and more and local speeds between 25 and 40 mph. The opportunity cost of travel time was estimated at 40% of the wage rate, a common and accepted practice in benefit estimation (Smith 1989; Fletcher *et al.* 1990). For respondents who did not indicate household income, income was estimated using a simple OLS regression that modeled income as a function of a constant term and education.

In order to determine the travel cost of substitute swimming locations, the same travel cost formulation used for the PES locations was applied to the substitute swimming locations. The average of travel costs to Long Island Sound, Block Island Sound, and the Atlantic Ocean, for each respondent, was used as the cost of travel to substitute sites and input into the travel cost model. Data on the number of trips to PES swimming locations comes directly from the survey questionnaire. The water quality variable used in the travel cost model is the sum of the probabilities that the PES swimming location would be ranked good and excellent. These probabilities are generated by the water quality model.

The results of the Poisson estimation are consistent with prior expectations, as presented in Table 7. As with the water quality model, all estimation was performed with the discrete choice model in LIMDEP™ (Greene 1995). The estimated coefficients indicate that the number of trips to a site decreases as the cost of travel to the site increases; trips increase when the cost of travel to a substitute site increases; trips increase the more times the individual participates in boating at the same location; and trips increase as water quality at the site increases. Each of the coefficients is significant at better than the 1% level. The Chi-squared statistic was calculated as 2 [restricted log-likelihood − unrestricted log-likelihood] and is also significant at better than the 1% level. The likelihood ratio index (ρ^2) was calculated as 1 − [unrestricted log-likelihood / restricted log-likelihood]. Following Greene (1995) marginal effects in the Poisson model are calculated as

$$\partial E\left[y \mid x\right]/\partial x = \lambda_i \beta .$$

The marginal effect of changes in water quality ranking calculated at its mean (0.5404) is 4.0325.

Alternative Model Specifications

A major concern for the Poisson model, often cited in the literature (Hellerstein 1991, Hellerstein and Mendelsohn 1993, and Haab and McConnell 1996) is the implied assumption that the mean and the variance are equal. A common alternative to the Poisson model is the negative binomial which relaxes the assumption of equality between mean and variance.

Table 11-7. Results of the Travel Cost Model (Poisson Regression)

Observations	995		Chi-Squared	1846.693	
Log Likelihood	-3707.225		D. Frdm.	4	
Res. L. Likelihood	-4630.572		Sig. Level	.000000	
Variable	Coefficient	Std. Err.	T Stat.	Sig. Level	Mean of X
Constant	-0.76984	0.16607	-4.636	0.00000	
Travel Cost	-0.028321	0.00210	-13.494	0.00000	23.29
Boating	0.0781	0.00157	49.757	0.00000	0.9246
Travel Cost to Substitute Locations	0.00978	0.00119	5.282	0.00000	31.74
Water Quality Rank	2.688	0.28209	9.529	0.00000	0.5404
Scale Factor	1.5002				
Marginal Effect for Water Quality Rank	4.0325				

However, Hellerstein and Mendelsohn (1993) show that the Poisson model provides a consistent estimate of the expected number of visits demanded as long as the estimator $\lambda = e^{x\beta}$ is the correct specification. Nonetheless, a negative binomial model was run on the data for the purpose of comparison with the Poisson model. A Tobit model was also run for the purpose of comparison, even though the continuous nature of the demand function modeled by the Tobit is known to be unsupported by the data. The results of these models are compared in Table 11-8

Table 11-8. Comparison Between Tobit, Poisson, and Negative Binomial Models

	Swimming Trip Estimates		
	Tobit	Poisson	Negative Binomial
Travel Cost Coefficient	-0.12973	-0.028321	-0.015051
(t-stat)	(-3.219)	(-13.494)	(-6.332)
Predicted Trips	2,413	2,056	9,162
Predicted – Observed Trips	357	0	7,106

The poor predictive power of the negative binomial in this case indicates that the Poisson model is the preferred choice. The Tobit model performs well in terms of predictive power and significance of coefficients, but is based on the false assumption of a continuous demand function.

The Environmental Fluid Dynamics Code Hydrodynamic Model

During 1985-1988, an unusually large and persistent brown tide algal bloom occurred. It caused serious damage to the bay scallop fishery as well as to other portions of the ecosystem, which demonstrated the need for a science-based nutrient management plan for the Peconic Estuary. The Brown Tide Comprehensive Assessment and Management Program (BTCAMP) was initiated in 1988 and included the development of a water quality model of Peconic

Estuary based on EPA's Water quality Analysis Simulation Program (WASP) model (TETRA TECH 1993).

Under the Peconic Estuary Program (PEP), the Environmental Fluid Dynamics Code (EFDC) three-dimensional hydrodynamic and water quality model has been developed and applied to the entire spatial area. This model was developed to comply with the requirements for the eutrophication model study of Peconic Estuary. The EFDC hydrodynamic model produced three-dimensional predictions of velocity, diffusion, surface elevation, salinity, suspended sediment and temperature on an intratidal time scale. The water quality model, an adaptation of the Corps of Engineers CE-QUAL-ICM (Cerco and Cole 1994), was integrated directly into EFDC and operates on a time step which is double that of the hydrodynamic model. In addition, the water quality model optionally interacts directly with a predictive sediment diagenesis model based on DiToro and Fitztrick (1993). Point-source and tributary (nonpoint source) loads were developed using water quality and flow monitoring data.

Two versions of the Peconic Estuary EFDC model have been developed under PEP: (1) a fine grid model having 772 computational elements in the horizontal and (2) a coarse-grid model having only 142 horizontal cells. Both models have two vertical layers.

The three major nutrients required by algae for growth are carbon, nitrogen, and phosphorus. Algal production is diminished or eliminated by the prolonged absence of one or more of the required nutrients. A rule of thumb is that phosphorus is the limiting nutrient in freshwater systems (Hecky and Kilham 1988) while nitrogen is limiting in estuarine and marine waters (Boynton *et al.* 1982). This presents a reasonable pattern for the Peconic Estuary system because nitrogen is of more importance as the limiting nutrient in the peripheral embayments and in the six main bays (TETRA TECH 2000).

Initially, using the coarse-grid version of the Peconic Estuary EFDC model, the eight-year verification period was simulated (TETRA TECH 2000). Comparisons of predicted and observed data for all parameters were considered to be reasonable with the coarse-grid model. Following calibration of the coarse grid model, the identical kinetic parameters and other conditions were input into the fine-grid model of Peconic Estuary and the model was verified for the 8 year verification period (October 1988 to October 1996).

The recurrence and magnitude of the winter and summer algal blooms were replicated reasonably well in all six main bays of the Peconic system. The seasonal and long-term trends of total phosphorus, total nitrogen, total organic carbon, and silica constituents were replicated reasonably well throughout the system.

The long-term sediment recovery scenario has provided a significant finding for estimation of economic benefits. Approximately eight years are required for the benthic sediments in Peconic Estuary to completely respond to changes in external loadings. Sediment flux rates reach approximately 80-90% of complete recovery after about three years of simulation. The delays in benefits implied by these science results are included in the economic results presented below.

The verified EFDC model of the Peconic Estuary can be used to evaluate the impacts of land use and pollution control scenarios on nitrogen, dissolved oxygen, light extinction, coliform bacteria, and other important parameters. Several viable standards and criteria are available so that model results can be weighted against numeric targets. For example, the recommended total nitrogen guideline of 0.45 mg/L (SCDHS 1998), based on measured violations of the New York State dissolved oxygen standard 5.0 mg/L, can be used to evaluate the extent of potential degradation at full build-out as well as the effectiveness of management scenarios. However, due to the sporadic, transient, and complex nature of dissolved oxygen depression, it is more desirable to utilize the nitrogen guideline and mean transect results to evaluate various management scenarios. Thus, we use nitrogen in evaluating the following management scenarios in the analysis.

Baseline conditions represent the status of water quality as of 1993 as measured through historic records. The sensitivity scenarios were run using the October 1992 to 1993 verification run conditions for many of the forcing conditions (*e.g.* atmospheric conditions, ocean boundary conditions, groundwater underflow and loads, and tributary flows and loads (except for Meetinghouse Creek). The Riverhead STP flow rate and nutrient loads were set to constant values of 0.75 MGD and 25.6mg/L total nitrogen, respectively. The base condition for Meetinghouse Creek was constant flow rate of 1.4 cfs and a total nitrogen concentration of 15mg/L.

Pastoral conditions reflect improvement of water quality that restores background or natural levels, *i.e.* levels expected with no anthroprogenic pollution. For example, pastoral conditions in the Meetinghouse Creek represent that the nitrogen concentrations in Meetinghouse Creek were set to pastoral conditions corresponding to 0.3mg/L total nitrogen; all other conditions were the same as in the base case.

Full build-out (the "Worst Case") assumes all developable lands are converted to residential use. In the aggregate, 40% of PES acreage is subject to development, although the fraction of developable land varies among the towns. Generally speaking, the full build out scenario has more significant implications for the eastern end of the PES (> 20%) as compared to the western end because more of this area is developable than the eastern parts.

The full build-out and pastoral conditions capture the extreme cases. Other management options considered include reduction of atmospheric deposition and restoration of pastoral conditions at Meeting House Creek, a former site of a heavily polluting duck farm. Generally speaking, the different policies have different effects on quality and these effects vary for different bays. Hence, the benefits can be expected to vary among policies and with the degree of strictness in their implementation.

While residential use contributes to nitrogen pollution through septic system releases, lawn fertilizers and other sources, agriculture generates twice the pollution per acre, mainly due to use of fertilizer. The full build-out condition is representative of the estimated nutrient loadings expected if the watershed surrounding the estuary is developed to its full potential. It was estimated that additional future total nitrogen loading under a full build-out scenario would be more modest in the western estuary (13% increase) than in the eastern estuary (greater than 20% increase). This is due to potential conversion of agricultural lands to low density residential land uses, which are generally less nitrogen-intensive. Virtually all of the TN increase would be in the form of residential TN loading except for the western estuary where there is the potential for industrial and commercial development.

Linking the Economic and Hydrodynamics Models

A full explanation of the details of the hydrodynamic model is beyond the scope of this chapter. It is sufficient to note that the model provides a dynamic simulation of nitrogen inputs, subsequent transport through the Peconic system, storage in and remobilization from the sediments, and degradation over time. Hence, when the model determines the effects of policy actions taken to control nitrogen inputs at a point, it simulates how nitrogen stocks change over space and time, until a new equilibrium is reached. The economic model identifies the associated changes in the economic value of recreational swimming, given the estimated water quality for each period (year). So, water quality changes spatially and temporally, and recreational demand shifts over space and time in response to these quality changes. A given scenario is assumed to simultaneously apply to all Bays.

RESULTS

The results show that benefits depend upon (1) the significance of water quality changes of each policy from the baseline and (2) the level of use of each Bay. Full build-out (the worst case) will have major consequences for water quality and for recreational uses and benefits for the PES, with the exception of Gardiners Bay. Gardiners Bay shows insignificant benefits because of high level of baseline water quality and little change associated with the various policies, due to the high degree of flushing from the open ocean. At the other extreme, achievement of pastoral conditions would result in substantial benefits, particularly for Flanders Bay and Great Peconic Bay, which have the highest nitrogen concentrations. Benefits are measured over a 25-year period using a discount rate of 7%.

Table 11-9. Present Value @ 7% for 25 years of Benefits to Swimmers in the PES from Improved Water Clarity due to Source Control of Nitrogen Pollution (millions of 1995 $)[a]

Scenario Case	Flanders Bay	Great Peconic Bay	Little Peconic Bay	Shelter Island	Gardiners Bay	Total
Pastoral	$4.9	$3	$2.2	$1.9	$0	$12
Full Build Out	-$1.9	-$1.4	-$1.2	-$5.4	$0	-$9.9
Pastoral Groundwater-Eastern Bays	$0.2	$0.5	$0.7	$0.8	$0.5	$2.7
Pastoral Groundwater-Western Bays	$0.2	$0	$0	$0	$0	$0.2
Pastoral Conditions-Meeting House Creek	$0	$0	$0	$0	$0	$0.0
50% Reduction in Atmospheric Deposition	$0.2	$0.5	$0.3	$0.2	$0	$1.2
20% Increase in Atmospheric Deposition	-$0.08	Negligible	Negligible	Negligible	Negligible	-$0.08

[a]All cases are changes as compared to the baseline. It is assumed that a given policy causes changes in water quality across all bays to occur simultaneously.

We have summarized the results from each policy alternative by bay in Table 11-9. Among all scenarios, the full pastoral scenario yielded the biggest change in the discounted sum of benefits, relative to a continuation of current conditions, which was $12 million in total discounted swimming benefits. A loss of $9.8 million results from the full build-out conditions, relative to a continuation of current conditions.

While the pastoral groundwater condition in eastern bays produced $2.7 million in total, the pastoral groundwater condition in western bays produced small benefits of $0.2 million only at Flanders Bay. The decrease in atmospheric deposition by 50% produced benefits of $1.3 million in total, but the increase in atmospheric deposition by 20% produced negligible negative impacts ($0.08 million) only on Flanders Bay. The pastoral conditions in the Meetinghouse Creek have no measurable benefit change at all. The reasons might be that Meetinghouse Creek is not a big contributor in terms of volume, since the last remaining duck farm, the former major contributor to nitrogen pollution, stopped direct discharge and started treating its waste water in 1988. An interesting finding is that reduction of atmospheric deposition by 50% would yield $1.2 million in benefits to swimmers in the PES.

SUMMARY AND CONCLUSIONS

Pollution of coastal waters by nutrient (phosphorus and nitrogen) over enrichment is a major cause of coastal water degradation throughout the United States (National Research Council, 2000; Boesch *et al.* 2001) and is an important issue in coastal waters virtually worldwide (Gren *et al.* 2000; KORDI 1996). The general consequences of eutrophication are widely recognized (*e.g.*, National Research Council 2000; Boesch *et al.* 2001), but controlling pollution sources can be expensive.

A rich literature examines the effects of water quality on recreational benefits using indirect or revealed preference methods, such as the travel cost model, the Random Utility Model, and hedonic model (*e.g.*, Braden and Koldstadt 1991; Freeman 1993; Diamantides 2001; Poor *et al.* 2000). However, few if any studies have been able to link multi-source pollution control to changes in ambient water quality at different sites, and to the resulting change in human behavior and benefits at these sites. In large part, this gap is due to lack of information on sources of pollution and the complicated and ambiguous effects of pollution source control policies on temporal ambient conditions, as noted.

The goal of this chapter is to provide an estimate of recreational benefits from alternative water quality policies to reduce a major nutrient in PES. As an initial effort, this chapter was restricted to benefits to swimmers from controlling nitrogen pollution, and focused on improvements of water clarity, which earlier research has found to be a key water quality attribute for users (Diamantides 2001; Poor *et al.* 2000; Freeman 1993). Using economic models—an ordered logit model, a count data and travel cost model—we linked the objective measures of water quality from sampling to the subjective perceptions of water quality and estimated recreational behavior by swimmers, which yielded changes in benefits. Exploiting this important link, several water quality policies being considered for the PES were analyzed. The consequences of these policies for water quality parameters relevant for benefits assessment were simulated by TETRA TECH (2000).

With these estimated water quality parameter (Secchi Disk Depth) results, estimates of benefits were given for each scenario. Among all scenarios, the full pastoral scenario yielded the biggest change in water quality, and in discounted benefits to swimming of $12 million. The full build out scenario resulted in the largest reduction in water quality and a loss in discounted benefits to swimming of $9.8 million. A decrease in atmospheric deposition by 50% produced benefits of $1.3 million in total, whereas the increase in atmospheric deposition by 20% produced negative impacts ($0.08 million) for Flanders Bay.

The ability to measure the changes in recreational benefits from alternative policy scenarios by linking the objective measure of water quality to the individual's subjective measure of water quality provides a basis to evaluate policy alternatives in economic terms. Given estimates of program costs, benefit-cost analyses can be carried out then for different policy alternatives.

Finally, we re-emphasize that only a subset of water quality parameters (water clarity) and prospective user benefits (to swimmers) have been considered. Hence, the estimates presented likely underestimate the total benefits from all uses.

REFERENCES

Beasley, S., W.G. Workman, and N.A. Williams. 1986. Amenity values of urban fringe farmland: A Contingent Valuation Approach. *Growth and Change* 17(4):70-78.

Besedin, E. 1996. Economics of Aquifer Protection in a Stochastic Environment. Unpublished Ph. D. Dissertation, Kingston: Department of Environmental and Natural Resource Economics, University of Rhode Island.

Bockstael, N.E., W.M. Hanemann, and C.L. Kling. 1987. Estimating the Value of Water Quality Improvements in a Recreational Demand Framework. *Water Resources Research* 23(5):951-960.

Boesch, D., R. Burroughs, J. Baker, R. Mason, C. Rowe and R. Seifert. 2001. Marine Pollution in the United States: Significant Accomplishments, Future Challenges. Pew Oceans Commissions. 51p.

Braden, J. and C.D. Kolstad, eds. 1991. *Estimating the Demand for Environmental Quality.* Amsterdam: North Holland Publishing Co. 370p.

Cerco, C.F. and T.M. Cole. 1994. Three-dimensional eutrophication model of Chesapeake Bay. Volume I: Main Report. Technical Report EV-94-4. U.S. Army Corps of Engineers, Waterways Experiment Station. May 1994.

Diamantides, J. 2001. Relating Objective and Subjective Measures of Water Quality in the Travel Cost Method: An Application to the Peconic Estuary System. Unpublished Ph.D. Dissertation, Kingston: Department of Environmental and Natural Resource Economics, University of Rhode Island.

Fletcher, J.J., W.L. Adamowitz, and T. Graham-Tomasi. 1990. The travel cost model of recreation demand: Theoretical and empirical issues. *Leisure Science* 12:119-147.

DiToro, D.M. and J.J. Fitzpatrick. 1993. Chesapeake Bay Sediment Flux Model. Contract Report EL-93-2. U.S. Army Corps of Engineers, Waterways Experiment Station. Vicksburg, MS.

Freeman, E.M.III. 1993. *Valuing Environmental and Natural Resource Services; Theory and Measurement.* Washington, D.C.: Resources for the Future. 516p.

----------------------------. 1995. The benefits of water quality improvements for marine recreation: A review of the empirical evidence." *Marine Resource Economics* 10(4)winter:385-406.

Greene, W.H. 1995. LIMDEP, Version 7.0: User's Manual. Economic Software: Bellport. NY.

Gren, I-M., K. Turner and F. Wulff. 2000. Managing a Sea: Ecological Economics of the Baltic Sea. London: Earthscan Publication, Ltd. 150p.

Grigalunas, T. and J. Diamantides. 1996. The Peconic Estuary System: A perspective on uses, sectors, and economic impacts. Revised final report prepared for the Peconic Estuary Program. Economic Analysis, Inc. Peacedale, Rhode Island.

Haab, T.C. and K.E. McConnell. 1996. Count data models and the problem of zeros in recreational demand analysis. *American Journal of Agricultural Economics* 78(February):89-102.

Halstead, J.M. 1984. Measuring the Nonmarket Value of Massachusetts Agricultural Land. *Journal of the Northeastern Agricultural Economics Council* 13(Apr):12-19.

Hellerstein, D. M. 1991. Using count data model in travel cost analysis with aggregated data. *Journal of Agricultural Economics* (August): 860-866.

Hecky, R. E., and P. Kilham. 1988. Nutrient limitation of phytoplankton in freshwater and marine environments: a review of recent evidences of the effects of enrichment. *Limnology and Oceanography* 33:796-822.

Hellerstein, D. and R. Mendelsohn. 1993. A theoretical foundation for count data models. *American Journal of Agricultural Economics* 75(August): 604-611.

Johnston, R.J., J.J. Opaluch, T.A. Grigalunas, and M.J. Mazzotta. 2001. Estimating amenity benefits of coastal farmland: Exploring differences between hedonic and contingent choice results. *Growth and Change* 32: 305-325.

Kline, J. and D. Wichelns. 1994. Using referendum data to characterize public support for purchasing development rights to farmland. *Land Economic* 70(2):223-233.

Korea Ocean Research and Development Institute. 1996. Final report for a study on formulating integrated coastal zone management system. Seoul, Korea. Korean Ministry of Construction and Transportation.

Morgan, C.L., J.S. Coggins, and V.R. Eidman. 2000. Tradable permits for controlling nitrates in groundwater. *Journal of Agricultural and Applied Economics* 32(2):249-258.

National Research Council. 2000. Clean Coastal Water: Understanding and Reducing the Effects of Nutrient Pollution. Washington DC: National Academy Press. 405p. http://books.nap.edu/books/0309069483/html/index.html

Opaluch, J.J., T. Grigalunas, J. Diamantides, M. Mazzotta, and Robert Johnston. 1999. Recreational and Resource Economic Values for the Peconic Estuary System. Final report prepared for the Peconic Estuary Program. Economic Analysis, Inc.: Peacedale, Rhode Island.

Ready, R. C., M. C. Berger, and G. C. Blomquist. 1997. Measuring amenity benefits from farmland: Hedonic pg vs. contingent valuation. *Growth and Change* 28(Fall):438-458.

Poor, P.J., K. Boyle, L. Taylor, R. Bouchard. 2000. Objective versus subjective measure of environmental quality in hedonic property-value models. Unpublished Marine Agricultural and Forest Experiment Station working paper.

Randall, A., Taylor, M.A. 2000. Incentive-Based Solutions to Agricultural Environmental Problems: Recent Development in Theory and Practice. *Journal of Agricultural and Applied Economics* 32(2)August:221-234.

Sherman, K., N. Jaworski and T. Smayda. 1996. *The Northeast Shelf Ecosystem: Assessment, Sustainability and Management.* Blackwell Science, Inc. Cambridge, Mass.

Shumway, S.E. 1990. A review of the effects of algal blooms on shellfish and aquaculture. *Journal of the World Aquaculture Society* 21(2):65-104.

Smith, V.K. 1988. Selection and recreation demand. *American Journal of Agricultural Economics* February:29-36.

Smith, V.K. 1989. Taking stock of progress with travel cost recreation demand methods: Theory and implementation. *Marine Resource Economics* 6:279-310.

Smith, V.K., W.H. Desvousges, and M.P. McGivney. 1983. Estimating Water Quality Benefits: An Econometric Analysis. *Southern Economic Journal* 50:422-437.

Suffolk County Dept. of Health Services, Office of Ecology. 1992. Brown Tide Comprehensive Assessment and Management Program. Riverhead, NY: Suffolk County Dept. of Health Services Office of Ecology. (November).

Suffolk County Dept. of Health Services, Office of Ecology. 1999. Nitrogen Loading Budget and Trends-Draft. Suffolk County: Dept. of Health, Peconic Estuary Program, Riverhead, NY: Suffolk County Dept. of Health Services (Jan. 15).

TETRA TECH. 2000. Three dimensional hydrodynamic and water quality model of the Peconic Estuary... draft final report. Riverhead, NY: Suffolk County Dept. of Health Services Office of Ecology. (June)

Wessells, C.R., C.J. Miller and P.M. Brooks. 1995. Toxic algae contamination and demand for shellfish: A case study of demand for mussels in Montreal. *Marine Resource Economics* 10:143-159.

Large Marine Ecosystems, Vol. 13
T.M. Hennessey and J.G. Sutinen (Editors)
© 2005 Elsevier B.V. All rights reserved.

12

Valuing Large Marine Ecosystem Fishery Losses Because of Disposal of Sediments: A Case Study

T. A. Grigalunas, J. J. Opaluch, M. Luo

INTRODUCTION

Background

Substantial economic pressures exist at ports throughout the Northeast Shelf Large Marine Ecosystem to carry out dredging in order to maintain or deepen channels, berths and turnaround basins to accommodate larger vessels. In all such cases, disposal is a major economic and environmental concern. While onshore disposal and beneficial uses are possible, in many cases such options are so costly that marine disposal is seen as the only viable option. Marine disposal, however, can cause losses to a variety of species through physical smothering, adverse effects of sediment plumes, and perhaps other effects. Both direct losses (smothering) and indirect losses through the food web could occur.

This chapter examines how governance decisions concerning marine disposal might be enhanced using economic information concerning the costs to fisheries because of marine disposal. We examine the costs to fisheries at alternative marine disposal sites, using dredging scheduled for the Providence River and Harbor as a case study. We consider costs to commercial and recreational fisheries using the methods and data presented below. Differences in transportation costs among the disposal sites are not considered.

Dredging of the Providence River will require disposal of some 5.1 million cubic yards (mcy) of clean sediment. This includes material dredged from the Providence River itself (3.1 mcy) (US Army Corps of Engineers 1998b), plus an additional 458,000 cubic yards requiring disposal from non-federal sources, such as petroleum docks and at marinas. In addition, disposal of 1.2 mcy of material in confined aquatic disposal (CAD) cells will generate an additional 1.5 mcy of clean material potentially to be disposed of at an open water site. Seven open-water locations are being considered as potential disposal sites, three in Narragansett Bay and four in Rhode Island Sound (Figure 12-1). Disposal of sediments will affect commercial and recreational fisheries, and the impact on fisheries is a factor considered in evaluating potential disposal sites. Hence, an important issue concerns the relative economic cost to fisheries due to disposal at each site.

Marine disposal of clean sediments will cause a temporary loss of commercial and recreational catch, and the effects can be quite complex. The number of fish of each species may differ among sites. Further, mortality to fish at a disposal site may cause lost short-run catch of adults offsite, if the affected species are mobile and would have been caught elsewhere (perhaps many miles away) if not lost at the disposal site. Long-term fishery losses also will be realized if, for example, mortality to juveniles because of disposal leads to an eventual loss in catch, perhaps many years in the future. Finally, lost catch due to ecological (food web) effects also may occur if disposal causes mortality to lower trophic biota consumed by species eventually harvested by commercial or recreational users. Each of these sources of lost catch due to sediment disposal can be expected to differ among potential disposal sites as a result of differences in fishery resources among sites. Any study of the full costs of marine disposal should attempt to consider all of these short- and long-term, and direct and indirect, effects.

Purpose and Scope

This chapter estimates the incremental costs to commercial and recreational fisheries in the Northeast LME from marine disposal of sediments from dredging of the Providence River and Harbor. Incremental costs are defined as the present value of annual lost benefits to users due to disposal of clean dredge material. The losses considered are (1) the lost annual value (economic rent) in commercial fisheries and (2) the lost annual user benefits (consumer surplus) to recreational anglers. These annual losses are measured over the period from mortality due to disposal through the time to recovery of the harmed resources.

Specific incremental costs of marine disposal to commercial and recreational fisheries considered in this chapter are those due to:

Short-term effects—lost catch during the disposal period from mortality to adult commercial and recreational species,

Long-term effects— lost catch over the recovery period following disposal,

Indirect effects— loss of biomass due to ecological effects, or food web effects, during the disposal period. Hereafter, these indirect effects are referred to as food web effects.

To estimate the economic costs from the short-term, long-term, and food web effects mentioned above, we use a cohort-type, bio-economic model. This model is described in the Concepts and Methods section and presented formally in *Appendix A*. Here we note that use of a cohort (or age-class) model is appropriate for this issue, for several reasons. A cohort model accounts for the fact that disposal of sediments will cause mortality to juveniles, larvae, and eggs, where the consequences for lost catch may not be realized for years and may occur offsite. A cohort model is appropriate, since otherwise this loss of sub-legal size fish might be omitted or would require use of an *ad hoc* framework. We also adapt the model to include a simplified food web component, generally following the framework used in Economic Analysis, Inc. (1987); Grigalunas *et al.* (1988) and updated in Applied Science Associates *et al.* (1994).

To implement the model, data is drawn from several sources, which are described later in the report. Any analysis of the consequences of disposal must address data issues. We recognize the considerable uncertainty surrounding some of the data, and information available for some variables is incomplete. For these reasons, extensive sensitivity ("what if") analyses are used in order to examine how the resulting estimates of incremental costs to fisheries change when alternative assumptions or data are employed. Thus, rather than present one estimate of the cost of disposal at a site, we show a range of estimates. Further, whenever judgments must be made, we adopt a conservative stance that serves to overstate the cost to fisheries from disposal. Lastly, we employ a transparent approach to allow readers to understand the methodology, assumptions and data used, and their strengths and limitations.

In this document, only the incremental costs of disposal at potential sites are considered; the effects of the actual dredging on fisheries are outside the scope of this effort. Also, only the economic consequences of open-water, marine disposal of clean sediments are assessed; disposal at non-marine sites, and disposal of any contaminated sediments are not within the scope of this report.[1] Further, transportation cost differences between sites are not considered.

Organization

The Concepts and Methods section below contains a non-technical explanation of the concepts and methodology used in order to facilitate access for interested readers who may not be specialists. The Data section describes the data needs, sources, and assumptions used to implement the methodology. In the fourth section, the results are presented and several sensitivity analyses are given. Detailed descriptions of the model and data are provided in appendices.

CONCEPTS AND METHODS

Economic Value

The coastal and marine waters and bottom sediments of Narragansett Bay and Rhode Island Sound can be viewed as natural assets which provide a range of valuable recreational and commercial services over time, if maintained. Among these services are the *in situ* natural "production" of commercial and recreational fisheries, production which takes on a *use value* when harvested as a result of effort by commercial and recreational fishers.

The value of fisheries ("economic rent") is measured by their worth in the market place, less what it costs to harvest them. For commercial fisheries, value is measured as the market value of landings, minus the cost of the effort used in harvesting the fish. Since the issue of concern is the effect on fisheries of dredge material disposal, we are interested in incremental effects caused by disposal—the reduced market value of landings less any cost savings from lower effort. However, we assume effort will not change in this case, for reasons given below.

First, as to the market value of landings, so long as the effect on annual commercial catch is very small as compared to total landings, or substitutes are readily available at the given

---

[1] An overview of issues associated with dredging and disposal of contaminated sediments can be found in Grigalunas and Opaluch (1989) and National Research Council (1989).

market price, the price of harmed species will not change. In this study, the annual effects for a given area are small in comparison to total landings, so that the market price for fish will not change and can be used to measure the value of changes in catch.

Second, regarding fishing effort, if fish prices do not change, and the overall effect on catch-per-unit-effort is small, fishing effort will not change, so no change in harvesting costs will occur. When price and effort do not change, the measure of lost economic rent is simply the ex-vessel (i.e., dockside) market price of the lost catch times the reduced catch for each harmed species, measured over the period from the beginning of disposal through full recovery of all affected populations. This is the estimate of economic value for commercial species we use in this chapter.

Recreational fisheries are not valued in markets, but similar concepts apply. The value of a change in recreational fisheries due to disposal of sediments is measured as the most anglers would pay to engage in recreational fishing, less what they actually pay for a fishing trip. Since the annual effect on fisheries is small, and lost catch is distributed over a wide area, we assume that recreational fishing effort also will not change. The value of recreational fishing to participants depends on the catch rate, among other factors, and hence lost recreational fishing in this case is measured as reduced value (willingness to pay) for a small reduction in catch-per-unit-effort (here, effort is a recreational trip). In sum, the value of the resultant change in catch rates (assuming constant effort) is the measure employed to estimate recreational fishery losses in this document.

We do not include so-called economic "multiplier effects" as an incremental cost in our base case analysis. This is because the use of naïve multipliers inappropriately assumes that the labor, fuel, and other variable inputs used in commercial fishing, or to support recreational fishing, have no alternative uses (*i.e.*, they have zero opportunity cost). This cannot be true; for example, if labor literally had no other uses, it would be possible to hire workers for a near-zero wage—which obviously makes no sense. Since variable inputs used in any economic activity, including fishing, have alternative uses, the assumption of zero opportunity cost underlying the use of naïve multipliers does not apply.[2]   Further, the percentage reductions in fish catch are spread over space and time and are so small that aggregate fishing effort is unlikely to change appreciably, as noted. Since effort is unlikely to change, it follows that expenditures for labor, fuel, ice, etc. often argued to give rise to multipliers also do not change. Still, as part of a "worst case" sensitivity analysis, we show how the results are affected when indirect economic ("multiplier") effects are considered.

Measurement of Changes in Value of Fisheries Due to Dredged Material Disposal

Overview of Bioeconomic Model

To estimate the incremental costs to commercial and recreational fisheries, we use a bio-economic model. The model is fairly involved; readers interested in the technical aspects of the model are referred to *Appendix A*. This section provides a non-technical explanation of

---

[2] When substantial unemployment exists, workers who otherwise would be unemployed may earn a net gain measured as the *difference* between the wage they receive and the value of their time in its next-best use (i.e., its "shadow value"). This net gain to labor will be much smaller than the total market wage.

the concepts and data used and steps involved in the analysis to allow interested readers to follow the logic and method in deriving estimates of fishery losses.

The model used is a Beverton-Holt cohort model (*e.g.* Ricker 1975). A cohort model accounts for each year class or size class for each species. Each year, legal-size members of all cohorts are subject to natural and fishing mortality; those fish surviving in a cohort grow and gain weight, and move on to the next age class, until they reach a maximum age at which time any surviving individuals die. The model also tracks eventual recruitment of eggs, larvae, and juveniles into the harvestable fishery.

A cohort model provides a very useful approach for assessing the consequences of disposal of sediments at marine sites. For example, a site might have a concentration of juveniles, but few legal-size fish or shellfish. For this site, if only lost catch of legal-size species is considered, disposal at the site might appear to cause negligible losses, when in fact mortality to juveniles might cause an eventual substantial loss in catch as juveniles would have reached a harvestable size. Hence, failure to account for loss of juveniles would understate losses. A cohort-type model is appropriate for this situation, where potentially significant adverse impacts would otherwise be missed, or would require a purely *ad hoc* approach.

Alternately, an area may have large numbers of species which have an important ecological role in the food web as prey, thereby indirectly supporting commercial or recreational fisheries. In some cases, the mortality at or around the disposal site due to loss of food fish may occur as lost catch many miles away, so that the link with the habitat site may not be apparent. Hence, omitting the food web would understate eventual losses to commercial and recreational fisheries of the predators of the lost food fish.[3]

The general categories of impacts considered by the model are illustrated in Figure 12-2 and elaborated upon below. Disposal of 5.11 mcy of clean sediments is presumed to kill all biota over the disposal area during the entire 18-month disposal period. The affected area is presumed to be the geographic extent of the mound created as a result of disposal. This mound can be many feet thick and has been estimated to cover 300 to 430 acres for sites in Narragansett Bay and 530 acres for all sites in Rhode Island Sound (Corps of Engineers 1998a).[4]

These assumptions serve to overstate costs for several reasons. Some species covered or contacted by sediments will not suffer complete mortality,[5] especially mobile species in the water column. Recovery may begin in some parts of the disposal area prior to completion of the 18-month disposal period. Also, dredged material will only impact a small area at a time (not the entire 500 or so acres) giving finfish time to move out of the area.

---

[3] This assumes that the non-commercial species that are part of the food web affected by disposal in a particular case are a limiting factor for commercial and recreational food stocks. If they are not a limiting factor in a particular case, then their loss will cause no reduction in commercial and recreational stocks.
[4] The area covered varies among sites due to differences in water depths and diffusion of material when deposited from barges, and different characteristics of the areas, *e.g.*, currents, tidal actions and slope. Our estimates of loss are for the size sites stated in the text.
[5] For example, buried quahogs may survive, if the covering by disposed sediments is not too deep (US Army Corps of Engineers 1999).

The bio-economic model estimates the annual commercial and recreational lost catch due to the deposited sediments. To estimate short-term lost catch, the fishing mortality rate (percent of harvestable fish caught) is multiplied by the biomass for each species times the length of the disposal period. Long-term lost catch is estimated through the period of recovery. Recovery is assumed to begin at the end of the 18-month disposal period. Full recovery varies by species in the model and can be up to 19 years, the maximum life of some species in the affected area. The recovery trajectory is calculated for each species using the age-class model described in *Appendix A*.

Lost catch due to food web effects also is included. To incorporate these effects, we adapt the approach used in the Natural Resource Damage Assessment Model for Coastal and Marine Environments (NRDAM/CME) published by the U.S. Department of the Interior (Economic Analysis, Inc. and Applied Science Associates, 1987; Applied Science Associates *et al.* 1994). In the NRDAM/CME, food web losses result when losses in lower trophic organisms occur. These losses pass up the food web, and ultimately cause losses of biomass to consumer species harvested and valued by people for commercial or recreational purposes. Following the NRDAM/CME, we assume a proportional relationship between food resources and consumer species. For example, if 50% of the food resources are lost (the assumption used in our base case), there is a 50% reduction in biomass of consumer species. This assumption is consistent with our intent to overstate losses, since it assumes that there exists no potential for consumer species to substitute other food resources opportunistically.

Two issues must be addressed when considering food web effects. One concerns how loss of biomass is distributed among consumer species, and the other is how to address onsite versus offsite food web losses. To distribute food web losses among species further up the food web, we follow the general approach used in the NRDAM/CME, as noted. The loss of each consumer species is proportionate to the biomass of each consumer of the food resource in question.[6]

The magnitude of food web effects depends upon the extent to which food at a site is consumed by species that remain at the site, or whether food moves offsite. If all food resources are consumed by species in the disposal area, losses in non-commercial food resources cause no additional loss to commercial and recreation fish beyond the direct effect. This is because we assume mortality of all biota on site due to disposal, as noted, so that the model already accounts for a total loss of all onsite consumer species. To include additional food web effects to onsite consumer species in this situation would double count losses.

However, loss of food resources could cause an additional adverse impact if food resources are exported to commercial and recreational consumer species residing offsite. In this case, loss of food as a consequence of disposal at a site would cause an ultimate reduction in the biomass and catch of commercial and recreational species elsewhere, since offsite predators experience loss of food. As noted above, we assume a 50% loss of food resources to offsite fisheries in our base case estimates.

---

[6] For this purpose, we assume that food web effects are distributed over all species in the affected area.

Overview of Data

To apply the model outlined above, considerable biological data and economic information are needed. Biological information needed includes the abundance of each species (in grams wet weight) at the disposal sites and species-specific parameters for growth in length, weight as a function of length, time to recruitment (*i.e.*, entry) to the legally harvestable fishery, maximum age for each species, natural mortality, and fishing (commercial and recreational) mortality. A more complete discussion is provided in the next section; a listing of parameters and data for each species is given in *Appendix B*.

Economic data required includes commercial prices and landings for affected species and estimates of non-market value and landings for recreationally harvested species. Of course, in many cases, as with flounder and tautog, a species is harvested by both commercial and recreational users. For these species, lost catch is allocated to commercial and recreational losses in proportion to the relative sizes of total commercial versus recreational catch. As noted, the data used is described more fully below and in *Appendix B*; commercial values for RI landings are presented in *Appendix C*.

DATA

Data Sources

The principal source of biological information is the Natural Resource Damage Assessment Model for Coastal and Marine Environments ("NRDAM/CME") published by the U.S. Department of the Interior (French *et al.* 1998). Information taken from this source includes, for each species group: average biomass per unit area, natural and fishing mortality, growth, weight as a function of length, years to recruitment, maximum life, and food web parameters (see *Appendix B*).

The advantages of using the NRDAM/CME to assess the consequences of sediment disposal are several. Principal among these are: (1) the scientists who developed the model provide a synthesis of a wealth of available biological data in a format that readily lends itself to the development of the cohort model used in this report; (2) the model is incorporated in federal regulation (43CFR11 Subpart D) and has been subject to an extensive public review process; (3) the NRDAM/CME has been used on numerous occasions by states and by the federal government so it has some degree of acceptance among public trustees. Of course, data in the NRDAM/CME are averages and hence are only approximations.

Another important source of biological data is the information summarized in the Draft Environmental Impact Statement (DEIS) prepared for the Providence River dredging project by the Corps of Engineers (1998a; 1998b). Key information from these sources includes abundance data of quahogs and catch-per-unit-effort (catch-per-tow or per-trap) data for fin fish and lobster for each potential disposal site. However, the catch-per-unit-effort (CPUE) data in the DEIS does not match the format needed (biomass per unit area) to implement our bioeconomic model, described below. Thus, we use data from the NRDAM/CME to provide estimates of biomass per unit area, and use the CPUE data from the DEIS to calibrate the data at each site, as discussed below.

Data concerning commercial fish prices was obtained from the National Marine Fisheries Service (US Department of Commerce 1999). For fishery categories that include multiple species (e.g., flounders), we use weighted averages, where the weights within a fisheries category are the relative shares of each species within the category. Obtaining a price for mollusks (quahogs) was complicated by the fact that different size clams have very different prices. For this species, a price index was constructed using data from a telephone survey of three dealers in the Narragansett Bay area. For recreational fishing values, we use benefit transfer—that is, we adapt estimates of recreational fishing values from the environmental and resource economics literature. In this case, we adopt the marginal value for recreational fishing ($10.16 per pound) given the NRDAM/CME.[7]

Abundance

Biomass data for affected species at potential disposal sites in Narragansett Bay and Rhode Island Sound are developed as follows. First, we use the estimates of biomass per unit area for species categories given in the NRDAM/CME. These data are stated as *kg per kilometer square* for Narragansett Bay and for Rhode Island Sound. For some species, abundance is the same for the two areas, while the abundance differs for other species.

Sampling data reported on in the COE DEIS found many species of fish at the disposal sites. However, a handful of commercial or recreationally important species dominates the sampling data: lobster, quahogs, flounder, tautog, and Atlantic herring. Hence, the bio-economic analysis focused on these species. Remaining species were aggregated into a category termed "All Other Species" for our assessment of economic costs.

The estimates of biomass given in the NRDAM/CME are average values per unit area, for species groups, for each season. Many species prefer different environment types, such as sand-mud bottom or rocky bottom; hence, average biomass for these species varies by type of environment as well as by season.

For most species, biomass data in the NRDAM/CME shows substantial variability from season to season within an environment type and location. For example, in the NRDAM/CME flounder abundance per unit area in Narragansett Bay is highest during the spring and lowest during the fall. Of course, mollusks (*e.g.*, quahogs) do not migrate, and hence their biomass per unit area in the model is the same for each season.[8]

Disposal of dredged sediments would take place over 18 months, as noted, so an issue is *which* seasonal biomass estimate from the NRAM/CME should be used for each species. In keeping with the conservative (*i.e.*, overstated) cost approach adopted in this document, we

---

[7] Detailed descriptions of the development of price indices for commercial and for recreational species are referred to Economic Analysis Inc. (1987) and Grigalunas, Opaluch and Tyrrell (1988).
[8] The use of average biomass figures per unit area is a convenient approach but is problematic with species that school.

use the *highest* seasonal biomass estimate given in the NRDAM/CME database to characterize each species.

We estimate total biomass for each site by multiplying biomass per unit area by the total affected area for each site. The area covered by the disposal mound of sediments has been estimated to range from 300 to 430 acres in Narragansett Bay and 530 acres in Rhode Island Sound (US Army Corps of Engineers 1998a; Oliver 1999, personal comm.).

As mentioned above, the NRDAM/CME uses averaged data for each species and environment type. The biomass data in the NRDAM/CME is the same for each site within the Narragansett Bay area and for each site within the Rhode Island Sound area. To reflect site-specific variations among potential disposal sites, the NRDAM/CME biomass estimate for each species was calibrated to reflect available site-specific fisheries data. For this purpose, we use data given in the DEIS (US Army Corps of Engineers 1998b) to develop calibration factors. Recognizing differences between the estuarine environment in Narragansett Bay and the open-ocean environment in Rhode Island Sound, two calibrations are done for each species, one for each area.

Specifically, the DEIS contains data on CPUE (catch-per-unit-effort) for various species of fish and shellfish. For each potential disposal site, several samples were taken, and the DEIS reports average CPUE for each species, at each site. In order to capture differences in species biomass across sites, we use the DEIS data on CPUE to calibrate the NRDAM/CME biomass per square kilometer, as follows. In keeping with our overstated-cost approach, we apply the biomass per square kilometer from the NRDAM/CME to the site with the *lowest* CPUE data (or abundance for quahogs) for each species from the DEIS. We then calibrate the NRDAM/CME data upward for other sites by multiplying by a calibration factor equal to the CPUE at the site divided by the *minimum* CPUE over all sites for that species.

For example, Table 12-1 contains the calculations for calibrating lobster populations for each site. The DEIS data for adult lobster catch per pot ranges from 0.27 for site 18 in Rhode Island Sound to 2.0 for site 158 in Narragansett Bay. The calibration factor for each site

Table 12-1. Calibrated Adult Lobster Populations for Each Site

Area	Site	DEIS Data (A)	Calibration Factor (B=A/Min A)	NRDAM Data (KG/KM^2) (C)	Calibrated Population (= B*C)
Narra-gansett Bay	3	.67	1.00	1,190	1,190
	157	1.7	2.54	1,190	3,019
	158	2.0	2.99	1,190	3,552
Rhode Island Sound	16	0.4	1.48	1,190	1,763
	18	0.27	1.00	1,190	1,190
	69A	0.5	1.85	1,190	2,204
	69B	0.48	1.78	1,190	2,116

equals the CPUE for that site divided by the minimum CPUE for lobster (0.27). These calibration factors are then multiplied by the NRDAM/CME data on lobster population per square kilometer to give the calibrated lobster population for each site. Calibrated populations range from 1,190 kg/sq. km for site 18 to 8,818 kg/sq. km for site 158.

In sum, we calibrate or adjust the abundance data from the NRDAM/CME using a factor, A_{ij}, for species i and potential site j, where A_{ij} is determined as:

$$A_{ij} = CPUE_{ij} / (Min \{CPUE_{ij} \, j = 1,...J \, |CPUE_{ij}{>}0\})$$

where $CPUE_{ij}$ is the average catch-per-unit-effort for species i at site j from the DEIS data.[9] In cases where the DEIS sampling found no members of a particular species, the site is imputed a value equal to the site with the lowest nonzero DEIS estimate. For example, the DEIS sampling found no quahogs in site 3. Thus, site 3 was given an imputed value equal to the lowest nonzero site in Narragansett Bay, site 157. The DEIS found no tautog in any of the Rhode Island Sound sites. In this case, the Rhode Island Sound sites are not calibrated; instead, we apply the NRDAM data to all four sites in the Sound.

Disposal is estimated to take place over 18 months. We presume that (1) all biomass is lost during the disposal period, (2) recovery begins when the disposal is complete, and (3) recovery of the impacted species follows the model outlined in *Appendix A*. As noted earlier, the period for complete recovery could be long for some species.

Economic data were adopted from the following sources. Commercial fish losses are valued using *ex-vessel* prices for lost catch. Fish prices are from the most recent (1998) *ex-vessel* values for Rhode Island from the National Marine Fisheries Service (*Appendix C*). For fish categories that include multiple species (*e.g.*, flounders), the commercial value is calculated as the weighted average of the commercial prices for each species, where the weights are the percentage of landings of each species (*e.g.*, winter flounder) relative to the total landings of the species group (all flounders). The value for lost recreational catch is adopted from the NRDAM/CME. This value represents the reduced value (consumer's surplus) by anglers due to a reduction in recreational catch. The discount rate used in our base case, 6.87 %, is the rate specified for use in Corps of Engineer projects.

RESULTS

Base Case

Using the model, data, and assumptions outlined above, estimates of the annual lost commercial and recreational catch for each potential site are derived. These results are considered the "base case." Results are given in a spreadsheet. Selected major findings are

---

[9] Note that DEIS data for quahogs is based on abundance estimates. Thus, for quahogs, calibration is carried out with the same methodology, except that the DEIS abundance at each site is used in place of the CPUE.

presented in Table 12-2, and summarized here, with all monetary values given in 1999 dollars.

Key base case results can be summarized as follows:

- Incremental costs range from $0.3 million for Site 16 in RI Sound to $2.04 million for site 158 in Narragansett Bay (see Table 12-2). These losses include short-term, long-term and food web effects.
- Losses for Narragansett Bay sites are considerably higher than for Rhode Island Sound sites. Differences between sites within Narragansett Bay, and differences between sites within Rhode Island Sound are small. The uncertainties of the data probably imply that differences among sites *within* each of the two areas are not significant.
- The largest loss is the "All Other Species" category for four out of seven sites. The significant size of losses attributed to this category is to be expected given that this "All Other Species" has the largest biomass of any category for each site.
- Tautog is the highest *single* species value for Narragansett Bay sites due to the large numbers of this species affected by disposal and the substantial use of tautog in recreational fisheries (which typically have a higher unit value than commercial catch). Lobster (for Sites 16 and 18) and flounder (for Sites 69A and 69B) are highest *single* species value for Rhode Island Sound sites (Table 12-2).
- Recreational losses are greater than commercial losses for potential sites in Narragansett Bay, while the opposite is the case for potential sites in RI Sound. This emphasizes the importance of including recreational effects in disposal evaluations.
- For all potential sites, long-term effects account for the highest share of costs, underscoring the significance of using a cohort-type framework (Table 12-4).
- For all sites, ecosystem (food web) effects are non-trivial—and significant in some cases (Table 12-4), pointing out the importance of incorporating these effects.

Table 12- 2. Lost Values of Commercial and Recreational Catch by Site and Species

	Site	Quahog ($000)	Lobster ($000)	Herring ($000)	Flounder ($000)	Tautog ($000)	All Other Species ($000)	Total ($000)
Narra-	3	$8.3	$42.7	$0.3	$67.9	$854.6	$694.7	$1,668.5
gansett	157	$5.8	$48.5	$0.1	$133.1	$447.2	$969.4	$1,604.1
Bay	158	$13.0	$58.6	$0.2	$155.3	$521.7	$1,131.0	$1,879.9
Rhode	16	$4.7	$96.2	$17.5	$27.1	$23.3	$87.4	$256.1
Island	18	$6.1	$59.6	$12.1	$41.7	$23.3	$147.0	$289.9
Sound	69A	$1.1	$71.5	$11.6	$110.0	$23.3	$184.1	$401.6
	69B	$1.5	$58.2	$1.4	$101.2	$23.3	$117.5	$303.1

Table 12-3. Allocation of Losses to Commercial and Recreational Uses

Area	Site	Commercial ($000)	Recreational ($000)	Total ($000)
Narra-	3	$501.2	$1,167.3	$1,668.5
gansett	157	$648.8	$955.3	$1,604.1
Bay	158	$765.3	$1,114.6	$1,879.9
Rhode	16	$182.7	$73.5	$256.1
Island	18	$183.7	$106.2	$289.9
Sound	69A	$250.5	$151.1	$401.6
	69B	$186.2	$116.9	$303.1

Table 12-4. Short Term, Long-term and Food Web Effects

Area	Site	Short Term ($000)	Long Term ($000)	Food Web ($000)	Total ($000)
Narra-	3	$241.4	$1,306.4	$120.7	$1,668.5
gansett	157	$311.4	$1,137.1	$155.7	$1,604.1
Bay	158	$367.0	$1,329.3	$183.5	$1,879.9
Rhode	16	$44.4	$189.5	$22.2	$256.1
Island	18	$54.7	$207.9	$27.3	$289.9
Sound	69A	$78.6	$283.7	$39.3	$401.6
	69B	$57.6	$216.7	$28.8	$303.1

Sensitivity Analysis

At the outset, several important sources of uncertainty were acknowledged. Recognizing this uncertainty, throughout this report assumptions were used that would serve to overstate costs. Nevertheless, an important issue concerns how the results given above might change if the assumptions used were altered?

A great many sensitivity results were generated. Here we describe a few important sensitivity analyses. In each case, only one factor at a time is varied (and all other base case data are maintained) so that the effect of the particular sensitivity analysis can be isolated. We then report results for a "worst case," which includes the effect of all sensitivity factors simultaneously.

Here we report sensitivity analyses that were carried out for the following factors:

- Impact area extends beyond disposal mound
- Delayed habitat recovery
- Use of a lower discount rate

- Different assumptions regarding food web losses

- Economic multiplier losses are included

- Combination of the above

For all sensitivity analyses, disposal costs are substantially higher for Narragansett Bay sites as compared to RI Sound sites, and costs do not vary substantially among sites within the Bay or among sites within Rhode Island Sound. The results for the sensitivity analyses are reported in Table 12-5 and each is discussed briefly below.

Impact area extends beyond disposal mound
This sensitivity analysis increases the area of impact by 25% to account for the possibility that there will be impacts outside of the disposal area. This results in a proportionate increase in damages at all sites, so damages range from $.42 million at site 16 to $2.55 million at site 158. Increasing the impact area by 25% leads to a 25% increase in damages at all sites, and so does not change the relative desirability of sites.

Delayed habitat recovery
The base case results assume that recovery begins as soon as the disposal period ends. This sensitivity analysis assumes a one-year period for habitat recovery, which delays recovery of all biota by one year. The delay of habitat recovery results in proportionately larger direct and food web effects, but does not change long term effects. Assuming a 1-year period for habitat recovery results in damages that range from $0.38 million at site 16 to $2.30 million at site 158. While there is an increase in impacts at all sites, no change in the relative sizes of impacts across sites occurs.

Lower discount rate
Lowering the discount rate from 6.87 % used in the base case to 3% results in larger damages at all sites. With a 3% discount rate, damages range from $377,600 (Site 16) to $2.28 million (Site 158). However, changing the discount rate does not affect the relative sizes of impacts across sites.

Food Resource Export
As a worst case analysis, we assume 100% export of food resources leads to a doubling of the food web effects relative to the base case which assumed a loss of 50% of export of food resources. With the assumed 100% offsite loss of food resources, damages range from $0.37 million (Site 16) to $2.24 million (Site 158). Once again, there is no change in the relative sizes of impacts across sites.

Economic Multipliers
Again, as a worst case, we assume that lost commercial fishing income due to dredge disposal leads to indirect economic losses in other sectors. These indirect economic losses are assumed to occur when commercial fishermen who receive slightly lower incomes each year spend less on local goods (e.g., food and entertainment), thereby lowering the income of owners and operators of these businesses (often referred to as induced effects). We also assume that reduced commercial fish landings lower the amount of fish received by seafood

wholesale and retailers leading to losses in income to owners and operators of these businesses.[10] These are extreme assumptions since (1) labor and other resources have alternative uses and (2) substitute fish will be available from other sources, especially given time to adjust to new circumstances. Hence, the multiplier sensitivity analysis is included only as a worst case.

All Sensitivity Conditions
Finally, as a worst case analysis we consider a case that incorporates all of the sensitivity conditions discussed above. Under this worst case analysis, losses range from $0.63 million at site 16 to $3.70 million at site 158. Once again, there is no change in the relative sizes of impacts across sites.

Table 12-5. Sensitivity Analyses for Total Commercial and Recreational Losses ($000)

Area	Site	Base Case Result ($000)	25% Increase in Size of Mound ($000)	1 Year Delay in Habitat Recovery ($000)	3% Discount Rate ($000)	100% Food Resource Export ($000)	All Sensitivity Conditions ($000)
Narra-	3	$1,668.5	$2,085.7	$1,803.2	$1,881.6	$1,789.3	$2,849.4
gansett	157	$1,604.1	$2,005.2	$1,814.2	$1,789.4	$1,759.8	$2,901.6
Bay	158	$1,879.9	$2,349.8	$2,128.3	$2,095.9	$2,063.4	$3,404.2
Rhode	16	$256.1	$320.2	$284.3	$295.5	$278.3	$462.7
Island	18	$289.9	$362.4	$326.2	$329.6	$317.2	$528.1
Sound	69A	$401.6	$502.0	$454.9	$453.8	$440.9	$735.1
	69B	$303.1	$378.9	$341.5	$342.6	$331.9	$550.7

SUMMARY AND CONCLUSION

Port-related dredging and marine dredge disposal can be important stresses on the productivity of the Northeast Shelf Large Marine Ecosystem. This paper provides a conceptual framework for estimating losses to fisheries and a case study, information which can contribute to selecting between different marine sites for disposal.

The conceptual framework involves the extension of a cohort-type age-class or size-class model to estimate lost catch over time due to short-term, long-term, and indirect (food web) effects of marine disposal. Dredging planned for the Providence River and Harbor was used as the case study. Seven different sites were assessed in Narragansett Bay and RI Sound. Biological data was taken from several sources and calibrated based on site-specific sampling using net tows and pots.

The main findings were that losses to fisheries are higher within the Bay than in RI Sound, although considerable differences exist between sites with respect to species affected, recreational versus commercial losses, and short- versus long-term effects. Several

---

[10] The estimate for induced effects is from Jin and Hoagland (1999); information for forward linkages is from Ascari and Grigalunas (1983).

sensitivity analyses show the effect on estimates of a longer recovery period, a larger affected area, a different discount rate, and different assumptions concerning food web effects. The sensitivity analyses show the overall changes in lost fisheries but do not change the relative costs to fisheries across areas. The results also illustrate how information from the natural and social sciences can be combined to contribute to governance issues for the NE Shelf Large Marine Ecosystem.

Appendix A

Modeling Losses in Commercial and Recreational Catch

Three components of fishery losses are modeled: short-term effects, long term effects and food web effects. We use standard methods (*e.g.*, Economic Analysis and Applied Science Associates 1987) for estimating each of these components of losses, along with assumptions that overstate losses. The short-term effects are reductions in catch during the disposal period from mortality of adult finfish and shellfish. All biomass in the disposal area is presumed lost over the disposal period, and short term lost catch is equal to the fishing mortality rate (F) times the adult biomass (B) integrated over the disposal period.

Long term effects are reductions in catch over the period of recovery of the population, including losses due to mortality of adult fish, juveniles, and young of the year. The long-term losses are determined by simulating recovery of the population following the disposal period using a Beverton and Holt age class model (*e.g.*, Ricker 1975). The model simulates the mortality and growth of each age class of the population for each species, and calculates the lost catch during the recovery period.

Food web effects are losses in commercial and recreational catch due to impacts on lower trophic food resources. Losses in lower trophic organisms are translated into lost biomass of commercial and recreational species, and consequent reductions in catch. To do this we follow French (1994) and use a proportionality rule for food web effects. An X % loss in food is assumed to lead to an X % loss in species ultimately harvested by commercial and recreational users

The Age Class Model

We use the Beverton and Holt approach to model number of individuals and weight of individuals for each age class. Prior to the age of recruitment to the fishery, the number of individuals in an age class declines due to natural mortality, and after recruitment, the number declines due to both natural and fishing mortality. So the number of individuals in age class t is:

$$N(t) = N(0)\ e^{-Mt}\ \text{for}\ t < t_R$$

where N(0) is the number of individuals in the initial age class, M is the instantaneous natural mortality rate and t_R is the time of recruitment to the fishery. The number of individuals in the same age class j years after recruitment to the fishery is

$$N(t_R+j) = N(t_R)\ e^{-(M+F)t} = N(0)\ e^{-M(t_R+j)-Fj}$$

The von Bertalanffy equation (Ricker 1975) is used to model the growth of individuals within an age class. Here, the length of an individual member of an age class is assumed to asymptotically approach the maximum length for the species, and the weight of individuals is calculated as a function of their length. Length is modeled as:

$$L_t = L_\infty\ (1\text{-}\exp(\text{-}k(t\text{-}t_0)))$$

where L_t is length at age class t, L_∞ is the maximum length for the species, k is the Brody growth coefficient, and t_0 is a constant. Weight of individuals is then determined as a function of length as:

$$W_t = a\ L_t^{\ b}$$

where W_t is weight at age t and a and b are constants. Catch is the fishing mortality rate times the biomass, and total catch from age class k through its remaining life span is:

$$C_k = \int_{t_k}^{t_{max}} F\ N(0)W_t\ e^{-Mt-F(t-t_R)}\ dt \qquad\qquad \text{for } t_k > t_R$$

where t_{max} is the maximum age for the species and:

$$C_k = \int_{t_R}^{t_{max}} F\ N(0)W_t\ e^{-Mt+F(t-t_R)} dt \qquad\qquad \text{for } t_k \le t_R.$$

Total lost catch for all age classes during the recovery period is:

$$C = \sum_{k=1}^{t_{max}} C_k$$

Recovery is presumed to start at the end of the disposal period. In the year following the end of disposal, the habitat is presumed to support a new initial age class, N(0). Recovery for each age class occurs in sequence over time as the initial age class matures and grows, with full recovery of the population occurring after t_{max} years. The loss during this recovery period is the long term effect, which is calculated by determining what would have been caught during the recovery period from age classes lost due to disposal.

Food Web Effect

The food web effect provides a measure of losses in commercial and recreational catch as a result of a loss of lower trophic organisms that function as food resources for commercial and recreational species. The food web loss in commercial and recreational species is estimated using a simple assumption that losses in biomass for consumer species are proportionate to losses of lower trophic production. However, no food web losses would occur if consumption of all food resources occurs on-site, since all consumer species within the disposal area are also lost due to disposal. So if all lower trophic production is consumed on site, the total loss of all consumer species negates food web losses, so that adding a food web effect to the direct effect double counts losses from disposal.

However, off-site food web losses may occur to the extent that food resources are lost to consumer species outside the disposal area. For example, consider food resources that are transported by currents. All production within the disposal area is lost, plus food resources are lost that are produced outside the disposal area but are transported into the area. Because no production is presumed to occur on site during disposal, production export from the disposal area is lost to adjacent areas. Thus, there may be a net loss in production outside the disposal area.

Consistent with our efforts to overstate losses, we assume a net loss in lower trophic production export of 50% of the food resources for the potential disposal sites. Our assumption of proportionate losses of consumer species implies that the off site food web effect equals 50% of the biomass in the disposal area for all commercial and recreational species. Food production is assumed to fully recover immediately following the end of the disposal period.

Appendix B

Listing of Biological Data

The USCOE DEIS provided the basic relative abundance of fishery species in the Narragansett Bay and Rhode Island Sound area. Using various methods described in the text and the fishery abundance data provided in the NRDAM, we calibrated the fishery abundance data, as listed in the next table.

Table B-1: Calibrated Biomass Abundance for the Disposal Sites (kg/km^2)

Sites		Quahog	Lobster	Atl. Herring	Flounder	Tautog	All Others
Narra-	3	743	1,190	1,714	3,128	4,112	20,685
gansett	157	743	3,019	1,200	8,795	3,084	41,369
Bay	158	1,439	3,552	1,200	8,795	3,084	41,369
Rhode	16	3,230	1,763	80,010	1,012	91	3,384
Island	18	4,186	1,190	55,529	1,559	91	5,692
Sound	69a	743	2,204	53,141	4,114	91	7,127
	69b	1,037	2,116	6,568	3,783	91	4,548

The biological parameters are taken from the NRDAM model. The "All Others" group is the
weighted average for all other species in the NRDAM model.

Table B-2. Biological Parameters for Fish Species in the Analysis

Species		F	M	L_∞ (cm)	k	t_0	$a*10^3$ (kg)	b (cm)	t_r	t_{max}
Quahog		0.3	0.1	14.6	0.253	0.734	0.0174	2.95	5	15
Lobster		0.25	0.1	25.3	0.063	0	0.75	3.04	5	15
Atl. Herring		0.22	0.4	20.8	0.59	0	0.0073	3.232	2	19
Flounder		0.25	0.1 1	59	0.34	0.05	0.0079	3.14	3	12
Tautog		0.38	0.3 2	83	0.16	0	0.0385	3	2	17
All Others	(NB)	0.38	0.3 2	83	0.16	0	0.0385	3	2	17
	(RIS)	0.26	0.3 3	44	0.28	0.557	0.075	3.0383	3	13

Appendix C

Table C-1. Commercial Values of Rhode Island Landings (1998)

Species	Weight (pounds)	Value
HERRING, ATLANTIC	34,321,527	$2,065,088.00
SKATES	11,858,301	$862,399.00
HAKE, SILVER	10,296,455	$3,486,898.00
GOOSEFISH	7,185,583	$4,110,272.00

MACKEREL, ATLANTIC	5,770,552	$1,626,324.00
BUTTERFISH	2,631,067	$1,457,574.00
MENHADEN, ATLANTIC	2,018,800	$101,020.00
SHARK, SPINY DOGFISH	1,760,324	$274,537.00
FLOUNDER,SUMMER	1,716,463	$3,924,671.00
FLOUNDER,WINTER	1,236,942	$1,536,438.00
HAKE, RED	1,221,018	$219,289.00
SCUPS OR PORGIES	792,530	$1,151,322.00
FLOUNDER,YELLOWTAIL	725,966	$909,609.00
BLUEFISH	570,740	$146,994.00
TILEFISH	553,485	$835,558.00
COD, ATLANTIC	416,782	$350,199.00
PLAICE, AMERICAN	199,011	$230,451.00
FLOUNDER,WITCH	196,972	$229,042.00
SEA BASS, BLACK	134,888	$256,656.00
HADDOCK	129,715	$149,854.00
BASS, STRIPED	94,663	$191,308.00
WEAKFISH	77,095	$49,885.00
LITTLE TUNNY	62,137	$6,817.00
FINFISHES,UNC GENERAL	61,781	$141,454.00
TUNA, BIGEYE	57,007	$158,558.00
DORY, AMERICAN JOHN	52,982	$14,266.00
TUNA, YELLOWFIN	51,816	$112,287.00
HAKE, WHITE	37,832	$18,013.00
BONITO, ATLANTIC	37,176	$24,947.00
SHAD, AMERICAN	33,590	$5,558.00
POLLOCK	33,514	$20,497.00
FLOUNDER,WINDOWPANE	32,225	$10,948.00
HAKE,ATLANTIC,RED & WHITE	29,988	$3,920.00
MACKEREL, CHUB	29,519	$4,428.00
CONGER EEL	25,604	$9,018.00
TUNA, ALBACORE	24,128	$17,715.00
TAUTOG	20,327	$21,796.00
SHARK, SMOOTH DOGFISH	15,896	$4,450.00
HAKE, OFFSHORE SILVER	11,654	$5,806.00
SEAROBINS	9,835	$1,938.00
REDFISH OR OCEAN PERCH	9,575	$4,275.00
SHARK, DOGFISH	8,710	$1,724.00
WOLFFISH, ATLANTIC	7,638	$2,670.00
DOLPHIN	7,366	$12,495.00
CUSK	5,572	$2,844.00
SHARK, UNC	4,535	$4,384.00
SEA RAVEN	4,137	$4,458.00
JACK, CREVALLE	2,799	$1,991.00
TUNA, SKIPJACK	2,634	$1,439.00
TUNA, BLUEFIN	2,344	$14,007.00
SCUP	2,239	$5,117.00
PERCH, WHITE	1,750	$2,317.00
POUT, OCEAN	1,470	$442.00
CUNNER	1,140	$1,285.00

KING WHITING	1,053	$557.00
TUNA, UNC	846	$425.00
LEATHERJACKETS	842	$408.00
SHARK, LONGFIN MAKO	830	$289.00
SHARK, SANDBAR	588	$231.00
SHARK, SHORTFIN MAKO	560	$570.00
SHARK, THRESHER	510	$140.00
HALIBUT, ATLANTIC	283	$489.00
STURGEONS	245	$220.00
FINFISHES,UNC FOR FOOD	215	$179.00
SHARK, DUSKY	183	$67.00
SHARK, PORBEAGLE	177	$163.00
DRUM, RED	165	$25.00
AMBERJACK	123	$55.00
MACKEREL, SPANISH	109	$66.00
COBIA	81	$41.00
MACKEREL,KING AND CERO	65	$33.00
GROUPERS	27	$34.00
LUMPFISH	24	$3.00
SHARK, SAND TIGER	21	$11.00
DRUM, BLACK	20	$24.00
TOADFISHES	5	$8.00
DEALFISH	3	$1.00

REFERENCES

Applied Science Associates, Inc., A.T. Kearney, Inc. and HBRS, Inc. 1994. *The CERCLA Type A Natural Resource Damage Assessment Model for Coastal and Marine Environments.* Technical Document Submitted to U.S. Department of the Interior.

Battelle, Inc. 1998. Shellfish sampling and site characterizations: Narragansett Bay and Rhode Island Sound potential disposal sites. In US Army Corps of Engineers, Draft Environmental Impact Statement, 1998b.

Economic Analysis, Inc. and Applied Science Associates, Inc. 1987. Measuring damages to coastal and marine natural resources. Technical Report Submitted to US Department of Interior.

Grigalunas, T.A. and C. Ascari. 1983. Estimation of income and employment multipliers for marine related activity in the Southern New England marine region. *Journal of the Northeast Agricultural Council* XI(1): 25 – 34 (Spr.).

Grigalunas, T.A., J. J. Opaluch, D. French and M. Reed. 1988. Measuring damages to marine natural resources from pollution incidents under CERCLA: Application of an integrated ocean systems/economic model. *Journal of Marine Resources* 5(1):1-21.

Grigalunas, T. A. and J. J. Opaluch.1989. Managing Contaminated Marine Sediments: Economic Considerations, in *Contaminated Marine Sediments: Assessment and Remediation*, Committee on Contaminated Marine Sediments, Marine Board, NRC, Washington, D.C.: National Academy Press.

Jin, D. and P. Hoagland. 1999. Economic activity associated with the Northeast large marine ecosystem: Application of an input-output approach. Marine Policy: Issues and Solutions *Proceedings* of Second Annual Korea-US Marine Policy Forum. Kingston: Korea-America Joint Marine Policy Research Center, University of Rhode Island.

National Research Council. 1989. *Contaminated Marine Sediments: Assessment and Remediation*, Committee on Contaminated Marine Sediments, Marine Board, NRC, Washington, D.C.: National Academy Press. Ricker, W.E. 1975. Computation and interpretation of biological statistics of fish populations. *Bull. Fish. Res. Board Can.* 191, 382 p.

U.S. Army Corps of Engineers. 1998. *Draft Environmental Impact Statement Providence River and Harbor Maintenance Dredging.*

------------------------------------ 1998b. *Draft Environmental Impact Statement, Appendix C: Biological Resources Providence River and Harbor Maintenance Dredging.*

U.S. Department of Commerce. 1999. *Fisheries Statistics of the United States.*

Part III:
The Role of Governance and Institutions

Large Marine Ecosystems, Vol. 13
T.M. Hennessey and J.G. Sutinen (Editors)

13

Emergence of a Science Policy-Based Approach to Ecosystem-Oriented Management of Large Marine Ecosystems

F. J. Gable

ABSTRACT

This article addresses interdisciplinary sustainable aspects of fisheries as linkages for the adaptive management of large marine ecosystems (LMEs). Natural and human-induced impacts on living marine resources are considered. Management and the ecological aspects of fish stock populations in the United States Northeast Continental Shelf ecosystem are examined for prospective and emerging "best practices" from a synthesis of the scientific literature. With the passage of the Oceans Act of 2000 (Public Law 106-256; *e.g.* Watkins 2002) in the United States, this article seeks to enhance the fostering of sustainability through natural and social science by forging linkages between the best available science practice and the "precautionary approach" that includes ecosystem considerations of fish stocks as component parts of a representative model LME.

INTRODUCTION

The concept of LMEs emerged from an American Association for the Advancement of Science (AAAS) selected symposium in the mid 1980s concerning variability and management of large marine ecosystems (Sherman *et al.* 1991; Alexander 1993). Rosenberg (2003) states that the "LME concept is helpful for thinking of the linkages of biological, chemical and physical factors of transboundary coastal ocean areas. Affecting any one part of the LME can have repercussions throughout the region. The LME provides a framework for thinking about potential impacts." The impacts on fisheries ecosystems including the biological, oceanographic and physical environment that supports commercial and recreational species within a specified management area and other economic activities such as sand and gravel mining, submarine telecommunications links, oil and gas energy development, marine transportation, contaminants disposal, recreational tourism and aquaculture, can occur at the scale of LMEs or may be localized in scope (Rosenberg 2003). Other examples may include "sector" analysis such as governance, for example, integrated coastal zone management (*e.g.* Juda 1999; Hennessey 1998; Juda and Hennessey 2001). Other more recent developments include the implications of climate change on marine fisheries science, policy, and management (Scavia *et al.* 2002; Healey 1990; Francis 1990).

ORIGINS OF OCEAN MANAGEMENT REGIMES

Following the September 1945 Truman Proclamations (nos. 2667 & 2668) in the United States concerning U.S. policy on natural resources of the subsoil and seabed of the continental shelf and on coastal fisheries, several ocean law measures were discussed and debated in a series of international fora. One of them, the Convention on the Continental Shelf, was agreed to in April of 1958 in Geneva (signed by the U.S. in June 1964). It contains 15 codified articles and developed actions. The 1958 Convention on the Territorial Sea and the Contiguous Zone was also codified at Geneva and was ratified by the U.S. Senate in 1961. This agreement contains 32 articles, including preexisting rules regarding international customary law, that provide a degree of precision and clarity not before then achieved. The Convention on the High Seas contains 37 mostly short articles and an Annex III (Convention on Fishing and Conservation of the Living Resources of the High Seas), also adopted in 1958. According to Merrell *et al.* (2001), "in 1958, the United Nations convened the first international conference of plenipotentiaries to examine the law of the sea, and to embody the results of its work in one or more international conventions. The 1958 conference produced four conventions that codified, to a great extent, customary law and brought international attention to the oceans." Years later, the Third United Nations Conference on the Law of the Sea (UNCLOS) began its substantive work in 1974, two years after the first U.N. Conference on the Human Environment in Stockholm (Emmelin 1972). UNCLOS III, consisting of 319 articles plus several annexes, was signed on 10 December 1982, was ratified by the requisite (60) countries and entered into force in November of 1994.

Domestically, the U.S. Congress enacted the Marine Resources and Engineering and Development Act of 1966 (Public Law 89-454) that created a blue ribbon executive-level commission on marine science activities later known as the Stratton Commission, named for the chairman of the 15 member panel. Their 1969 report, *Our Nation and the Sea,* reviewed the status of American ocean policy and provided specific recommendations for improving ocean science and ocean management practice. One of the major outcomes from those recommendations was the creation of the National Oceanic and Atmospheric Administration (NOAA est. October 1970; Nelson 1969).

Three noteworthy actions pertinent to ocean management were (1) the Presidential Proclamation of 27 December 1988 (No. 5928) in accordance with international law as reflected in the applicable provisions of the 1982 United Nations Convention on the Law of the Sea and customary international law extending the U.S. territorial sea to 12 nautical miles; (2) The Presidential Proclamation (5030) of 10 March 1983, establishing the Exclusive Economic Zone (EEZ) of the U.S. designated sovereign rights over natural resources out to 200 nautical miles from the baseline from which the breadth of the territorial sea is measured in accordance with international law; (3) The Oceans Act of 2000 (P. L. 106-256; effective 20 January 2001), was passed by Congress with the task of reviewing the importance of American oceans and marine resources and formulating a "scientifically based strategy for protecting and sustaining our oceans." It was recognized that the strategy required a coordinated and comprehensive national ocean policy (Watkins 2004). Ecosystem based management is one ocean governance approach advocated by the U.S. Ocean Commission (Watkins 2004).

LARGE MARINE ECOSYSTEMS (LMES): NATURAL ECOLOGICAL AREAS TO FOSTER REGIONAL FISHERIES MANAGEMENT AND SCIENCE ARRANGEMENTS

Large marine ecosystems (LMEs) are regions of ecological unity of ocean space extending from river basins and estuaries to the outer margins of continental shelves and seaward boundaries of coastal current systems (Griffis and Kimball 1996). A combination of ecological criteria including unique bathymetry, hydrography, productivity and trophic relationships characterize LMEs (Sherman 1989). LMEs are areas yielding 90 percent of the annual catch of global marine fisheries (Garibaldi and Limongelli 2003; Sherman and Duda 2001). The LME approach considers accommodating human utilization of its resources while maintaining ecosystem integrity (Sherman 1995,1994) .

The Northeast Shelf ecosystem extends over 260,000 km^2 of the US Northeast Continental Shelf. Four subareas have been identified within the LME boundary—Gulf of Maine, Georges Bank, Southern New England, and the Mid-Atlantic Bight (Sherman *et al.* 1996)

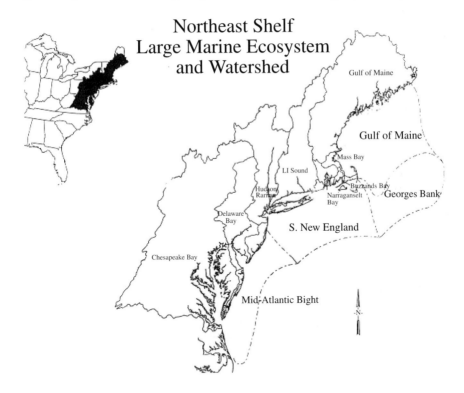

Figure 13-1. The Northeast Shelf Large Marine Ecosystem

Initiated in 1983, as a response to the Third United Nations Conference on the Law of the Sea (UNCLOS), was the Southeast Asian Project on Ocean Law, Policy and Management (SEAPOL). It was designed to promulgate a network of regional specialists in ocean development and management as a part of the Law of the Sea. These regional specialists selectively incorporated information on coral, mangrove and soft-bottom benthic communities in the coastal living resources project (English *et al.* 1988). They noted that "science is a central issue in any attempt to manage LMEs" and saw clearly that "the management of LMEs involves political, socio-economic, scientific and technical aspects." The ASEAN Coastal Living Resources Project was an early example of multidisciplinary and multinational approaches to the management of LMEs (English *et al.*1988).

In Australia and New Zealand the LME approach was selected as a means for introducing an ecosystem-based approach to the assessment and management of marine resources. Done and Reichelt (1998) emphasize that in the Oceania LME, fishery management focuses on optimization of catch-per-unit-effort (CPUE) for targeted commercial species along with bycatch and discard minimization. Integrated within the LME approach, as utilized in Australian and New Zealand jurisdictional waters, there is also a focus on both coastal zone and watershed catchment management. Here, emphasis for coastal zone management (CZM) is directed toward habitat protection for both catch and bycatch species (prohibited and non-specified species bycatch) as well as water quality maintenance. The reduction of polluted land-based runoff into surface waterways that drain towards the shore is the principle scope of emphasis for watershed catchment management (Done and Reichelt 1998). Thus, in Oceania, the quest for resource sustainability may best be achieved through the combination of management effort directed towards coastal habitats and catchment watersheds as well as the fishery (Done and Reichelt 1998).

In the eastern Atlantic Ocean the governments of the Republic(s) of Angola, Namibia and South Africa, in their desire to manage development and protect for future use the Benguela Current LME in an integrated and sustainable manner, committed themselves to establishing the "Benguela Current Large Marine Ecosystem" (BCLME) program with specific ecosystem-based actions, principles and policies (O'Toole 2002). The reasons for the establishment of the BCLME, included (a) significant transboundary implications of unsustainable practices of harvesting of living marine resources (fish stocks), (b) increasing habitat degradation and alteration which may have contributed to the increased incidence of harmful algal blooms, as well as (c) inadequate governance capacity to assess and monitor ecosystem status and trends, either nationally or regionally. An original Strategic Action Program (SAP) was adopted by signature of government ministers at the end of February of 2000 in the spirit of the United Nations Conference on Environment and Development (Rio Declaration) and Agenda 21 principles (O'Toole 2002).

The BCLME program was established as an international body under the terms and conditions of the Third United Nations Convention on the Law of the Sea and international customary law principles (see *e.g.* Belsky 1989, 1985). At the outset, for example, The United Nations Development Programme (UNDP) is represented on the Interim Benguela Current Commission for the initial five year BCLME program development phase. Original start-up funding was secured from the Global Environment Facility (GEF) in partnership with UNDP, and scientific and technical assistance from the National Oceanic and Atmospheric Administration (NOAA) of the United States, and ocean science agencies in Germany, Norway and France.

In another eastern Atlantic regional setting north of the BCLME, according to Ukwe *et al.* (2003, p.219) the countries of the Gulf of Guinea littoral "adopted an integrated and holistic approach using the LME concept to manage sustainably the environmental and living resources of the region." The Gulf of Guinea LME project began in 1995 with a pilot project initiative by six Gulf of Guinea countries. The project was focused on biodiversity conservation and water pollution control. Ministers representing the six countries responsible for the LME project signed the Accra Declaration as an expression of support for international cooperation in fostering sustainable management practices (Ukwe *et al.* 2003). Donor agency funding was secured via the GEF with implementation provided through the UNDP in concert with the U.N. Industrial Development Organization (UNIDO) with technical support from NOAA/National Marine Fisheries Service (NMFS) and the U.N. Environment Programme. "The project is anchored in the concept of LMEs as geographic units for improving the assessment and management of marine resources" (Ukwe *et al.* 2003). The overriding goals of the ongoing Guinea Current LME Strategic Action Plan (SAP) are biological diversity and the control of aquatic pollution with regard to restoring and sustaining the health of the living marine resources of the region. Ukwe *et al.* (2003) mention four specific objectives for the Guinea Current LME using the LME five module approach (see Figure 13-2). These are governance capacity building, ecosystem management database development, living marine resource assessment and long-term monitoring and protection strategies. The modular approach to the assessment and management of LMEs is presently being undertaken by a growing number of the world's developing coastal nations (Duda and Sherman 2002). The "predominant variables" for any given LME may be different even from those of its neighbor, depending upon the results of issue prioritization based on consensus reached through a transboundary diagnostic analysis (Sherman and Skjoldal 2002). Taking a holistic ecosystem approach, the LME concept "highlights the interrelationships of the different variables of each system and encourages cooperative dialogues across traditional disciplinary boundaries" (Knauss 1996).

ECOSYSTEM-ORIENTED MANAGEMENT AS A LINK FOR FOSTERING SUSTAINABLE FISHERIES

Recently, at the United Nations General Assembly in New York, "resolutions" have been crafted for adoption by member nations to apply by 2010 the "ecosystem approach" to the conservation, management and exploitation of highly migratory (pelagic) and "straddling" fish stocks (Jahnke 2003). Resolution A/57/L.49 concerning a number of fisheries issues was introduced by the United States of America through Ambassador Mary Beth West, the then Deputy Assistant Secretary of State for Oceans and Fisheries, to the fifty-seventh session of the General Assembly on 10 December 2002 (West 2003). Resolution A/57/L.50 regarding the conservation and management of straddling fish stocks and highly migratory fish stocks was also introduced at the same time by Ambassador West. In her remarks before the General Assembly, she indicated that "the fisheries draft resolutions are an assemblage of current ocean issues drawn from the priorities and interests of Member States." And, "they represent consensus... in making the oceans safe and healthy environments for sustainable development." Ambassador West's statement also contained an emphasis on the agreed-to Johannesburg World Summit on Sustainable Development Plan of Implementation adopted on 4 September 2002. She remarked that the "Plan calls on the world community to establish

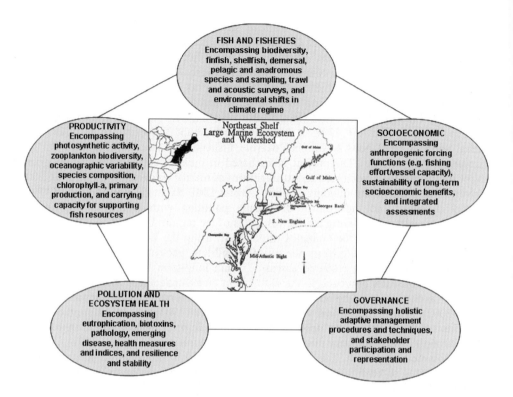

Figure 13-2. A five module LME strategy developed for assessing and analyzing ecosystem-wide changes in support of improved adaptive management decision practices (adapted from Sherman 2000 and Gable 2003).

by 2004, a regular United Nations process for global reporting and assessment of the state of the marine environment based on existing regional assessments. The Plan calls on the world community to elaborate regional programs of action and to improve links with strategic plans for the sustainable development of coastal and marine resources" (see: UNGA 2002; Jahnke 2003). Thus, the prescribed benefits of an in-place LME approach to living marine resources conservation biology and management can be seen at work in the international arena (Alexander 1999; Belsky 1985). The introduced "Resolution (on Oceans and the Law of the Sea, A/57/L.48) similarly calls upon States to develop national, regional and international programmes aimed at halting the loss of marine biodiversity. The United States welcomes this emphasis on integrated regional approaches to oceans issues." While at the podium, Ambassador West went on to state, "in that context (regarding integrated, regional approaches to ocean issues), we would like to bring to this body's attention the White Water to Blue Water oceans partnership initiative currently being planned for the Caribbean... it aims for an integrated approach to the management of freshwater watershed and marine ecosystems." "We hope it might serve as a successful model for similar efforts in other regions of the world." Moreover, "the United States also looks forward to collective efforts to establish an interagency coordination mechanism on oceans and coastal issues within the United Nations system" (UNGA 2002).[1]

Specifically, the written draft resolution A/57/L49 introduced by Ambassador West notes also, with particularity, "the importance of implementing the principles elaborated in Article 5 of the Provisions of the United Nations Convention on the Law of the Sea of 10 December 1982 relating to the Conservation and Management of Straddling Fish Stocks and Highly Migratory Fish Stocks (entered into force on 11 December 2001), including *ecosystem considerations* in the conservation and management of straddling fish stocks and highly migratory fish stocks." Draft resolution A57/L.49 as adopted (now known as 57/142) "encourages all States to apply by 2010 the ecosystem approach... and supports continuing work under way at the Food and Agricultural Organization of the United Nations (FAO) to develop guidelines for the implementation of *ecosystem considerations* in fisheries management..." (UNGA 2003).

Similarly, Resolution 57/141 Oceans and the Law of the Sea (formerly draft A/57/L/48) "calls upon States to promote the conservation and management of the oceans in accordance with Chapter 17 of Agenda 21 (*i.e.*, Earth Summit, Rio De Janeiro, June 1992; *e.g.* Garcia and Newton 1994) and other relevant international instruments, to develop and facilitate the use of diverse approaches and tools, including the ecosystem approach, the elimination of destructive fishing practices, the establishment of marine protected areas (MPAs; see also Presidential Executive Order 13158 of 26 May 2000 on MPAs of the U.S.) consistent with international law and based on scientific information, including representative networks by 2012 and time/area closures for the protection of nursery grounds and periods, proper coastal and land use and watershed planning, and the integration of marine and coastal areas management into key sectors." In Section XI Marine Environment, marine resources and sustainable development of said Resolution 57/141 of 12 December 2002 calls upon States "to improve the scientific understanding and assessment of marine and coastal ecosystems as a fundamental basis for sound decision-making through the actions identified in the Johannesburg Plan of Implementation, including that of relevant data collection of the marine environment" (UNGA 2003).

In November of 2003 an analogous resolution was adopted by the General Assembly. Demonstrating a pattern of agreement by the international community towards sustainable fisheries, another marine-oriented instrument was placed on the table at the U.N. General Assembly, reaffirming its resolutions, *inter alia*, 57/142 and 57/143 of 12 December 2002 (see above). Draft resolution A/58/L.18 was on the agenda at the fifty-eighth session in New York. After a successful roll-call adoption of the "sustainable fisheries... and related instruments resolution" (adopted as RES/58/14 on 24 November 2003), there was affirmation that in seeking responsible fisheries in the marine ecosystem (Section IX) there is the encouragement for Member States to apply by 2010 the ecosystem approach. This ecosystem

---

1 At the 74th Plenary Meeting of the U.N. General Assembly, on 12 December 2002, 132 nations were in favor of A/57/L.48 adopted as Resolution A/57/141, only one against (Turkey) and two abstaining (Colombia and Venezuela). In their affirming statement before the General Assembly, Japan through Ambassador Akamatsu, explained its position prior to voting on the three resolutions. In essence, Japan decided to associate itself with "the consideration of the ecosystem in the conservation and management of marine living resources" regarding draft resolution A57/L.49. By voice vote draft resolution A/57/L.49 was adopted as resolution 57/142. Similarly, draft resolution A/57/L.50 was adopted as resolution 57/143. See: http:www.un.org/ga/57pv.html online; available 10 November 2003.

approach and its relevant guidelines, in part, developed by FAO (Rome, Italy) would provide for the "implementation of ecosystem considerations in fisheries management." Resolution 58/14 of 2003 also "notes with satisfaction" the activities of the World Bank housed Global Environment Facility (GEF) aimed at "promoting the reduction of bycatch and discards in fisheries activities." Discards add to the effect of fishery landings, for example, "a mid-1990's assessment suggested that about 25 percent of marine catch is discarded" (Hanna 1999). Moreover, the GEF supports projects that apply the LME approach to ocean stewardship of living marine resources (Duda and Sherman 2002). The flexible LME approach can aid in achieving sustainable fisheries by addressing ecosystem considerations like: fishing overcapacity, large-scale pelagic drift-net fishing, fisheries bycatch and discards, aid in accomplishing sub-regional and regional cooperation in fostering responsible fisheries in the marine ecosystem, as well as address capacity-building and cooperation as it relates to science policy technical assistance and financial aid mechanisms (UNGA 2004).

Garcia (1994) suggests that, although U.N. General Assembly resolutions are not legally binding, they can have pertinent political significance; he notes their resolutions in the early 1990s on large-scale pelagic driftnets. "A U.N. General Assembly resolution may have an effect wider than that of a recommendation (its legal status) in revealing what State practice is, or pointing to what States might be willing to accept." He also indicates that the "precautionary principle" is no more than a non-binding norm, operating within the framework of particular agreements, but it "may be on its way to becoming part of customary international law" (see also Belsky 1985). As regards "good governance" for the environment, West (2003) emphasizes the promotion of "sound science-based decision-making" within legal, programmatic, and regulatory frameworks while stating that "changes in marine and coastal systems can undermine the basic economic and environmental services provided by the oceans." She also writes that "when it comes to the coastal environment, however, we have learned that regional approaches are often most effective" (West 2003). The large marine ecosystem (LME) paradigm provides just such an effective approach internationally and/or domestically in the U.S. The LME approach or initiative provides and promotes science-based decision making for the ocean and coastal activities, especially in the realm of commercial fisheries science policy. The LME modular assessment approach (Figure 13-2) is an improved science-based application to best practices integrated coastal management (*e.g.* West 2003; Ajayi *et al.* 2002; Done and Reichelt 1998).

CUSTOMARY INTERNATIONAL LAW AND THE MANAGEMENT OF LARGE MARINE ECOSYSTEMS

While the adoption of ocean affairs related resolutions by the Member States of the United Nations General Assembly demonstrate a willingness to move towards ecosystem-based fisheries management (as a tenet of adaptive management), more importantly "this acceptance may be emerging into customary rules of international law which promote consideration of total ecosystems and the establishment of standards for those systems" (Belsky 1985).[2]

---

[2] From Belsky (1985) "at a certain point, a series of state practices, codified in treaties or working agreements, and supported by the writings of legal scholars and the acknowledgement or acceptance of the world community, passes from mere examples of national action to a customary norm of international law. A total ecosystem approach to conservation and management of resources could become binding customary international law via this route."

Knecht (1994) recognized "that the use of the ecosystem approach in dealing with large marine ecosystems is already close to becoming international law." "Soft laws essentially are statements of international cooperation, usually in the form of an international treaty or agreement, which are not binding on (all) States but have the capacity to promote evolving notions of customary law, they have great importance in the evolution of customary law" (MacDonald 1995). He points out that customary international law consists of 'rules' and 'norms,' written and unwritten, that may or may not find expression in treaties. Precisely because of its informal nature, he says, customary law is central to international dialogue, often forming the basis on which to find common ground in international disagreements (MacDonald 1995). Alexander (1999) postulates that "the articles of the 1982 United Nations Conference on the Law of the Sea (UNCLOS) generally support the principles of ecosystem management for living marine resources. Most indications now point toward a general acknowledgement of the benefits of integrated ecosystem management in the world's oceans and seas." The objectives of UNCLOS are parallel to those of LME management (Alexander 1999). Moreover, Cole (2003) asserts that "there have been structural changes in fisheries decision-making, notably a transformation from a state-led approach towards multi-leveled decision-making procedures due to key developments in, *inter alia*, international law." Further, she says that "there have been considerable shifts in authority dealing with fisheries regulation and a new, distinct, global structure is emerging in essence attributed to globalization" (Cole 2003).

To that effect, the European Community has recently enacted reforming legislation for its Member States proscribing a "road map" towards their Common Fisheries Policy. The Council of the European Union, a regional body of Member States, enacted Council Regulation (EC) No. 2371/2002 of December 20th 2002 on the conservation and sustainable exploitation of fisheries resources under the Common Fisheries Policy (COEU 2002). This regulation is binding in its entirety and directly applicable in all Member States. The Europeans seem to be taking a "cue" from the United States by establishing Regional Advisory Councils (Article 31) to enable fisherfolk and other stakeholders to network their local knowledge and experience concerning diverse conditions throughout European Community jurisdictional waters. Note that the advisory councils are not designed to be independent management bodies with the authority to make decisions (Gray and Hatchard 2003). The eight regional fishery management councils structured in the U.S.A do have such authority.

The scope and objectives of EC No. 2371/2002 (Article 2(1)) include the provision to "aim at a progressive implementation of an ecosystem-based approach to fisheries management." Included here is the "good governance" objective of a "decision-making process based on sound scientific advice which delivers timely results. Broad involvement of stakeholders at all stages of the Common Fisheries Policy from conception to implementation" is another objective under the "principles of good governance" (Article 2(2)). Specifically, the Regional Advisory Councils were established to "contribute to the objectives of Article 2(1), that is, an "ecosystem-based approach to fisheries management" and in particular to advise the European Commission on matters of fisheries management in relation to prescribed sea areas or fishing zones.

Under the heading "conservation and sustainability, Article 5(3) Recovery Plans" and Article 6 (3) "Management Plans" "may cover either fisheries for single stocks or fisheries exploiting a mixture of stocks, and shall take due count of interactions between stocks and fisheries."

Therefore, objectives or aims of the European Commission's "new" approach to fisheries management refocuses policy towards a long-term view to fostering higher yield sustainable fisheries while moving towards an ecosystem-based approach to fisheries management. Curiously under Article 3 "definitions," none was provided for what is meant by an ecosystem-based approach.

Considerable progress has been made in recent years in developing ecosystem-based approaches to large marine ecosystems (Larkin 1996). Belsky (1999) argues that prevention of harm and 'rational and equitable use' mean that resources and uses must be studied and managed in a comprehensive manner, focusing on the large marine ecosystems in which resources exist." As such, according to Probert, "the concept of large marine ecosystems (LMEs) is now widely accepted" (Probert 2002). Belsky adds that "the evolution of the marine ecosystem approach from preferred policy to binding (international) customary law is demonstrated by the United Nations Convention on the Law of the Sea (UNCLOS 1982), which came into effect in November, 1994 (Belsky 1999). "The movement towards an ecosystem-approach is best represented by the Convention on the Conservation of Antarctic Marine Living Resources (CCAMLR) which was ratified in 1982. This treaty represents the first attempt to develop and apply an ecosystem management approach (English *et al.* 1988)." And, as described earlier here, LMEs are now a part of "best practices" international customary law (see: Belsky 1985; Juda and Hennessey 2001; Duda and Sherman 2002). Knecht (1994) argues that "the ocean governance process and policymakers need to take account of goals and principles emerging at the international level since these are likely to play a role in shaping future national ocean governance schemes" (see also Costanza *et al.* 1998).

ECOSYSTEM CONSIDERATIONS: THE FORMULATION OF A BEST PRACTICES LME APPROACH

"There is a need to enhance the conservation objectives of fisheries management plans to include explicitly ecosystem considerations. It is often difficult to separate out the effects of fishing from other anthropogenic influences (*e.g.*, pollution, habitat modification) and from natural environmental variability. This is particularly the case in nearshore ecosystems" (Gislason *et al.* 2000). Internationally, Wagner (2001) affirms that the recent Reykjavik Declaration of Responsible Fisheries in the Marine Ecosystem (October 2001) includes "ecosystem considerations in fisheries management that provide a framework to enhance management performance." These "considerations" incorporate increased attention to predator-prey relationships and to an understanding of the impact of human activities as well as the role of habitat and factors affecting ecosystem stability and resilience, among others (Table 13-1).[3]

---

[3] From the New England Fishery Management Council Executive Committee Minutes of January 9, 2004 (correspondence and reports document #10 of January 27-29, 2004) one of the discussion "problem statements" was the need for advice on ecosystem management principles. "What is ecosystem-based management? We need to know what it is before it is applied to our plans." Further, from the minutes it was mentioned that…" "descriptive knowledge should be a first step. Understanding ecosystems is a difficult but right thing to do." In addition, one stated draft goal was to "improve ecosystem-based management integration of habitat and bycatch, and interactions between fishery management plans" (FMPs).

Table 13-1. Elements for consideration in an ecosystem-oriented fisheries management approach for the Northeast United States Continental Shelf Large Marine Ecosystem (LME).*

Definition NOAA's three strategic objectives under the mission Goal 1: (1) "Protect, restore, and manage the use of coastal and ocean resources through ecosystem-based management. Protect, restore and manage the use of our ocean, coastal, and Great Lakes resources, (2) Protect, restore, and manage species and their habitats listed under the Endangered Species and Marine Mammal Protection Acts, (3) Manage and rebuild fisheries to population levels that will support economically viable and sustainable harvests.

ECOSYSTEM-BASED MANAGEMENT
Because we recognize that our three strategic objectives are scientifically, socially, and economically interdependent, we are improving our science, management, and regulatory processes to support comprehensive, integrated ecosystem-based management of our coastal, ocean, and Great Lakes resources. We will invest in improved understanding of ecosystems, identification of regional ecosystems, development of ecosystem health indicators, and new methods of governance to establish the necessary knowledge, tools, and capabilities to fully implement ecosystem-based management. (from New Priorities for the 21st Century: NOAA's Strategic Vision 2004)

Objective The basic ecosystem consideration is a precautionary approach to extraction of fish resources to provide and ensure the intergenerational sustainability of ecosystem goals, services and socioeconomic benefits by establishing appropriate reference points and/or sustainability indicators for restoring and maintaining the fish and fisheries produced by this ecosystem

Goals
1. Maintain ecosystem productivity and biodiversity consistent with multiple spatial scales, natural evolutionary and ecological processes, including dynamic change and variability.
2. Maintain and restore habitats essential for fish and their prey, that is, "those waters and substrates necessary to fish for spawning, breeding, feeding or growth to maturity" (Fluharty 2000)
3. Maintain system sustainability and sustainable yields of fisheries resources for human consumption at a rate or level of fishing mortality that does not reduce the capacity of a fishery to produce the maximum sustainable yield on a continuing basis (Fluharty 2000).
4. Maintain the concept that humans are integral components of the ecosystem.

Guidelines
1. Integrate ecosystem-oriented management through interactive partnerships among the states and regulatory agencies, stakeholders, public, regional and international organizations.
2. Utilize peer reviewed ecological models as an aid in understanding the structure, function, and dynamics of the Northeast Shelf ecosystem.
3. Utilize best available science research and monitoring to validate a "best practices" ecosystem approach for sustainable uses of fishery resources.
4. Use precaution when faced with uncertainties to minimize risk; management decisions should err on the side of resource conservation.

Assumptions
1. Ecosystem-oriented management is an adaptive process which requires periodic evaluation preferably on an annual basis for refining and incorporating updated scientific information as it becomes available.
2. Ecosystem-oriented management requires temporal scales that transcend human generations.

Understanding
1. "The ecosystem is considered to be a unit of biological organization made up of all of the organisms in a given area interacting with the physical environment so that a flow of energy leads to characteristic trophic structure and material cycles within the system" (Odum 1969).
2. Science policy, management measures that are consistent with an ecosystem-oriented strategy include precautionary-conservative catch (allocation) limits, comprehensive monitoring and enforcement, and additional ecosystem considerations that are based on scientific research and advice (Witherell, *et al.* 2000).

*Adapted from concepts described by NOAA 2004, Sinclair and Valdimarsson 2003, Witherell *et al.* 2000, Fluharty 2000, Sherman and Duda 1999a,1999b, Witherell 1999, Slocombe 1998, MacKenzie 1997, Haeuber 1996, Schramm and Hubert 1996, Odum 1969.

In the U.S., the North Pacific Fishery Management Council (NPFMC) utilizes as ecosystem consideration indicators: physical oceanography indices (*e.g.*, temperature and decadal regime shifts); habitat (*e.g.*, groundfish bottom trawling effort by subregion, closed areas to trawling, and biota bycatch by all gears in habitats of particular concern (HAPC's)); target groundfish (*e.g.* total biomass, total catch by subregion, groundfish discards including target species discards, recruitment by subregion); fleet size – analogous to humans as a part of the ecosystem – (*e.g.,* total number of vessels actually fishing); forage (*e.g.* forage species such as herring et al., bycatch by subregion); <u>other species</u> (*e.g.* spiny dogfish, various shark species, jellyfish and prohibited, other, and nonspecified species bycatch – example(s) of prohibited bycatch include halibut mortality, herring, crab and salmon species, among others); <u>marine mammals</u> (*e.g.* seals, sea lions); <u>seabirds</u> (*e.g.* population trends and bycatch as well as breeding chronology and species productivity); and, <u>aggregate indicators</u> (such as possible regime shifts and trophic-level food web catch by subregion). All of these categories come under the rubric of ecosystem considerations (Livingston 2001, 1999) at an LME scale whether in the Gulf of Alaska or the U.S. Northeast Continental Shelf (Sherman and Skjoldal 2002).

Regarding precautionary and conservative catch limits, the North Pacific Fishery Management Council (NPFMC) mandates that "all fish caught in any fishery (including bycatch), whether landed or discarded are counted towards the TAC for that stock" (Witherell *et al.* 2000). As a further management precautionary approach it is assumed that there is a 100 percent mortality for all discards regardless if some fish actually survive (Witherell *et al.* 2000). Species are discarded by a fishing vessel because they are either unwanted "economic discards" or they are regulatory "prohibited species" (Witherell *et al.* 2000). In the North Pacific, a "best practices" approach requires a "comprehensive and mandatory observer programme," requires 100 percent coverage on any vessel more than 49m in length overall (Witherell *et al.* 2000). While this has been adopted as a "best practice" to provide limits on bycatch and discards, it does not necessarily address "ecosystem concerns" (Witherell *et al.* 2000).

Other "best practices" (see also Sainsbury and Sumaila 2003) used in the American waters of the North Pacific for limits on bycatch and discards include certain gear restrictions. For example, to prevent ghost fishing and reduce bycatch of non-target species, gillnets for groundfish are prohibited (Witherell *et al.* 2000). Further, the NPFMC "adopted an improved retention and utilization programme for all groundfish target fisheries. Beginning in 1998, 100 percent retention of pollock and Pacific cod was required, regardless of how or where it was caught" (Witherell *et al.* 2000). By 2004, the NPFMC expects that for most regulated species, the discard rate will be about five percent (Witherell *et al.* 2000). It is a plausible way to manage commercial fisheries while incorporating ecosystem considerations.

Ecosystem considerations also may translate to specific concerns in a given LME or subarea. Examples of these concerns may entail harvest rate(s) fishery effects on species composition. Significant differences exist in the rate of harvest of groundfish species in the New England Region. Some are harvested close to their acceptable biological catch [F(abc)] levels while other species are taken at variable lower levels. Some trawl fisheries are constrained by bycatch limitations for prohibited species (*e.g.*, yellowtail flounder) and commercial landings prices for flatfish. As witnessed in the Northeast United States Continental Shelf LME (Sherman *et al.* 1996; Sherman and Skjoldal 2002) shifting or resulting high biomasses of predator species (*e.g.* dogfish and skates) can have substantial impacts on the trophodynamics of the marine ecosystem and shift the species assemblages. Disproportionate harvest rates require constant analysis for lasting season-to-season implications on the commercial groundfishery. "Fish populations on Georges Bank changed from dominance by commercially important groundfish species to less desirable species such as dogfish and sandlance. Concurrent with a decline in the desirable groundfish from overfishing were increases in pelagics (herring, mackerel) and elasmobranches (spiny dogfish, skates)" (Boehlert 1996).

LMEs and the Precautionary Approach to Fisheries Management

Witherell *et al.* (2000) emphasize that for the North Pacific, "the basic ecosystem consideration is a precautionary approach to extraction of fish resources." They suggest that the "precautionary principle was developed over the past 10 years as a policy measure to address sustainability of natural resources in the face of uncertainty" (*e.g.* Kinzig *et al.* 2003; Hilborn 1987). One of their main hypotheses concerning integrating ecosystem considerations in fisheries management is that "if fisheries are managed sustainably using a precautionary approach, it is likely[4] that the overall ecosystem processes, ecosystem integrity, and biodiversity are also protected to some degree" (Witherell *et al.* 2000; also Table 13-2).

Table 13-2. An ecosystem approach requires new thinking about how marine ecosystems are defined, and how problems and solutions are framed.

- LMEs are regions of ocean space encompassing coastal areas from river basins and estuaries on out to the seaward boundary of continental shelves and the seaward margins of coastal current systems. They are relatively large regions on the order of 200,000 km^2 or larger, characterized by distinct bathymetry, hydrography, productivity, and trophically dependent populations.
- Management scales are nested in a multiple spatial and temporal application of 5-module multisectoral suites of indicators (productivity, fish and fisheries, pollution and health, socioeconomics and governance) extending in scale from coastal community level to seaward extensions of economic interests to the full extent of the large marine ecosystem.
- Ecosystem categories of threat, level of threat, and distance from optimum condition can be combined to rank ecosystems at risk. Ranks should be based on a review of quantitative information by a scientific panel with stakeholder participation. Ranks can be employed to plan and prioritize management regulatory agency action for ecosystems at various levels of present and future risk.
- The UN Food and Agricultural Organization (FAO) looks for an ecosystem approach to fisheries, balanced with diverse societal objectives and taking into account variabilities in biotic, abiotic and human components of ecosystems and their interactions, and applying an integrated approach to fisheries within ecologically meaningful boundaries.
- Holling (1996) advocates that, "at a minimum, the goal of ecosystem management is understanding to reduce uncertainties, action to maintain or restore resilience (*i.e.* the ability of a system to absorb change and variation without flipping into a different state where the variables and processes controlling structure and behavior suddenly change) as insurance for the unknown, and creation of incentives for maintaining sustainable systems."

Witherell (1999) mentions that specific ecosystem consideration(s) chapters have been prepared as supplementary information in select annual stock assessment and fishery

---

4 From *inter alia*, Mahlman (1997), Easterling *et al.* (2000), and the U.S. Global Change Research Program (http://www.usgcrp.gov/), quantitative terminology used in global (climate) change and policy (and applicable here) includes:
- "virtually certain projections" = > 99% probability or chance of being true
- "very likely" or "very probable" = 90 to 99% or 9 out of 10 probability or chance of being true
- "likely" or "probable" = 67 to 90% or 2 out of 3 chances of being true
- "possible" = 33 to 66% probability or chance of being true
- "unlikely" or "some chance" = 10 to 33% probability or chance of being true
- "very unlikely" or "little chance" = 1 to 10% probability or chance of being true
- "improbable" = < 1% probability or chance of being true.

evaluation reports. In addition, the North Pacific Fishery Management Council (NPFMC) established an Ecosystem Committee in 1996 to suggest possible ecosystem-oriented approaches into the fishery management process (*e.g.*, hosting workshops, meetings and informal discussions) whereby the Committee utilized the scientific literature to identify elements and prospective principles of ecosystem-oriented management (see Tables 13-1 and 13-2). Witherell (1999) stresses that the NPFMC and the National Marine Fisheries Service have used a precautionary approach, incorporated as part of ecosystem considerations, by: (a) relying on scientific research and advice, (b) conservative catch quotas, (c) comprehensive monitoring and enforcement, (d) bycatch controls, (e) habitat conservation areas, and (f) additional ecosystem considerations. The NPFMC also incorporates select marine protected areas (MPAs) as a tool for managing bycatch and habitat protection as well as time/area closures (see Presidential Executive Order 13158 of 26 May 2000).

Additional ecosystem considerations result from the impacts of fishing gear on habitat and ecosystems. Of numerous articles on this subject in the open scientific literature, most research has been on trawl gear. Though not the focus of this article, bottom trawls, as well as other gear types can alter the benthic structure, sediments and nutrient cycling in certain situations (Witherell *et al.* 1997). Now banned pelagic drift nets or "ghost fishing" created bycatch discard issues as well as marine debris problems. Climatic changes are another consideration. Related to oceanic temperature conditions are year class strengths of commercially important species (*e.g.* Sainsbury *et al.* 2000). Herring and cod appear to respond favorably with strong year classes with the onset of warm current regimes. Declines in stocks may be seen, however, for other finfish (Witherell 1998; Mountain 2002; Fogarty 2001). More "retrospective" ecosystem change research on this topic might prove valuable when trying to prepare optimal yield (OY) and maximum sustainable yield (MSY) figures from biomass estimates for a commercial species. Witherell (1998) writes about the occurrence, on a decadal or longer frequency in the North Pacific Ocean, of shifts between warm and cool periods and the compelling links between ocean conditions and living marine resources production. These decadal oscillations in the ocean are characterized as "regime shifts" (Steele 1998).

Other ecosystem-oriented management approaches include the NPFMCs adopted regulation prohibiting a directed fishery for select forage fish that are found to be important prey for higher trophic level species (such as groundfish) (Witherell *et al.* 2000). These authors discuss continuing progress towards ecosystem-based management that the NPFMC is trying to fulfill. Elements for consideration in introducing ecosystem-based management for the Northeast U.S. Continental Shelf LME have been crafted to foster dialogue (Table13-1). The approach is grounded in elements and principles of ecosystem-based management identified in the existing scientific literature. The approach provides a definition for fisheries ecosystem-based management as well as a presentation on objectives, goals, guidelines, assumptions and understanding.

The Precautionary Principle/Approach: Control Rules and Reference Points

Richards and Maguire (1998) profess that the "precautionary approach is now embodied in several international agreements, including the United Nations Straddling Fish Stocks and Highly Migratory Fish Stocks Agreement and the voluntary FAO Code of Conduct for Responsible Fisheries. Article 6 of the "Straddling Stocks" Agreement, which was ratified by the requisite number of countries as of 11 December 2001, and was thus incorporated into the Law of the Sea Treaty, provides "the essence of the precautionary approach whereby 'States shall be more cautious when information is uncertain, unreliable or inadequate. The absence of adequate scientific information shall not be used as a reason for postponing or failing to take conservation and management measures' and improved methods are required for dealing with risk and uncertainty" (Richards and Maguire 1998).

Stock-specific reference points provide the principle mechanism for applying the precautionary approach for harvest management strategies for developed fisheries. The "Straddling Stocks" Agreement, in Article 6, provides that signatory States "shall determine, on the basis of the best scientific information available, stock-specific reference points and the action to be taken if they are exceeded. Two types of reference points are identified: limit reference points set boundaries which are intended to constrain harvesting within safe biological limits within which the stocks can produce maximum sustainable yield while target reference points are intended to meet management objectives" (Richards and Maguire 1998). "Reference points have been generally defined in terms of the fishing mortality rate F and expressed as targets rather than limits. Although reference points have been applied mainly in the context of biological science, economic or social reference points could and should also be developed and adopted" (Richards and Maguire 1998). Basically, "the status of an ecosystem can be assessed" according to Link *et al.* (2002), and it is "not novel to assess the status of single species fish stocks." For the assessment and management of "large marine ecosystems," lessons from single species stock assessment, environmental impact assessment (EIA), and ecological risk assessment tools and procedures provide appropriate management decision criteria (Link *et al.* 2002).

For fisheries management, tools to achieve ecosystem objectives – gear restrictions, closed areas and seasons, including MPAs, quotas and bycatch limits and restrictions on days-at–sea, are the same as those already in use to achieve single species related conservation objectives (Gislason *et al.* 2000). These are also referred to as input-output controls and technical measures. "The similarity between single-species fisheries management and an ecosystem approach should not come as a surprise" (Sissenwine and Mace 2003). Rosenberg (2002) discusses control rules stating they "essentially relate management action to control the fishing mortality rate to the status of the resource in terms of biomass or some other measure. A control rule provides a framework for pre-agreed management actions as called for in the precautionary approach. Uncertainty in the status of the resource can be included explicitly through the specification of management targets to be achieved on average and management thresholds that should never be exceeded." "Control rules leave little room for negotiation and consideration of issues such as (stock) rebuilding timeframes and allocation between States, groups or gear types" (Rosenberg 2002). Generally, these were designed by marine scientists before the managers had provided any precautionary management systems of their own. Rosenberg (2002) concludes that "the mechanistic approach of control rules to implementation of precautionary management may be hindering agreement on conservation restrictions, simply because it leaves so little room for negotiation."

Concerning the implementation of the precautionary approach domestically, Rosenberg (2002) indicates that "the Sustainable Fisheries Act of 1996 carries forward many of the ideas of the precautionary approach with regard to preventing overfishing, the use of reference points, reducing bycatch and protecting habitat." And, "the burden of proof continues to be on managers to prove that restrictive measures are essential rather than to show that harvesting can be safely allowed." Therefore, reference points, to establish targets or thresholds for defining overfishing, are used to implement precautionary management in the USA and maximum sustainable yield (MSY) remains as a standard reference point. Garcia (1994) theorizes that "in a way, MSY could be considered a measure of the maximum assimilative capacity of the stock." "The need to reduce fishing pressure has resulted in (control) rules that do not allow fishers to shift from one fishery to another as easily as in the past" (Rosenberg 2002). Thus, the need for an LME ecosystem-based approach to living resources biomass allocation in an adaptive management environment is necessary to foster sustainable yields. Rosenberg (2002) laments that "as Regional Administrator for the National Marine Fisheries Service, I found it hard to understand all the rules and changes, and the fishermen certainly found it equally hard."[5]

In a recent synopsis, the European Community on 20 December 2002, addressing the conservation and sustainable exploitation of fisheries resources under the Common Fisheries Policy—Council Regulation LEC No. 2371/2002—noted in the *Official Journal of the European Communities* dated 31 December 2002, entered into force on 1 January 2003, that member states of the Community "shall apply the precautionary approach in taking measures designed to protect and conserve living aquatic resources, to provide for their sustainable exploitation and to minimize the impact of fishing activities on marine ecosystems. It should aim at a progressive implementation of an ecosystem-based approach to fisheries management"[Article 2(1)]. Article 3(i) states that the "precautionary approach to fisheries management means that the absence of adequate scientific information should not be used as a reason for postponing or failing to take management measures to conserve target species, associated or dependent species and non-target species and their environment. Precautionary reference points are biological reference points and are designed to mark the boundary between acceptable risks and unacceptable risks." Further, Article 5 (3) and Article 6 (3) require that "recovery plans and management plans," respectively, should be drawn-up on the basis of the precautionary approach. Article 6 (2) shall include conservation reference points, which under Article 3 (k) are "means values of fish stock population parameters (such as biomass or fishing mortality rate) used in fisheries management, for example, with respect to an acceptable level of biological risk or desired level of yield." Three types of reference points are typically considered including limit reference points (means values of fish stock population parameters such as biomass or fishing mortality rate), which should be avoided because they are associated with unknown population dynamics, stock collapse or impaired recruitment (Art. 3 (j)), precautionary or buffer reference points and target reference points. Thus a precautionary approach has been linked to "best practices" for living marine resource capture and exploitation actions and it is therefore incumbent upon countries to apply it through customary international law and practice.

Sinclair and Valdimarrson (2003) argue that "a first step in moving towards ecosystem-based fishery management is to identify and describe the different ecosystems and their boundaries, and then to consider each as a discrete entity for the purposes of management. Thereafter, ecosystem management objectives must be developed. The central objective of ecosystem-

---

[5] From the NEFMC Executive Committee Minutes of January 9, 2004 under the discussion section on "problem statement" the present NOAA/NMFS Regional Administrator stated that "ecosystem management means something different to everyone; at some point all must agree as to what it means ... [we] need to have better linkage with everything we're doing and not working separately." This article provides a baseline dialogue on the meaning of large marine ecosystem-oriented fisheries management (see Table 13-1).

based fishery management is to obtain optimal benefits from all marine ecosystems in a sustainable manner." These authors suggest that "once the objectives have been identified and agreed upon, it is necessary to establish appropriate reference points and/or sustainability indicators... which must be based on the best scientific evidence available." The general principles utilized in conventional single-species management will still apply regarding achieving objectives in suitable ecosystem-based fisheries management strategies.

Governance Issues for Ecosystem-based Fisheries Management

To Sissenwine and Mace (2003) the "precautionary approach means that, when in doubt, err on the side of conservation." Further, they state that "an ecosystem approach for responsible fisheries management requires taking into account trophic interactions in a precautionary fishing mortality rate strategy which they define as, "geographically specified fisheries management that takes account of knowledge and uncertainties about, and among, biotic, abiotic and human components of ecosystems, and strives to balance diverse societal objectives" (see also Table 13-1). Interdisciplinary science employing ecosystem considerations, along with developing local and regional institutions and frameworks that can integrate scientific information into socioeconomic and political decisions, are needed (Boesch 1999; Botsford *et al.* 1997).

Imperial (1999) asserts that ecosystem-oriented management "needs to develop low-cost mechanisms to facilitate communication, make decisions, and resolve conflicts between scientists, agency officials, interest groups, and the public in order to minimize information asymmetries (*e.g.* Table 13-1). This may be one reason why many ecosystem-based management programs utilize collaborative approaches to decision-making." "Like many other government programs, ecosystem-based management is the result of an evolutionary process of experimentation, goal definition and redefinition, and the search for appropriate implementation strategies" (Imperial 1999).

Domestically, Griffis and Kimball (1996) suggest that the regional marine fishery management councils "appear to have the breadth of responsibility and adequate structure needed for stakeholder input and involvement in decision making... some Councils have functioned better than others and there are lessons to be learned from both the successes and failures." Murawski (2000) emphasizes that in the U.S., "current management is characterized as being concerned with 'conservation of the parts' of systems, as opposed to the interrelationships among them." He suggests that "there is no specific ecosystem analogue to single-species definitions of overfishing." "For the Northeast Shelf, the decline in the groundfish resource, combined with restrictive management directed to that component, has resulted in the predictable scenario of serial depletion. The practice of allowing many species to remain outside any management control until they show signs of overfishing encourages excess depletion (*e.g.* Hagfish) and serial depletion, and exacerbates bycatch problems" (Murawski 2000). He reiterates that "situations such as those existing off the northeast USA could benefit greatly from a more formal mechanism to incorporate ecosystem perspectives (*i.e.*, considerations or interactions) in the development of management goals and conservation measures" (Murawski 2000).[6] "Ecosystem approaches, whether

---

[6] NEFMC member Erik Anderson of New Hampshire at the November 4, 2003 Council meeting in Peabody, Massachusetts put forward a motion stating that population parameters used for groundfish would incorporate "ecosystem interactions" as a condition of (stock) status determination. This was the first Council meeting where

implemented as perspectives on traditional overfishing paradigms or through explicit ecosystem-based definitions, require research and advisory services not typically provided by fish stock assessment science. Nevertheless, additional ecosystem monitoring and research is necessary with increased emphasis on species interactions, diversity and variability – at various temporal and spatial scales" (Murawski 2000). He suggests that "*ecosystem considerations* may increasingly be used to modify regulations intended primarily to conserve high-value species, to address bycatches (*e.g.* sea turtles and marine mammals are of significant concern), predator-prey demands and the side-effects of fishing effort" (Murawski, 2000). Griffis and Kimball (1996) argue that a main ingredient of ecosystem approaches to resource management includes defining sustainability and making it the primary goal or objective.

CONCLUSION

Management agencies "continue to struggle with the problem of how to define in operational terms, let alone implement, an ecosystem-based framework for managing fisheries" (Hall, 2002). It appears that facets of the LME paradigm, in combination with the precautionary approach, have taken root in international custom and law. An LME *policy orientation* paradigm (Gable 2003), provides a view of fisheries professionals as participants involved in decision making streams and actions over time that collectively determine what truly happens to fishery stocks within an identifiable marine ecosystem.[7]

REFERENCES

Ajayi, T., K. Sherman, and Q. Tang, 2002. Support for Marine Sustainability Science. *Science* 297 (5582):772.

Alexander, L.M.. 1999. Management of large marine ecosystems: A Law of the Sea based governance regime. In: *The Gulf of Mexico Large Marine Ecosystem: Assessment, Sustainability, and Management,* Kumpf, H., K. Steidinger, and K. Sherman, (eds.), Blackwell Science. 511-515.

---

an ecosystem-oriented approach was voted on and passed unanimously by the members. John Boreman, Director of the NMFS/NEFSC (Woods Hole, MA) said, when asked by vice-chairman Thomas Hill to elaborate, that …"what we're going to be looking at is just all species, the whole ecosystem. So, it's all trophic interactions and competitors, predators, prey, anything that would affect the population growth of a given species." The word ecosystem is an "expression of carrying capacity in the system that may be less than the biomass target. If there are factors out there in the ecosystem that limit the carrying capacity of that particular species, we should know about that, or we should include that factor in the analysis." Erik Anderson's use of the term "ecosystem interaction(s)" is equivalent to the use of "ecosystem considerations" found here. (Source: notarized written transcription of the audiographic tape dated 5 March 2004 and received 2 April 2004 by the NEFMC at pages 158-161).

[7] Acknowledgments: funding for this study was provided through a competitive contract from NOAA/NMFS/NEFSC Narragansett, Rhode Island Lab contract P.O.No. EA133F-03-SE-0707. Thanks to Kenneth Sherman, NMFS Office of Marine Ecosystem(s) Studies for comments on an earlier version of the manuscript as well as to Professor Tim Hennessey (URI) and Phil Logan (NMFS, Woods Hole, MA.) for discussions on aspects of the study and other anonymous reviewers. Portions of the project are a spin-off of my *marine affairs* Ph.D. and graduate program of study at the University of Rhode Island.

Alexander, L.M. 1993. Large marine ecosystems: A new focus for marine resources management. *Marine Policy* 17 (3):186-198.

Belsky, M.H. 1999. Using legal principles to promote the "health" of the ecosystem. In: Kumpf, H., K. Steidinger, and K. Sherman, eds. *The Gulf of Mexico Large Marine Ecosystem: Assessment, Sustainability, and Management,* Blackwell Sciences. 416-430.

Belsky, M.H. 1989. The ecosystem model mandate for a comprehensive United States ocean policy and law of the sea. *San Diego Law Review* 26 (3): 417-495.

Belsky, M.H. 1985. Management of large marine ecosystems: Developing a new rule of customary international law. *San Diego Law Review* 22 (4):733-763.

Boehlert, G.W. 1996. Biodiversity and the sustainability of marine fisheries. *Oceanography,* 9 (1): 28-35.

Boesch, D.F. 1999. The role of science in ocean governance. *Ecological Economics* 31 (2): 189-198.

Botsford, L.W., J.C. Castilla, and C.H. Peterson. 1997. The management of fisheries and marine ecosystems. *Science,* 277 (5325):509-515.

Cole, H. 2003. Contemporary challenges: Globalisation, global interconnectedness and that 'There *are not* plenty more fish in the sea': Fisheries, Governance and Globalization: Is There a Relationship? *Ocean and Coastal Management* 46 (1-2):77-102.

Costanza, R. and 15 others. 1998. Principles for sustainable governance of the oceans. *Science* 281 (5374):198-199.

Council of the European Union (COEU). 2002. Council Regulation (EC) No. 2371/2002 of 20 December 2002 on the Conservation and Sustainable Exploitation of Fisheries Resources Under the Common Fisheries Policy. *Official Journal of the European Communities* 358:59-71. Online available, November 22, 2003 @ http://europa.eu.int/.

Done, T.J., and R.E. Reichelt. 1998. Integrated coastal zone management and fisheries ecosystem management: Generic tools and performance indices. *Ecological Applications* 8 (1) Supplement:110-118.

Duda, A.M., and K. Sherman. 2002. A new imperative for improving management of large marine ecosystems. *Ocean & Coastal Management* 45 (11-12):797-833.

Easterling, D.R., G.A. Meehl, C. Parmesan, S.A. Changnon, T.R. Karl, and L.O. Mearns, 2000. Climatic extremes: Observations, modeling, and impacts. *Science* 289 (5487):2068-2074.

Emmelin, L. 1972. The Stockholm conferences. *Ambio,* 1 (4):135-140.

English, S.A., R.H. Bradbury, and R.E. Reichelt. 1988. Management of large marine ecosystems – A multinational approach. In: *Proceedings of the 6th International Coral Reef Symposium, Australia,* Vol. 2:369-374.

Fluharty, D. 2000. Habitat protection, ecological issues, and implementation of the sustainable fisheries act. *Ecological Applications* 10 (2):325-337.

Fogarty, M.J. 2001. Climate variability and ocean ecosystem dynamics: Implications for sustainability.   In   Steffen,W., J. Jaeger, D.J. Carson, and C. Bradshaw, eds. *Challenges of a Changing Earth: Proceedings of the Global Change Open Science Conference,* Amsterdam, Holland, 10-13 July, 2001,   Ch. 4, Springer. 27-29.

Francis, R.C. 1990. Climate change and marine fisheries. *Fisheries* 15 (6):7-9.

Gable, F.J. 2003. A practice-based coupling of the precautionary principle to the large marine ecosystem fisheries management concept with a policy orientation: The Northeast United States Continental Shelf as a case example. *Coastal Management* 31 (4):435-456.

Garcia, S.M. 1994. The precautionary principle: Its implications in capture fisheries management. *Ocean & Coastal Management* 22 (2):99-125.

Garcia, S.M., and C.H. Newton. 1994. Responsible fisheries: An overview of FAO policy developments (1945-1994). *Marine Pollution Bulletin*, 29 (6-12):528-536.

Garibaldi, L., and L. Limongelli, 2003. Trends in Oceanic Capture and Clustering of Large Marine Ecosystems. FAO Fisheries Technical Paper 435. 71p.

Gislason, H., M. Sinclair, K. Sainsbury, and R. O'Boyle. 2000. Symposium overview: Incorporating ecosystem objectives within fisheries management. *ICES Journal of Marine Science*, 57 (3):468-475.

Gray, T. and J. Hatchard. 2003. The 2002 reform of the common fisheries policy's system of governance—Rhetoric or reality? *Marine Policy* 27 (6):545-554.

Griffis, R.B., and K.W. Kimball. 1996. Ecosystem approaches to coastal and ocean stewardship. *Ecological Applications*, 6 (3):708-712.

Haeuber, R. 1996. Setting the environmental policy agenda: The case of ecosystem management. *Natural Resources Journal* 36 (1):1-28.

Hall, S.J. 2002. The Continental Shelf benthic ecosystem: Current status, agents for change and future prospects. *Environmental Conservation* 29 (3):350-374.

Hanna, S.S. 1999. Strengthening governance of ocean fishery resources. *Ecological Economics* 31 (2):275-286.

Healey, M.C. 1990. Implications of climate change for fisheries management policy. *Transactions of the American Fisheries Society* 119(2):109-118.

Hennessey, T. M. 1998. Ecosystem management: The governance approach. In Soden, D., B.L. Lamb and J.R. Tennert, eds. *Ecosystems Management: A Social Science Perspective*, Kendall-Hunt, Dubuque, Iowa, Ch. 2, 13-29.

Hilborn, R. 1987. Living with uncertainty in resource management. *North American Journal of Fisheries Management* 7 (1):1-5.

Holling, C.S.1996. Surprise for science, resilience for ecosystems, and incentives for people. *Ecological Applications* 6 (3):733-735.

Imperial, M.T. 1999. Institutional analysis and ecosystem-based management: The institutional analysis and development framework. *Environmental Management* 24 (4):449-465.

Jahnke, M. ed. 2003. United Nations activities: Environmental policy decisions. *Environmental Policy and Law* 33 (1):2-13.

Juda, L. 1999. Considerations in developing a functional approach to the governance of large marine ecosystems. *Ocean Development & International Law* 30 (2):89-125.

Juda, L., and T. Hennessey. 2001. Governance profiles and the management of the uses of large marine ecosystems. *Ocean Development & International Law* 32 (1):43-69.

Kinzig, A. and 20 others. 2003. Coping with uncertainty: A call for a new science-policy forum. *Ambio* 32 (5):330-335.

Knauss, J.A. 1996. The Northeast Shelf Ecosystem: Stress, Mitigation, and Sustainability Symposium – Keynote Address. In Sherman, K., N.A. Jaworski and T.J. Smayda, eds. *The Northeast Shelf Ecosystem: Assessment, Sustainability, and Management.* Blackwell Science, Cambridge, Massachusetts, USA. 21-29.

Knecht, R.W. 1994. Essay: Emerging international goals and principles and their influence on national ocean governance. *Coastal Management* 22 (2):177-182.

Larkin, P.A., 1996. Concepts and Issues in Marine Ecosystem Management. *Reviews in Fish Biology and Fisheries,* 6 (2), pp. 139-164.

Link, J.S., J.K.T. Brodziak, S.F. Edwards, W.J. Overholtz, D. Mountain, J.W. Jossi, T.D. Smith, and M.J. Fogarty. 2002. Marine ecosystem assessment in a fisheries management context. *Canadian Journal of Fish and Aquatic Science* 59 (9):1429-1440.

Livingston, P. ed. 2001. Ecosystem Considerations for 2002. North Pacific Fishery Management Council, Anchorage, Alaska, November, 2001. Online Available, September 29th, 2003 @ http://www.fakr.noaa.gov/npfmc/ecosystem/ecobased.htm .

Livingston, P ed. 1999. Ecosystem Considerations for 2000. North Pacific Fishery Management Council, Anchorage, Alaska, November, 1999. Online Available, September 29th, 2003 @ http://www.fakr.noaa.gov/npfmc/ecosystem/ecobased.htm .

MacDonald, J. M. 1995. Appreciating the precautionary principle as an ethical evolution in ocean management. *Ocean Development & International Law* 26 (3):255-286.

Mackenzie, S.H. 1997. Toward integrated resource management: Lessons from the ecosystem approach from the Laurentian Great Lakes. *Environmental Management* 21 (2):173-183.

Mahlman, J.D. 1997. Uncertainties in projections of human-caused climate warming. *Science* 278 (5342):1416-1417.

Merrell, W.J., M.H. Katsouros, and J. Bienski. 2001. The Stratton Commission: The model for a sea change in national marine policy. *Oceanograph*, 14 (2):11-16.

Mountain, D.G. 2002. Potential consequences of climate change for the fish resources in the Mid-Atlantic region. *American Fisheries Society Symposium* 32:185-194.

Murawski, S.A. 2000. Definitions of overfishing from an ecosystem perspective. *ICES Journal of Marine Science* 57(3):649-658.

Nelson, B. 1969. Marine commission invokes NOAA, urges refitting of nation's ark. *Science* 163 (3864):263-265.

NOAA. 2004. New Priorities for the 21st Century: NOAA's Strategic Vision. USDOC, NOAA Strategic Planning, Silver Spring MD.

Odum, E.P. 1969. The srategy of ecosystem development. *Science* 164 (3877):262-270.

O'Toole, M. 2002. Benguela Current Large Marine Ecosystem Programme Strategic Action Programme. (brochure) updated November 2002. Windhoek, Namibia. 23p.

Probert, P.K. 2002. Ocean science and conservation – Catching the wave. *Aquatic Science: Marine and Freshwater Ecosystems* 12 (2):165-168.

Richards, L.J., and J-J. Maguire. 1998. Recent international agreements and the precautionary approach: New directions in fishery management science. *Canadian Journal of Fisheries and Aquatic Science* 55(6):1545-1552.

Rosenberg, A.A.. 2003. Multiple uses of marine ecosystems. In Sinclair, M. and G. Valdimarsson , eds. *Responsible Fisheries in the Marine Ecosystem.*189-196.

Rosenberg, A.A. 2002. The precautionary approach in application from a manager's perspective. *Bulletin of Marine Science* 70 (2):577-588.

Sainsbury, K and U.R. Sumaila. 2003. Incorporating ecosystem objectives into management of sustainable marine fisheries, including 'best practice' reference points and use of marine protected areas. In M. Sinclair and G. Valdimarsson, eds. *Responsible Fisheries in the Marine Environment*, FAO and CABI Publishing. 343-360.

Sainsbury, K.J., A.E. Punt, and A.D.M. Smith. 2000. Design of operational management strategies for achieving ecosystem objectives. *ICES Journal of Marine Science* 57(3):731-741.

Scavia, D. and 13 others. 2002. Climate change impacts on U.S. coastal and marine ecosystems. *Estuaries*, 25 (2):149-164.

Schramm, H.L., Jr., and W.A. Hubert. 1996. Ecosystem management: Implication for fisheries management. *Fisheries* 21:6-11.

Sherman, K. 2000. Why regional coastal monitoring for assessment of ecosystem health? *Ecosystem Health* 6 (3):205-216.

Sherman, K. 1995. Achieving regional cooperation in the management of marine ecosystems: The use of the large ecosystem approach. *Ocean & Coastal Management* 29 (1-3):165-185.

Sherman, K. 1994. Sustainability, biomass yields, and health of coastal ecosystems: An ecological perspective. *Marine Ecology Progress Series* 112:277-301.

Sherman, K. 1989. Biomass yields of large marine ecosystems. *Ocean Yearbook* 8, University of Chicago Press. 117-137.

Sherman, K. and H.R. Skjoldal, eds. 2002. *Large Marine Ecosystems of the North Atlantic: Changing States and Sustainability.* Elsevier Science, Amsterdam, Netherlands. 449p.

Sherman, K. and A.M. Duda. 2001. Toward ecosystem-based recovery of marine biomass yield. *Ambio* 30 (3):168-169.

Sherman, K. and A.M. Duda. 1999a. Large marine ecosystems: An emerging paradigm for fishery sustainability. *Fisheries.* 24 (12):15-26.

Sherman, K. and A.M. Duda. 1999b. An ecosystem approach to global assessment and management of coastal waters. *Marine Ecology Progress Series* 190:271-287.

Sherman, K., N.A. Jaworski, and T.J. Smayda, eds. 1996. *The Northeast Shelf Ecosystem: Assessment, Sustainability, and Management.* Blackwell Science. 564 pp.

Sherman, K., L. M. Alexander, and B. D. Gold, eds. 1991. *Food Chains, Yields, Models, and Management of Large Marine Ecosystems.* (AAAS) Westview Press, Boulder, Colorado. 320p.

Sinclair, M., and G. Valdimarsson, eds. 2003. Towards ecosystem-based fisheries management. In *Responsible Fisheries in the Marine Ecosystem.* FAO and CABI Publishing, 393-403.

Sissenwine M.P., P.M. Mace. 2003. Governance for responsible fisheries: An ecosystem approach. In Sinclair, M. and G. Valdimarsson, eds. *Responsible Fisheries in the Marine Environment,* FAO and CABI Publishing. 363-390.

Slocombe, D.S. 1998. Defining goals and criteria for ecosystem-based management. *Environmental Management* 22(4):483-493.

Steele, J.H. 1998. Regime shifts in marine ecosystems. *Ecological Applications* 8(1) Supplement: 33-36.

Ukwe, C.N., C.A. Ibe, B.I. Alo, and K.K. Yumkella. 2003. Achieving a paradigm shift in environmental and living resources management in the Gulf of Guinea: The large marine ecosystem approach. *Marine Pollution Bulletin* 47 (1-6):219-225.

U.N. General Assembly, (UNGA) 2002. Verbatim Records of the Plenary Meetings of the General Assembly, Fifty-Seventh Session, 71st Plenary Meeting, 10 December 2002, A/57/PV.71 & the 74th Plenary Meeting, 12 December 2002, A/57/PV.74. Online Available, November 10th, 2003 @ http://www.un.org/ga/57/pv.html .

U.N. General Assembly, (UNGA) 2003. Resolution(s) Adopted by the General Assembly, Fifty-Seventh Session. A/RES/57/141 Oceans and the Law of the Sea, & A/57/142 Large-scale Pelagic Drift-net Fishing, Unauthorized Fishing in Zones of National Jurisdiction and on the High Seas/Illegal, Unreported and Unregulated Fishing, Fisheries By-catch and Discards, and Other Developments, & A/RES/57/143 Agreement for the Implementation of the Provisions of UNCLOS of 10 December 1982 Relating to the Conservation and Management of Straddling Fish Stocks and Highly Migratory Fish Stocks, February, 2003. Online Available, November 10th, 2003 @ http://www.un.org/ga/57/ .

U.N. General Assembly, (UNGA) 2004. Fifty-Eighth Session, Agenda Item 52(b), Resolution 58/14, Sustainable Fisheries... and Related Instruments. Document A/RES/58/14 (A/58/L.18), January 21st 2004, Online Available February 9th, 2004 @ http://www.un.org/58 .

Wagner, L., ed. 2001. Summary of the Reykjavik Conference on Responsible Fisheries in the Marine Ecosystem. *Sustainable Developments* 61 (1). 11p.

Watkins, J.D. 2004. Sustaining our oceans: A public resource, a public trust. *Oceanography*, 17 (1):102-106.

Watkins, J. D. 2002. U.S. Commission on Ocean Policy. *Oceanography* 15 (4):4-6.

West, M.B. 2003. Improving science applications to coastal management. *Marine Policy* 27 (4):291-293.

Witherell, D., C. Pautzke and D. Fluharty. 2000. An ecosystem-based approach for Alaska groundfish fisheries. *ICES Journal of Marine Science*, 57 (3):771-777.

Witherell, D. 1999. Incorporating ecosystem considerations into management of Bering Sea groundfish fisheries. In *Ecosystem Approaches for Fisheries Management*. Alaska Sea Grant College Program, AK-SG-99-01, Fairbanks. 315-327.

Witherell, D. ed. 1998. Ecosystem Considerations for 1999. North Pacific Fishery Management Council, Anchorage, Alaska, November, 1998. Online Available, September 29th, 2003 @ http://www.fakr.noaa.gov/npfmc/ecosystem/ecobased.htm .

Witherell, D. and 11 others. 1997. Ecosystem Considerations for 1998. North Pacific Fishery Management Council, Anchorage, Alaska, November, 1997. Online Available, September 29th, 2003 @ http://www.fakr.noaa.gov/npfmc/ecosystem/ecobased.htm .

Large Marine Ecosystems, Vol. 13
T.M. Hennessey and J.G. Sutinen (Editors)
© 2005 Elsevier B.V. All rights reserved.

14

Applications of the Large Marine Ecosystem Approach Toward World Summit Targets

Alfred Duda and Kenneth Sherman

WORLD SUMMIT TARGETS FOR THE COASTAL OCEAN

Continued over-fishing in the face of scientific warnings, fishing down food webs, destruction of habitat, and accelerated pollution loading – especially nitrogen export – and the resulting significant degradation to coastal and marine ecosystems of both rich and poor nations, have given rise to the emergence of an ecosystem-based approach to the assessment and management of marine resources and environments. Fragmentation among institutions, international agencies, and disciplines, lack of cooperation among nations sharing marine ecosystems, and weak national policies, legislation, and enforcement all contribute to the recognition of a new imperative for adopting ecosystem-based approaches to managing human activities in these systems in order to avoid serious social and economic disruption. The Global Environment Facility (GEF) has been approached by developing countries in growing numbers for assistance in securing the futures of their shared Large Marine Ecosystems (LMEs). GEF supported processes are now being used to assist those countries in adopting a science-driven, ecosystem-based approach to the management of human activities affecting coastal and marine ecosystems and linked freshwater basins. At risk are renewable goods and services valued at $10.6 trillion per year. In recognition of threats to coastal ocean sustainability, the 2002 Johannesburg World Summit on Sustainable Development participating heads of state reached agreement for implementing a Plan of Implementation (POI) on several specific ecosystem-related targets. The targets for improving conditions in coastal ocean water included a POI for achievement of "substantial" reductions in land-based sources of pollution by 2006; introduction of the ecosystems approach to marine resource assessment and management by 2010; designation of a network of marine protected areas by 2012; and the maintenance and restoration of fish stocks to maximum sustainable yield (MSY) levels by 2015.

Earlier international agreements fell short of attaining the goals of the 1992 UNCED process for oceans, in part because they were designed around sectoral themes such as pollution; sewage; waste disposal; fisheries; biodiversity; or global climate change that fail to link international and local problems in a cross-sectoral strategic approach applicable for the particular priorities of that LME and its coastal area. They remain thematic and have encouraged narrowly focused institutions to develop. To bridge this gap, the GEF, its UN partner agencies, and other organizations including IUCN, IOC of UNESCO and NOAA, have joined together to support developing countries committed to the introduction of ecosystem-based projects to assess and manage marine resources in an emerging global movement toward the WSSD targets.

Developing country officials responsible for coastal and marine resources have understood the ramifications of the declining status of their marine ecosystems and the link to land-based activities that has been so difficult to foster. Across Africa, Asia and the Pacific, Latin America and the Caribbean, and in Eastern Europe, country officials have been cooperating with the GEF to reverse the decline of their marine ecosystems, testing methods for restoring once abundant biomass in order to sustain growing populations of coastal communities and to conserve highly fluctuating systems to ensure continued benefits for future generations. Since the early 1990s, these nations have approached the GEF, its implementing agencies, and other executing agencies like the UN Industrial Development Organization (UNIDO) for assistance in restoring and protecting sustainable use of their LMEs.

GLOBAL SCOPE OF GEF-LME PROJECTS

Table 14-1 lists the LME projects that have been approved by the GEF or are under preparation with GEF funding. The approved GEF-LME projects include developing nations or those in economic transition as well as other OECD countries, since the living resources, the pollution loading, or the critical habitats have transboundary implications across rich and poor nations alike. A total of $650 million in project costs from the North and South is currently being invested as of May 2004 in 10 LME projects in 72 countries with $225 million in GEF grant finance. An additional 7 LME projects are under preparation involving 54 different nations. Involved with these GEF LME projects are 121 different countries. With OECD countries involved that share LMEs with the GEF recipient nations, expectations are that reforms will take place in both the North and the South in order to operationalize this ecosystem-based approach to managing human activities in the different economic sectors that contribute to place-specific degradation of the LME and adjacent waters. Each of the GEF-LME projects assists participating countries in moving toward the World Summit targets with the implementation of their ecosystem-based projects having objectives consistent with the Summit POI for ocean coastal waters.

The significant annual global biomass yields of marine fisheries from ecosystems in the GEF-LME Network of 44.8% provides a firm basis for moving toward the Summit goal for introducing an ecosystem-based assessment and management approach to global fisheries by 2010, and fishing MSY levels by 2015. Even now there is an international instrument supported by most coastal nations that could have immediate applicability to reaching Summit fishery goals. The FAO Code of Conduct for Responsible Fishery practice of 2002 argues for moving forward with a "precautionary approach" to fisheries sustainability, given a situation wherein available information can be used to recommend a more conservative approach to fish and fisheries total allowable catch levels (TAC) than has been the general practice over the past several decades. Based on the decadal profile of LME biomass yields from 1990 to 1999 (Garibaldi and Limongelli 2003), it appears that the yields of total biomass and the biomass of 11 species groups of 6 LMEs have been relatively stable or have shown marginal increases over the decade (Sherman 2003). The yield of marine biomass for these 6 LMEs was 8.1 mmt, or 9.5 percent of the global marine fisheries yield in 1999 (Sherman, this volume).

Table 14-1. Countries Participating in GEF/Large Marine Ecosystem Projects

Approved GEF Projects	
LME	Countries
Gulf of Guinea (6)............................	Benin, Cameroon, Côte d'Ivoire, Ghana, Nigeria, Togo
Yellow Sea (2)...................................	China, Korea
Patagonia Shelf/Maritime Front (2)..............	Argentina, Uruguay
Baltic (9)...	Denmark, Estonia, Finland, Germany, Latvia, Lithuania, Poland, Russia, Sweden
Benguela Current (3)............................	Angola, Namibia, South Africa
South China Sea (7)..............................	Cambodia, China*, Indonesia, Malaysia, Philippines, Thailand, Vietnam
Black Sea (6).......................................	Bulgaria, Georgia, Romania, Russia*, Turkey, Ukraine
Mediterranean (19)................................	Albania, Algeria, Bosnia-Herzegovina, Croatia, Egypt, France, Greece, Israel, Italy, Lebanon, Libya, Morocco, Slovenia, Spain, Syria, Tunisia, Turkey*, Yugoslavia, Portugal
Red Sea (7)...	Djibouti, Egypt* Jordan, Saudi Arabia, Somalia, Sudan, Yemen
Western Pacific Warm Water Pool-SIDS[a] (13)...	Cook Islands, Micronesia, Fiji, Kiribati, Marshall Islands, Nauru, Niue, Papua New Guinea, Samoa, Solomon Islands, Tonga, Tuvalu, Vanuatu

Total number of countries: 70*

GEF Projects in the Preparation Stage

Canary Current (7)..	Cape Verde, Gambia, Guinea, Guinea-Bissau, Mauritania, Morocco*, Senegal
Bay of Bengal (8).................................	Bangladesh, India, Indonesia*, Malaysia*, Maldives, Myanmar, Sri Lanka, Thailand*
Humboldt Current (2).............................	Chile, Peru
Guinea Current (16)..............................	Angola*, Benin*, Cameroon*, Congo, Democratic Republic of the Congo, Côte d'Ivoire*, Gabon, Ghana*, Equatorial Guinea, Guinea*, Guinea-Bissau*, Liberia, Nigeria*, São Tomé and Principe, Sierra Leone, Togo*
Gulf of Mexico (3)................................	Cuba, Mexico, United States
Agulhas/Somali Currents (8).....................	Comoros, Kenya, Madagascar, Mauritius, Mozambique, Seychelles, South Africa*, Tanzania
Caribbean LME (23)..............................	Antigua and Barbuda, The Bahamas, Barbados, Belize, Colombia, Costa Rica, Cuba*, Grenada, Dominica, Dominican Republic, Guatemala, Haiti, Honduras, Jamaica, Mexico*, Nicaragua, Panama, Puerto Rico[b], Saint Kitts and Nevis, Saint Lucia, Saint Vincent and the Grenadines, Trinidad and Tobago, Venezuela

Total number of countries: 51*

*Adjusted for multiple listings

[a] Provisionally classified as Insular Pacific Provinces in the global hierarchy of LMEs and Pacific Biomes (Watson *et al.* 2003).
[b] A self-governing commonwealth in union with the United States

The countries bordering these six LMEs—Arabian Sea, Bay of Bengal, Indonesian Sea, Northeast Brazil Shelf, Mediterranean Sea and the Sulu-Celebes Sea—are among the world's most populous, representing approximately one-quarter of the total human population. These LME border countries are increasingly dependent on marine fisheries for food security and for national and international trade. Given the risks of "fishing-down-the-food-chain," (Pauly and Christensen 1995; Pauly *et al.* 1998) it would appear opportune for the stewardship agencies responsible for the fisheries of the bordering countries to consider options for mandating precautionary total allowable catch levels during a period of relative biomass stability.

Evidence for species recovery following a significant reduction in fishing effort through mandated actions is encouraging. Following management actions to reduce fishing effort, the robust condition of the U.S. Northeast Shelf ecosystem with regard to the average annual level of primary productivity ($350gCm^2.yr$), stable annual average levels of zooplankton (33 cc/100m^3), and a relatively stable oceanographic regime (Sherman *et al.* 2002), contributed to: (1) a relatively rapid recovery of depleted herring and mackerel stocks, with the cessation of foreign fisheries in the mid-1970s; and (2) initiation of the recovery of depleted yellowtail flounder and haddock stocks following a mandated 1994 reduction in fishing effort (Sherman *et al.* 2002).

Three LMEs remain at high risk for fisheries biomass recovery expressed as a pre-1960s ratio of demersal to pelagic species—the Gulf of Thailand, East China Sea and Yellow Sea (Sherman 2003). However, mitigation actions have been initiated by the People's Republic of China toward recovery by mandating 60 to 90 day closures to fishing in the Yellow Sea and East China Sea during summer months (Tang 2003). The country-driven planning and implementation documents supporting the ecosystem approach to LME assessment and management practices can be found at www.iwlearn.org.

FEATURES AND EARLY RESULTS OF LME PROJECTS

Through the GEF LME projects, countries are testing methods to demonstrate how integrated management of oceans, coasts, estuaries, and freshwater basins can be implemented though an ecosystem-based approach. See Figure 1-3 this volume for a description of the TDA and SAP processes of the LME approach. It is noteworthy that non-recipient OECD countries also share these LMEs or are located in contributing basins such as Germany and Austria in the Danube Basin draining to the Black Sea. Emphasizing the global situation in which both the developed and developing nations must cooperate in order to reverse the continuing degradation of coastal and marine ecosystems, a total of 18 non-recipient, developed States are collaborating with the GEF recipient States in those LME projects on the particular high priority concerns relevant for each water body—some focused on the depletion of fisheries, others on habitat restoration and protection, and still others on the reduction of pollution from land-based sources.

Danube/Black Sea Basin LME

An example where the Global Program of Action for reduction of land-based sources of pollution (GPA) concerns prevail is the case of accelerated eutrophication of the Danube Delta and Black Sea LME from excessive levels of nitrogen loading. A series of small GEF projects for the Danube and Dnipro River basins and the states of the Black Sea LME since the early 1990s have been programmed to focus on reducing nitrogen loadings from the 17 contributing nations. Following successful completion of the TDA and SAP processes in the mid and late 1990s for the Black Sea LME (Black Sea Environment Programme 1996), and the Danube Basin (Environmental Programme for the Danube River Basin 1999), political commitments were achieved for nutrient reduction and abatement of persistent toxic substances being released from hotspots.   Reforms in policy, laws, institutions, and investments are now being supported by GEF in each country for nitrogen abatement from the agriculture, municipal, and industrial sectors. Billions of dollars of water quality investments are being mobilized through EU accession, agriculture pollution is being reduced, and wetlands are being restored in the upstream basins to serve as nutrient sinks to protect the LME.

A GEF Strategic Partnership is in place for 2001-2006 with all 3 GEF implementing agencies (UNEP, UNDP, World Bank) to assist the ·17 collaborating nations. Through the GEF recommended strategic processes, political commitments have been agreed to among the states (including nutrient reduction action by the Danube basin states of Austria and Germany, supported by national funding). The Partnership among the 3 GEF agencies, donors, and the 17 States is now bringing coordinated support and benefits to the transboundary basin and its linked marine environment under the Bucharest Convention and the Istanbul Convention and has fostered an adaptive management approach.   Community and NGO participation is fostered with extensive small grants programs for mobilizing support for hotspot cleanup. GPA Protocols to the conventions are to be adopted, codifying country commitments to action, and a fisheries convention is to be negotiated by the 6 Black Sea states to adopt an ecosystem-based management approach.   This is the largest GEF international waters initiative of its kind and is intended to serve as a test of whether a more comprehensive level of participation by GEF and streamlined sub-project approvals can leverage significant environmental improvements for a large LME and its drainage basin.

Red Sea LME

The Red Sea and Gulf of Aden LME represents another example of all three GEF implementing agencies working together to assist the collaborating States in a modest, catalytic project with GEF finance being just a small part of a much larger effort in different economic sectors funded through other sources that help protect the unique coral reefs of the sea.   Formulation of the Red Sea SAP (Regional Organization for the Conservation of the Environment of the Red Sea and Gulf of Aden 1998) was initiated in 1995 and was the first one completed under the GEF Operational Strategy in 1997. The processes of formulating the TDA and the SAP played an important role in uniting the countries under their previously adopted regional seas convention, the Jeddah Convention.   The SAP identifies actions needed to protect the uniquely fragile coral reefs, sea grass beds, and mangroves of the Red Sea coast. An array of actions supported includes development and implementation of Integrated Coastal Management (ICM) plans for specific coastal areas and the development of Marine Protected Areas (MPAs).

The Red Sea project was programmed with a complementary GEF international waters project for the pollution hotspot of the Gulf of Aqaba in Jordan to accompany World Bank assistance.   The reefs in the Gulf are the northern most warm water-type coral reefs on Earth and the 17 kilometer marine protected area, shared by Israel and Jordan, serves as an example of how developed and developing countries may work together jointly to sustain their valuable coastal and marine resources.   The marine park serves as a haven for fish and contributes to the repopulation of other areas subject to exploitation.   The use of MPAs is an essential management component of LMEs in order to conserve biomass and biodiversity. The project also assisted Jordan to develop a modern environmental management institution as part of its economic development processes in areas to protect the sensitive reefs from excesses of tourism, pollution discharges, and industrial development. The institution is now more stringent in its requirements to protect the marine ecosystem than the rest of the country.

Western Pacific Warm Pool Marine Ecosystem

While not strictly an LME, the Western Pacific Marine Ecosystem is the life blood of Pacific SIDS economies with its rich tuna fisheries. Its island archipelagos represent an agglomeration of a number of LMEs. Heads of States of the 13 PACSIDS adopted their GEF SAP (South Pacific Regional Environment Programme 1997) in September 1997 and began implementation of their GEF/UNDP international waters project thereafter. While a number of components were involved including community-based fisheries management, ICM, and interventions addressing their water supplies, an important component included GEF support to the countries through the Forum Fisheries Agency, which included the establishment of a regional convention on conservation, management, and sustainable use of their highly migratory fish stocks. A commission is being established to oversee a more ecosystem-based approach to management, known as the "Convention on the Conservation and Management of Highly Migratory Fish Stocks of the Western and Central Pacific Ocean." The GEF assistance helped level the playing field among the Pacific SIDS as they negotiated the Convention with Asian, North American and European nations. Following 7 sessions of what was known as the MHLC process (Sydnes 2001a and 2001b), the Convention was signed in September 2000 and is the first agreement to be successfully negotiated on the basis of the 1995 UN Fish Stocks Agreement.

Mediterranean LME

In the Mediterranean project, GEF assistance resulted in a SAP for land-based sources of marine pollution being adopted by all 20 nations under their Barcelona Convention (Mediterranean Action Plan 1999), with enforceable commitments to action on pollution reduction for specific pollutants with specific timetables and targets—the first such commitments to action in the program's 20 year history. GEF played a catalytic role in its transition from a research focus to an on-the-ground implementation focus. The 8 non-recipient nations must also adhere to the pollution reduction timetables as the SAP process operationalized their GPA Protocol under their regional seas convention and expanded the collaboration from just the saltwater to the basins draining to the sea. UNEP and the World Bank are assisting the Mediterranean countries according to their comparative advantages. UNEP is assisting in the more controversial processes of developing a TDA and SAP for living resources and their critical habitats, processes that will take a number of years to complete in conjunction with the review of the EU Common Fisheries Policy. The World Bank is assisting with feasibility studies for high priority bankable investments that will help the states implement their Mediterranean SAP for land-based sources of pollution.

South China Sea LME

The South China Sea project with UNEP has been programmed in conjunction with two other GEF international waters projects to fit programmatically in the attempt to restore and protect the globally significant coral reefs, sea grass beds, mangroves, and wetlands of the LME and its coast. The Mekong Basin project with its valuable delta receives GEF assistance through the World Bank, while the complementary hotspot remediation demonstration activities conducted through the GEF/UNDP/IMO program entitled Partnerships for Environmental Management of the Seas of East Asia (PEMSEA) are also an integral part of GEF's programmatic approach. While the South China Sea project undertakes collective strategic processes for developing a more ecosystem-based approach to management through production of a TDA (East Asian Seas Coordinating Unit 2000) and SAP, PEMSEA has supported a number of complementary local demonstrations of ICM since 1996 that are well-known throughout the ICM community (Thia-Eng 1998).

Of global policy significance has been the GEF/PEMSEA assistance to the Government of the Philippines as it developed the Manila Bay Declaration and Manila Bay Coastal Strategy in its part of the South China Sea. This complementary initiative is multi-jurisdictional in nature, with respective national governments, provinces in the drainage area, and the large municipalities of Manila. It represents a GPA-equivalent of an SAP for the contributing freshwater basin that is enacted in the framework of coastal sustainable development. The political declarations have been adopted at the highest level and represent a decade-long commitment to action.

Patagonia Shelf LME

Two international waters projects cover the Patagonia Shelf LME in Uruguay and Argentina. The Plata Maritime Front area is subject to management under a commission and a bilateral treaty. The remainder of the LME is in Argentina and suffers from land-based pollution from hotspots as well as extreme amounts of over-fishing recently brought about through agreements with the EU and Asian distant fleets. As noted by UNEP (2001), depletion of the ecosystem as a result of trade distortions and EU subsidies was rapid with the fishery lasting but 10 years with modern equipment of the EU and Asian fleets. UNDP is assisting the countries with the highly polluted and over-fished Maritime Front. The World Bank is assisting Argentina with two loans (one for land-based pollution abatement and another related to reforms in the fishery sector), to which GEF has added an incremental amount of grant funding toward restoration and protection of the marine biodiversity. The projects are under implementation.

COMPREHENSIVE LME DEMONSTRATION PROJECTS AND PROJECT MODULES

Four of the LME project areas involve testing comprehensive attempts at resolving complex and interlinked ecosystem problems: the Guinea Current, the Benguela Current, the Yellow Sea, and the Baltic Sea LMEs. A five-module approach to the assessment and management of LMEs has been proven to be useful in other LMEs and is being applied in these four areas to test its utility. The approach is customized to fit the situation within the context of the TDA and SAP processes for the groups of nations sharing the particular LME based on available information and capacity. These processes are critical to integrate science into management in a practical way and to establish governance regimes appropriate for the particular situation. The five modules (productivity, fish/fisheries, pollution/ecosystem health, socio-economics, and governance) are in the process of being adapted to four of the Comprehensive LME Demonstration projects. The first four models support the TDA process while the governance module is associated with a periodic updating of the Strategic Action Program or SAP. Adaptive management regimes are encouraged through periodic assessment processes (TDA updates) and updating of SAPs as gaps are filled.

Gulf of Guinea Pilot Project

The GEF is supporting coastal nations all along the western coast of Africa in the establishment of ecosystem-based assessment and management of their coastal environments and resources. Included among the projects was the pilot phase of the Guinea Current LME project from 1995 to 1999. The six participating countries—Benin, Cameroon, Ghana, Ivory Coast, Nigeria, and Togo—have used the GEF Grant to strengthen national infrastructure in staffing positions and engaging government support. The long-term objective of the project is to restore and sustain the health of the Guinea Current Large Marine Ecosystem and its living resources, particularly with regard to biological diversity, coastal habitats, and the control of water pollution.

Project participants included: networks of national environmental protection agencies and departments, public health administrations, sewage work authorities, industries, and universities/research institutions in the participating countries. Non-governmental organizations (NGOs) and community based organizations (CBOs) have been very active particularly as they related to public awareness and environmental education aspects. In order to provide the necessary focus, national focal point agencies and national focal point institutions were designated. National and regional experts were designated to support the monitoring and assessment module of the project at the national and regional level. The capacity of national institutes and experts was reinforced through the supply of appropriate equipment and by a series of workshops aimed at standardizing methodological approaches in the aforementioned components. Activity groups on specific topics (productivity, fish and fisheries, pollution monitoring, socioeconomics, and governance) were convened regularly to discuss the progress made and problems encountered, and to undertake joint assessments.

At the international level, UNDP served as the implementing agency, UNIDO as the executing agency and UNEP as a co-operating organization. The United States Department of Commerce through its National Oceanic and Atmospheric Administration (NOAA) provided technical support particularly in capacity building initiatives in addition to in-kind contribution to the funding of the project. Other United Nations and non-United Nations Agencies such as the Intergovernmental Oceanographic Commission (IOC 2000) or UNESCO, IMO, FAO, and IUCN provided guidance at specific stages in project implementation.

Actions included joint identification of major transboundary environmental and living resources management issues and problems, and adoption of a common regional approach, in terms of strategies and policies for addressing these priorities in the national planning process at all levels of administration, including local governments. Among the successfully completed activities is a cooperative survey of the bottom fish stocks using a chartered Nigerian vessel with representatives of each of the participating countries taking part in the trawling and data reporting operations. Funds were used to complete a report on the major multidecadal shifts in the abundance of fish stocks in the ecosystem, caused principally by environmental perturbations affecting the annual upwelling cycle and temperature regime of the ecosystem.   In addition to the cooperative trawl survey, surveys of the plankton community to address the carrying capacity of the Gulf of Guinea for supporting sustainable fisheries were conducted at six-week intervals using plankton recorder systems deployed from large container vessels transiting the region.   The samples are being processed in a plankton center established with GEF funds in Tema, Ghana in collaboration with the Sir Alister Hardy Foundation of the U.K.

Forty region-wide workshops attended by nearly 900 participants were held on the key transboundary concerns, including: pollution monitoring, ecosystem productivity studies; natural resource management and planning;   development of institutional capacities (including administrative and legal structures); and data and information management and exchange.   The pilot project for the Guinea Current LME established intra and international networks of scientific institutions and non-governmental organizations, with a total of more than 500 scientists, policy makers and other participants (making it the African continent's single largest network for marine and coastal area management), to undertake studies on ecosystem degradation, to assess living resource availability and biodiversity, and to measure socioeconomic impacts of actions and non-actions.   The capacity of the networks has been reinforced through the supply of appropriate equipment and by a series of group training workshops aimed at standardizing methodological approaches around five project modules: (1) productivity;   (2) fish and fisheries;   (3) pollution and ecosystem health;   (4) socioeconomics; and (5) governance.

Restoration of mangrove areas has been initiated along the coast; assessments of the principal sources of coastal pollution have been initiated.   National State of the Marine Environment Reports were issued as "initial assessments," encompassing published and unpublished data and including policy options and past interventions.   Plans for the management of transboundary coastal resources were completed by each of the countries (Adam 1998, Ibe 1998, Ibe et al. 1998, Mondjanagni et al. 1998 (2 citations), CEDA 1997).   Several studies have suggested options for increasing the long-term sustainability of coastal resources and increasing socioeconomic benefits to the people of the region (Ibe et al. Proceedings 1998). A detailed assessment of the nature and quantities of urban wastes and sewage and the present status of their management was completed.   With due recognition of the ongoing government efforts with the World Bank to implement master plans for urban wastes and sewage management, the project focused with municipal authorities on novel and low technology options, such as the use of settling pits in Ghana for sewage treatment in small communities and the sorting of domestic wastes prior to disposal as a means of encouraging recycling and reuse.   In addition, a parallel effort was made to develop strategies and policies to encourage reduction, recycling, recovery and reuse of industrial wastes.   One such initiative, now at the pilot stage in Ghana, is the establishment of the waste stock exchange management system. This concept, which has been enthusiastically embraced by manufacturing industries in Ghana and has as a slogan "one person's waste, another person's raw material," holds considerable promise as a cost effective waste management tool.

An Accra Declaration has been signed by the Environmental Ministers from each of the six countries indicating joint commitment for taking steps to promote the long-term sustainability of the Gulf of Guinea resources (Ibe 1998 Newsletter).   The ministers of the environment, fisheries, tourism, energy, mining, and finance of the six countries engaged in the first phase of the Guinea Current project agreed with counterpart ministers of ten neighboring countries along the coastal margins of the ecosystem, to extend the project in phase two from Guinea Bissau on the northwest part of the coast to Angola in the southwest. Phase two is presently focused on development of an expanded transboundary diagnostic analysis (TDA) and strategic action plan (SAP), in collaboration with the GEF, United Nations Industrial Development Organization (UNIDO), the U.N. Development Programme, NOAA, and IUCN.

Benguela Current LME Project

The GEF is supporting an ecosystem-based project requested by the governments of Angola, Namibia and South Africa for the "Integrated Management, Sustainable Development, and Protection of the Benguela Current Large Marine Ecosystem (BCLME)." The project is focused on sustainable management and utilization of living marine resources, mining and environmental variability, ecosystem forecasting, management of pollution, ecosystem health and protection of biological diversity, and capacity strengthening. Within an overall ecosystem approach, specific actions have been agreed upon through a series of meetings between stakeholders and government representatives. During a 12-month planning period, the three countries reached consensus on a strategic approach for the project, based on the preparation of a Transboundary Diagnostic Analysis (TDA) and a Strategic Action Plan (SAP). With regard to the fish and fisheries of the BCLME, the countries agreed to establish a regional structure to: (1) conduct transboundary fish stock and ecosystem assessments; (2) evaluate transboundary resource and environmental linkages; and (3) provide advice to the three governments based on the assessment results. They agreed to conduct joint surveys and assessments of shared fish stocks over a five-year period beginning in 2002 as a demonstration of the benefits to each of the countries of joint assessments for compiling baseline data and validating survey and assessment methodology.

The countries are establishing an Interim Benguela Current Commission (IBCC) to strengthen regional cooperation. The IBCC is to be supported by a project coordinating unit and by advisory groups. Within a period of five years it is expected that the IBCC will become a fully functioning Benguela Current Commission (BCC) with a supporting secretariat. The BCC is to serve as the organization for harmonizing technical issues including fishing gear, mesh size and type, data compatibility, and assessment methodology. Cooperative assessments of non-exploited species will also be made. Effort will be directed by the BCC to develop a viable mariculture policy for the three countries. Cooperative analyses of the socioeconomic consequences of harvesting methods will be undertaken by the IBCC, with a view to appropriate intervention within the framework of improving sustainable use of the BCLME resources, and in compliance with the FAO Code of Conduct for Responsible Fishing. In addition to fisheries, the IBCC will develop a regional framework for enhancing consultations, for the purpose of mitigating the negative impacts of marine mining particularly with regard to any potential or actual conflicts among fisheries and coastal and offshore diamond/gold mining and oil and gas exploration and/or production.

Among the principles adopted by the IBCC are: (1) the concept of sustainable development shall be used in a way that does not destroy the integrity of the BCLME ecosystem, or otherwise foreclose on options for use and enjoyment by future generations; (2) the precautionary principle where appropriate, shall be applied, preventative measures being taken when there are reasonable grounds for concern that an activity may increase the potential hazards to human health, living marine resources or marine ecosystems, may damage amenities, or interfere with other legitimate uses of the sea, even when there is no conclusive evidence of a causal relationship between the activity and the effects and by virtue of which greater caution is required when information is uncertain, unreliable or inadequate; and (3) the use of economic and policy instruments that foster sustainable development shall be promoted through, *inter alia*, the implementation of economic incentives for introducing environmentally friendly technologies, activities and practices; the phasing-out of subsidies which encourage the continuation of non-environmentally friendly technologies, activities and practices; the introduction of user fees and the 'polluter pays' principle. Environmental, ecosystem, and human health considerations shall be included into all relevant policies and sectoral plans, especially those concerning marine industrial development, fisheries, mariculture and marine transport.

The structure of the Interim Benguela Current Commission (IBCC) and terms of reference of the advisory groups to the Commission for fisheries, environment, pollution, legal affairs, and data exchange have been approved at the ministerial level in the participating countries.

The Yellow Sea LME Project

Notable progress has been made in the introduction of the ecosystem-based management and assessment activities for the Yellow Sea LME (YSLME), by ministerial representatives of China and South Korea serving together in a joint steering committee for a GEF-sponsored International Waters project. The project is being carried out in collaboration with the UNDP and other international partners including NOAA and IUCN. The Yellow Sea LME is an important global resource. This international water-body supports substantial populations of fish, invertebrates, marine mammals, and seabirds. Many of these resources are threatened by both land and sea-based sources of pollution and habitat loss resulting from extensive economic development in the coastal zone, and by the unsustainable exploitation of natural resources (primarily overfishing). Additionally, there is significant international shipping traffic through the waters of the Yellow Sea, with associated threats from spills and collisions with marine mammals.

In the western Yellow Sea, pollution sources include wastewater from Qingdao, Dalian, and Lianyungang port cities; oil discharged from vessels and ports; and oil and oily mixtures from oil exploration. More than 100 million tons of domestic sewage and about 530 million tons of industrial wastewater from coastal urban and rural areas are discharged into the nearshore areas of the Yellow Sea each year. The major pollutants carried by sewage and waste water are oils, mercury, cadmium, lead, COD, and inorganic nitrogen.

The eastern Yellow Sea has significant pollution in the shallow inlets of its southern coastline where the many islands prevent mixing with open ocean water and red tides persist. Demersal species used to be the major component of the resources and accounted for 65 to 90 percent of annual total catch. The resource populations of demersal species such as small yellow croaker, hairtail, large yellow croaker, flatfish, and cod declined in biomass by more than 40 percent when fishing effort increased threefold from the early 1960s to the early 1980s. Shifts in species dominance and biodiversity in the Yellow Sea are significant. The dominant species in the 1950s and early 1960s were small yellow croaker and hairtail, while

Pacific herring and chub mackerel became dominant during the 1970s. Some smaller-bodied, fast-growing, short-lived, and low-value fish (*e.g.*, *Setipinna taty*, anchovy, scaled sardine) increased markedly in about 1980 and have taken a prominent position in the ecosystem resources thereafter. As a result, some larger-sized and higher trophic level species were replaced by smaller-bodied and lower trophic level species, and the resources in the Yellow Sea declined in quality. About 70 percent of the biomass in 1985 consisted of fish and invertebrates smaller than 20 cm, and the mean body length in the catches of all commercial species was only 12 cm while the mean body length in the 1950s and 1960s exceeded 20 cm. The trophic levels in 1985 and in the 1950s were estimated to be 3.2 and 3.8, respectively. Thus it appears that the external stress of fishing has affected the trophic structure of the Yellow Sea ecosystem.

Aquaculture is a major use of the coastal waters of the Yellow Sea. Mariculture is commonly practiced in all coastal provinces of China, and it is most advanced in Shandong and Liaoning provinces. The total yield of invertebrate mariculture of ROK in 1997 was 301,873 metric tons (MT) representing 29.7 percent of ROK's total mariculture production (1,015,134 MT), including 200,973 MT of oysters (20 percent) and 63,572 MT of mussels (6.3 percent).

Offshore oil exploration has been successful in the Chinese and North Korean (DPRK) portions of the Yellow Sea. In addition, the sea has become more important with the growth in trade among its bordering nations. The main Chinese ports are Shanghai, Lu-ta, Tientsin, Qingdao, and Chin-huang-tao; the main South Korean (ROK) port is Inchon, the outport of Seoul; and that for the DPRK is Nampo, the outport for P'yongyang. Tourism is an industry in its infancy in both China and Korea. Several sites of picturesque beauty around the coastlines of these countries could be promoted as tourist attractions. As access to China and Korea becomes easier for foreign visitors, the tourist industry will expand. The granite mountains of the western Liaoning coast in China and the islands and swimming beaches of ROK, in particular Cheju Island, will be in even greater demand.

The Yellow Sea is an international water-body and many of its problems can be solved only through international cooperation. The management of the Yellow Sea is especially complicated in that it is surrounded by nations that share some aspects of their historical and cultural background, but differ in internal political systems, external political and economic alignment, and levels of economic development. For the future of the Yellow Sea, it is thus imperative for the coastal nations to realize the importance of regional cooperation. There are currently several agreements for bilateral regulation or development of the Yellow Sea and East China Seas, but none of them are binding on all the coastal nations; nor is any nation a party to all the agreements. Of global policy significance has been the GEF/PEMSA assistance to the Government of China as it developed the Bohai Sea Declaration for the internal sea connected to the Yellow Sea. This initiative is multi-jurisdictional in nature and involves the national government, provinces in the drainage area, and the large downstream municipality of Tianjin. It represents the national equivalent of strategic action program enacted in the framework of coastal sustainable development. The political declaration represents a decade-long commitment to on-the-ground action that will total billions of dollars of investments and policy/legal/institutional reforms to reduce coastal degradation. Such commitments are unprecedented in GEF-recipient countries, and they are quite similar to the United States' Chesapeake Bay Basin cleanup program that has been at work for two decades in coastal restoration.

The principal activities to be operationalized within the framework of the YSLME project are listed below. The activities include measurements of stock size and primary productivity for carrying capacity determinations for capture fisheries, mariculture and pollution assessments. Other activities involve the assessments of fish stocks and establishment of total allowable catch quotas for fish. A China-Korea forum for annual determination of TAC levels, based on the results of joint bottom trawl and acoustic surveys, will be introduced. Budgets have been provided for improving analyses of socioeconomic benefits in relation to short-term and long-term resource sustainability options. Consideration will also be given by both countries to the optimization of management actions for all shared marine resources. A bilateral China-Korea Project Coordination Unit (PCU) has been established to oversee the project for both countries.

Baltic Sea Regional Project

As late as 1950, the Baltic Sea was still regarded as environmentally "healthy;" its ecological deterioration has been caused in recent years by an increase of point source industrial and non-point source agricultural pollutants, degradation of the coastal zone and non-sustainable use of living marine resources. The natural vulnerabilities have been seriously aggravated by anthropogenic causes of environmental change and degradation. These problems of the Baltic Sea are transboundary in nature, and difficult to address on an individual country basis. The need to address the management of agricultural inputs into international waters, improve coastal zone management and adopt a sustainable management of living marine resources has been highlighted in the "Baltic Sea Joint Comprehensive Environmental Action Program (JCP)," prepared under the coordination of the Helsinki Commission by a broad-based task force. The JCP was adopted as the strategic action program for the region by the Ministers of Environment in 1992, and was updated and strengthened in 1998. HELCOM prepares assessments of transboundary trends and impacts in the form of Pollution Load Compilations and Periodic Assessments which support implementation of the JCP. The JCP recognizes the need to use an ecosystem-based management approach that recognizes the freshwater, coastal and marine resources as a management continuum. This GEF Project responds to the need to address regional transboundary issues and to establish a coordinated approach to ecosystem-based management, in order to alleviate burdens from anthropogenic impacts and meet the objectives of the JCP. In fact, for the first time, this project has all three international commissions with responsibilities in the Baltic working together. In addition to HELCOM, the Baltic Sea Fisheries Commission and the International Commission for the Exploration of the Sea are collaborating in the GEF project to address overfishing, the loss of genetic resources of valuable fisheries in the LME, and contaminants that bioaccumulate to pose ecosystem and human health threats (see ECOPS).

The Baltic Sea ecosystem and its catchment area have a range of ecotones and biological diversity. The brackish waters of the Baltic Sea contain a mixture of marine and freshwater species. The coastal areas serve as spawning, nursery, and feeding areas for several species of fish. Baltic 21 statistics have indicated that the fishery industry contributes significantly to the regional and local economy, and sustenance fishing is critical to the social and economic welfare of the coastal communities in the eastern Baltic. Major coastal and marine transboundary issues prevail due to current land, coastal and marine practices. They include: (i) changes in the productivity of the near coastal and offshore waters from eutrophication; (ii) unsustainable condition of fish stock yields; and (iii) the degraded condition of coastal water quality from pollution, harmful algal blooms, multiple ecological disturbances, and contaminant loading.

The Project components are based on the Large Marine Ecosystem (LME) concept and include integrated land, coastal and open sea activities to strengthen the local and regional capacity to achieve sustainable ecosystem management of the Baltic Sea resources. Sustainable management will improve ecosystem health while providing social and economic benefits to farming, coastal and fishing communities and sectors such as businesses and tourism. The Project introduces jointly planned and implemented multi-national monitoring and assessment surveys that facilitate local cooperation and coordination, and use of innovative methodologies for assessing the changing state of the ecosystem and development of effective strategies for the management of these shared resources. Component activities provide the mechanisms to meet these objectives through improving coastal and open sea monitoring and assessment practices, understanding the carrying capacity of the coastal and open sea ecosystem, and promoting sustainable fishery practices.

The Project supports activities in the coastal near shore environment of the Eastern Baltic Sea and in selected adjacent sections of the open sea environment. In general, the coastal near shore activities and monitoring network will correlate with land-based coastal and associated demonstration activities addressing land-based agricultural inputs to coastal and open sea waters. Improving coastal zone management are critical for management of the Baltic Sea ecosystem. The JCP highlights the management of agriculture inputs and the coastal areas of the Baltic as priority issues. The agricultural element of the Component will (i) test administrative and organizational mechanisms (regional and local) and provide advice and support to the farming community; (ii) assess farmers' interest in and willingness to pay for improving their environmental management practices; (iii) assist farmers to lower both the risk and barriers that currently hinder the adoption of new practices; and (iv) provide support for small-scale environmentally responsible agricultural investments.

The Project partially finances investment costs for on-farm environmental facilities, operating expenses of local implementers, equipment recommended by the farm management plans, and recurrent costs for local capacity building. The coastal zone management element of the Component covers the following: (i) focuses on the role that can be played by local communities in sustainable management of coastal resources; (ii) links activities in the demonstration watershed to activities being taken on the coast; (iii) supports implementation of previously prepared management plans; and (iv) assists local communities to overcome barriers to the adoption of new planning and management methods in these sensitive areas. The Project will partially finance costs for management activities, small-scale investments and demonstration activities and selected costs for local capacity building as well as encourage the three commissions to work together.

EXCESSIVE NUTRIENT LOADING OF OCEAN COASTAL WATERS

A common thread regarding degradation of LMEs in GEF projects is the large number of eutrophication cases. More and more, GEF receives requests for interventions in LMEs for such eutrophication concerns. Nitrogen over-enrichment has been reported as a coastal problem for two decades, from the southeast coast of the US as described by Duda (1982) twenty years ago, to the Baltic and other systems (Helsinki Commission 2001). More recent estimates of nitrogen export to LMEs from linked freshwater basins are summarized in Figure 14-1 as adapted from Kroeze and Seitzinger 1998. These recent human-induced increases in nitrogen flux range from 4-fold to 8-fold in the U.S. from the Gulf of Mexico (Jaworski *et al.*1999) to the New England coast, while no increase was documented in areas with few agricultural or population sources as in Canada (Howarth *et al.* 2000).

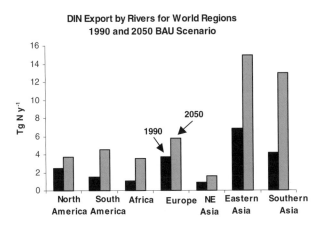

Figure 14-1. Model-predicted nitrogen (dissolved inorganic N) export by rivers to coastal systems in 1990 and in 2050 (based on a business-as-usual [BAU] scenario). Figure modified from Kroeze and Seitzinger 1998.

In European LMEs, recent nitrogen flux increases of from 3-fold in Spain to 4-fold in the Baltic and 11-fold in the Rhine basin draining to the North Sea LME have been recorded. Duda and El-Ashry (2000) described the origin of this disruption of the nitrogen cycle from the "Green Revolution" of the 1970s as the world community converted wetlands to agriculture, utilized more chemical inputs, and expanded irrigation to feed the world. As noted by Duda (1982) for the Southeast estuaries of the US, and Rabalais (Rabalais *et al.* 1999) for the Gulf of Mexico, much of the large increase in nitrogen export to LMEs is from agricultural inputs, both from the increased delivery of fertilizer nitrogen as wetlands were converted to agriculture and from concentrations of livestock as shown Duda and Finan (1983) for eastern North Carolina, where the increase in nitrogen export over the forested situation ranged from 20-500 fold in the late 1970s. Industrialized livestock production in the last two decades increased the flux, the eutrophication, and the oxygen depletion even more, as reported by the NRC (2000). The latest GESAMP Assessment (2001) also identified sewage as a significant contributor to eutrophication in drainages from large cities. Atmospheric deposition from automobiles and agricultural activities may also contribute depending on proximity to sources.

GEF is being asked more frequently by countries to help support the agreed upon incremental cost of actions that reduce such nitrogen flux. Actions range from assisting in development of joint institutions for ecosystem-based approaches for adaptive management described in this paper to on-the-ground implementation of nitrogen abatement measures in the agricultural, industrial, and municipal sectors and breaching of floodplain dikes so that wetlands recently converted to agriculture may be reconverted to promote nitrogen assimilation. The excessive levels of nitrogen contributing to coastal eutrophication constitute a new global environment problem that is cross-media in nature. Excessive nitrogen loadings have been identified as problems in the following LMEs that are receiving GEF assistance: the Baltic Sea, the Black Sea, the Adriatic portion of the Mediterranean, the Yellow Sea, the South China Sea, the Bay of Bengal, the Gulf of Mexico, and the Plata Maritime Front/Patagonia Shelf. In fact, preliminary global estimates of nitrogen export from freshwater basins to coastal waters were assembled by Seitzinger and Kroeze (1998) as part of a contribution to better understanding LMEs. Included in Figure 14-2, adapted from Kroeze and Seitzinger (1998), these preliminary estimates of global freshwater basin nitrogen export are alarming for the future sustainability of LMEs. Given the expected future increases in population and fertilizer use, LMEs may be, without significant N mitigation efforts, subjected to a future of increasing harmful algal bloom events, reduced fisheries, and hypoxia that further degrades marine biomass and biological diversity.

SUSTAINING MOMENTUM CREATED IN 121 COUNTRIES

An increasing number of developed and developing countries, now totaling 121 around the world, are concerned enough with the degraded condition of their coastal and marine ecosystems to collaborate on GEF-LME projects. Ministerial level commitments to ecosystem-based approaches for assessment and management may ultimately lead to establishing joint adaptive management regimes in support of the global objectives of Chapter 17 of Agenda 21, the Jakarta Mandate of the Convention on Biological Diversity (CBD), the United Nations Convention on Law of the Sea (UNCLOS), the GPA, and the regional seas agreements that countries have signed. It appears that an important corner has been turned by these countries toward a focused global effort to restore biomass and biological diversity to coastal oceans, as concerned governments understand the poverty reduction and security enhancement that accompanies more sustainable management regimes. The GEF international waters focal area has played a catalytic role through its emphasis on joint, integrated management of Large Marine Ecosystems, their coastal assets, and linked river basins. Through tests of these approaches, countries are starting to establish practical, science-based management regimes that address in collective and ecosystem-oriented ways the themes and programs under Agenda 21 and other global instruments.

While many of the multi-country-driven LME initiatives supported with GEF grant funding have just started, and while national and regional reforms in progress will take a number of years to achieve, several lessons are becoming evident for the world community to consider in reversing the decline of its coastal oceans. A geographic approach based on the LMEs of the world, their adjacent coastal areas and linked freshwater contributing basins (where appropriate), is likely to overcome the limits of more thematically directed activities to address global environmental problems as separate and distinct problems (*e.g.* fisheries, sewage, sediment, contaminants). In this manner, the different stresses that are important to each specific area can be addressed jointly through processes that result in collective national actions in different economic sectors where needed. Processes such as the TDA and the SAP foster multi-stakeholder dialogue, inter-ministerial dialogue, and a discourse with the science community in unraveling complex situations so they can be divided into priority components for more effective management than is now in general practice. Fragmented, thematic, single purpose agency programs are just not able to harness stakeholder involvement sufficiently to drive needed reforms, compared to geographic-based initiatives.

The assessment and management cycle based on the five modules in the TDA and SAP processes, fosters an adaptive management approach through the establishment of monitoring and evaluation indicators that are periodically measured by the nations and tracked over time for reporting to stakeholders and the GEF. GEF agencies have fostered participation of multi-level, multi-country, national-interministerial institutions, involving local government and local communities, for the buy-in and adoption of reforms. The geographic nature of LME areas is conducive for harnessing stakeholder participation and gaining political commitments to change. Thematic programs which are not place-based cannot garner real commitments for change in economic sectors without mobilizing local stakeholders as driving forces for reforms (Sherman and Duda 1999). The national-interministerial committee established in each country to operationalize reforms and programs is particularly important to achieve the practical integration of needed actions in different economic sectors. However, GEF was designed to play a minor, catalytic role and new partnerships are needed to sustain the momentum that has been created.

New LME partnerships

Now, at the beginning of this new century, a global common understanding is emerging in recognition of the accelerated degradation of coastal and marine ecosystems. The decline is not just a problem of developing nations but is also driven by over-consumption from developed nations. The $60 billion annual trade in fisheries makes those nations a stakeholder in LMEs of the South in addition to their own LMEs. Indeed, rich countries now acknowledge the need to adopt many reforms as well, not only for their degraded marine waters but also to provide a safety net to conserve the marine waters of developing nations that are exploited for global commerce. The $15 billion in annual fishing subsidies represent a powerful driving force for depletion, and reforms in those countries are just as essential as the reforms needed in developing nations. Many developed nations share LMEs with developing nations, and the GEF has shown that they can work together in adopting an ecosystem-based approach for joint assessment and management purposes.

If the spiraling degradation of coastal and marine ecosystems is to be reversed so that these ecosystems continue to provide both livelihood benefits to coastal communities and foreign exchange for governments, drastic reforms are necessary. Competing global programs, competing donor interests, competing priorities of international finance institutions also face an imperative to collaborate in harmony if the early momentum catalyzed through the GEF is to be sustained. Donor organization assistance and international agency support for the strategic, country-driven reforms being identified through LME projects need to be delivered in a coordinated and sequenced manner, to build capacity of nascent institutions that must learn to implement difficult reforms. New geographic-based partnerships are necessary to ensure the completion of the reform processes and the North is an essential member of those partnerships.

Perhaps most importantly, the GEF-LME projects are demonstrating that holistic, ecosystem-based approaches to managing human activities in LMEs, their coasts, and their linked watersheds are critical, and provide a needed place-based area within which to focus on the many benefits to be gained from multiple global instruments. Instead of establishing competing programs with inefficiencies and duplication, which is the norm now, LME projects foster action on priority transboundary issues—across instruments in an integrated manner—across UNCLOS, Chapter 17 of Agenda 21, the Jakarta Mandate of the CBD, the GPA and its pollution loading reductions, and in dealing with inevitable adaptation issues under UNFCCC. In fact, this ecosystem-based approach, centered around LMEs and the participative processes for countries undertaking to build political and stakeholder commitment and interministerial buy-in, can serve as the way ahead, consistent with Chapter 17, to reverse the degradation of marine ecosystems .

The adaptive management framework resulting from the iterative application of the GEF Operational Strategy allows for sequential capacity building, technology introduction, and investments to an ecosystem-based group of nations by the world community so that this collective response to global conventions and other instruments can be achieved in a practical manner. However, if international finance institutions, bilateral donors, and agencies cannot work collaboratively, in partnership with client countries that have identified their needs for reforms and investments, continued fragmentation and duplication will serve as a barrier to reversing the accelerated depletion of coastal and marine ecosystems. The 5 modules, including the results of joint surveys across LMEs for transparency of information, capacity building and technology transfer, ensure that management institutions are engaged with the science community in joint efforts developed in conjunction with stakeholders. In this way, ecological surprises of the future such as those generated by fluctuating climate may be able to be handled by the joint institutions and may have a chance to insulate the poor communities that are the first to suffer the adverse effects of inadequate management efforts.

This growing number of country-driven commitments, as fostered by the GEF, and the global imperative to change because of the degraded condition of the global coastal ocean, provides an unprecedented opportunity for accelerating the transition to the sustainable use, the conservation, and the development of coastal and marine ecosystems. The social, economic, and environmental costs of inaction are just much too high for multilateral and bilateral institutions and international agencies not to support the fledgling efforts of 121 countries, trying to implement Chapter 17 of Agenda 21 by focusing on specific, shared LMEs. A new partnership on ecosystem-based approaches to assess and jointly manage LMEs and linked watersheds is urgently needed to restore biomass and diversity. Such a partnership is needed to broaden and deepen reforms and investments triggered by the initial GEF catalytic action LME by LME, and to involve both developed and developing nations that have a stake in each particular LME and linked watershed. Momentum must not be lost, or the result may be irreversible damage to coastal and marine ecosystems, to the livelihoods and security of poor communities depending on them, and to the economy of many coastal nations.

The WSSD target for introducing ecosystem-based assessment and management practices by 2010 is likely to be met by the 121 countries constituting the existing LME network. It is unlikely that the Summit target for maintaining and restoring fishery resources to maximum sustainable yield (MSY) levels by 2015 will be met. However, progress is being made in the recovery of depleted fish stocks through mandated reductions in fishing effort (Sherman *et al.* 2002). With regard to the target for control and reduction of land-based sources of pollution, considerable additional effort will be required to achieve "substantial reductions in land-based sources of pollution by 2006." However, good progress has been made in designated Marine Protected Areas (MPAs) within the GEF-LME project Network.

REFERENCES

Adam K.S. Towards integrated coastal zone management in the Gulf of Guinea: a framework document. Les Editions du Flamboyant, Cotonou (Bénin). 1998. 85p.

Black Sea Environment Programme. 1996. Strategic action plan (SAP) for the rehabilitation and protection of the Black Sea. Istanbul. 29pp.

CEDA (Center for Environment and Development in Africa). Coastal profile of Nigeria. Ceda, Cotonou (Bénin). 1997. 93p.

Duda A.M. 1982. Municipal point sources and agricultural nonpoint source contributions to coastal eutrophication. *Water Resources Bulletin* 18(3):397–407.

Duda A.M. and M.T. El-Ashry. 2000. Addressing the global water and environmental crises through integrated approaches to the management of land, water, and ecological resources. *Water International* 25:115–26.

Duda A.M. and D.S. Finan. 1983. Influence of livestock on nonpoint source nutrient levels of streams. *Transactions of American Society of Agricultural Engineers* 26(6):1710–6.

Duda, A.M. and K. Sherman. 2002. A new imperative for improving management of large marine ecosystems. *Ocean & Coastal Management* 45:797-833.

ECOPS, ESF, European Commission DG XII, and Institute for Baltic Sea Research. Joint Baltic Sea Ecosystem Studies: A Science Plan for an Interdisciplinary Ecosystem Analysis in the Baltic Sea. Report of 4 international Baltic Sea Workshops: 18–19 March 1993, 4–10 November 1993, 6–8 June 1994, and 22–25 January 1995. HELCOM. Helsinki. 1995.

Garibaldi, L. and L. Limongelli. 2003. Trends in oceanic captures and clustering of large marine ecosystems: Two studies based on the FAO capture database. FAO Fisheries Technical paper 435. Food and Agriculture Organization of the United Nations. Rome. 71p.

GEF (Global Environment Facility). 1995. GEF Operational Strategy. Washington, DC: Global Environment Facility

GESAMP (IMO/FAO/UNESCO-IOC/WMO/WHO/IAEA/UN/UNEP Joint Group of Experts on the Scientific Aspects of Marine Environmental Protection) and Advisory Committee on Protection of the Sea. 2001.Protecting the oceans from land-based activities – Land-based sources and activities affecting the quality and uses of marine, coastal, and associated freshwater environment. Rep. Stud. GESAMP. No. 71, 162p.

Helsinki Commission. Environment of the Baltic Sea Area 1994–1998. 2001. Baltic Sea Environment Proceedings No. 82A. Helsinki. 23p.

Howarth R, D. Anderson , J. Cloern, C. Elfring , C. Hopkinson, B. Lapointe, T. Malone, N. Marcus, K. McGlathery, A. Sharpley, D. Walker. 2000. Nutrient pollution of coastal rivers, bays, and seas. *Issues in Ecology* (7) Ecological Society of America. Fall 2000. 15pp.

Ibe C, editor. Perspectives in Integrated Coastal Areas Management in the Gulf of Guinea. Center for Environment and Development in Africa (Ceda), Cotonou (Bénin). 91p.

Ibe C, Regional Coordinator. GOG LME Newsletter, No. 8, 10/97 - 3/98. Centre de Recherches Oceanologiques, Abidjan, Cote d'Ivoire. 1998. 28p.

Ibe C, A.A. Oteng-Yeboah, S.G. Zabi , D. Afolabi , editors. 1998. Integrated environmental and living resources management in the Gulf of Guinea: The Large Marine Ecosystem Approach. Proceedings of the first Symposium on GEF's LME Project for the Gulf of Guinea. Abidjan 26-30 January 1998. Institute for Scientific and Technological Information, CSIR. Accra, Ghana. 274+ p.

Ibe C and S.G. Zabi, eds. 1998. State of the coastal and marine environment of the Gulf of Guinea. Center for Environment and Development in Africa (CEDA), Cotonou (Bénin).

IOC (Intergovernmental Oceanographic Commission). 2002. IOC-IUCN-NOAA Consultative Meeting on Large Marine Ecosystems (LMEs). Fourth Session, Paris, France, 8–9 January 2002. UNESCO.

IOC (Intergovernmental Oceanographic Commission). IOC-IUCN-NOAA Consultative Meeting on Large Marine Ecosystems (LMEs). Third Session, Paris, France, 13-14 June 2000. IOC-IUCN-NOAA/LME-III/3. IOC Reports of Meetings of Experts and Equivalent Bodies, Series 162. 20 p.

Jaworski NA. 1999. Comparison of nutrient loadings and fluxes into the US Northeast Shelf LME with the Gulf of Mexico and other LMEs. In: Kumpf H., K. Steidinger, K. Sherman, editors. 1999. *The Gulf of Mexico Large Marine Ecosystem: Assessment, Sustainability, and Management.* Malden: Blackwell Science, Inc. 704p. 360-371.

Kroeze C. and S.P. Seitzinger. 1998. Nitrogen inputs to rivers, estuaries and continental shelves and related nitrous oxide emissions in 1990 and 2050: A global model. *Nutrient Cycling in Agroecosystems* 52:195–212.

Lubchenco J. 1994. The scientific basis of ecosystem management: framing the context, language, and goals. In: Zinn, J. and M.L. Corn, editors. Ecosystem Management: Status and Potential. 103rd Congress, 2d Session, Committee Print. U.S. Government Printing Office, Superintendent of Documents. 33–39.

Mondjanagni A.C., K.S. Adam, P. Langley. Côte d'Ivoire - Profil environnemental de la zone côtière. Centre pour l'environnement et le développement en Afrique (Ceda), Cotonou (Bénin). 1998. 87p.

Mondjanagni A.C., K.S.Adam , P.Langley. 1998. Profil de la zone côtière du Bénin. Centre pour l'environnement et le développement en Afrique (Ceda), Cotonou (Bénin). 93p.

NRC (National Research Council). 2000. Clean Coastal Waters: Understanding and reducing the effects of nutrient pollution. National Academy Press, Washington, DC.

NRC (National Research Council). 1999. Sustaining Marine Fisheries. National Academy Press, Washington, DC.

NOAA (National Oceanic and Atmospheric Administration). 1993. Emerging Theoretical Basis for Monitoring the Changing States (Health) of Large Marine Ecosystems. Summary Report of Two Workshops: 23 April 1992, National Marine Fisheries Service, Narragansett, Rhode Island, and 11–12 July 1992, Cornell University, Ithaca, New York. NOAA Technical Memorandum NMFS-F/NEC-100.

Pauly D., Christensen V. 1995. Primary production required to sustain global fisheries. *Nature*, 374:255–7.

Pauly D., Christensen V, Dalsgaard J, Froese R, Torres F, Jr. 1998. Fishing down marine food webs. *Science* 279:860–3.

Rabalais N.N., R.E. Turner, W. J. Wiseman, Jr. 1998. Hypoxia in the Northern Gulf of Mexico: Linkages with the Mississippi River. In: H. Kumpf, K. Steidinger and K. Sherman, editors. *The Gulf of Mexico Large Marine Ecosystem: Assessment, Sustainability, and Management.* Blackwell Science, Inc. Malden, MA. 297–322.

Seitzinger S.P., C. Kroeze. 1998. Global distribution of nitrous oxide production and N inputs to freshwater and coastal marine ecosystems. *Global Biogeochemical Cycles* 12:93–113.

Sherman, K. 2003. Physical, biological and human forcing of biomass yields in large marine ecosystems. ICES CM2003/P:12.24-27 presented September 2003, Tallin, Estonia.

Sherman, B. 2000. Marine ecosystem health as an expression of morbidity, mortality, and disease events. *Marine Pollution Bulletin* 41(1–6):232–54.

Sherman, K. 1994. Sustainability, biomass yields, and health of coastal ecosystems: an ecological perspective. *Marine Ecology Progress Series 112:277–301.*

Sherman, K., J. Kane, S. Murawski , W. Overholtz , A. Solow. 2002. The U.S. Northeast Shelf Large Marine Ecosystem: Zooplankton trends in fish biomass recovery. In: Sherman, K. and H.R. Skjoldal, eds. *Large Marine Ecosystems of the North Atlantic: Changing States and Sustainability.* Amsterdam: Elsevier Science. 195-215. 449p.

Sherman, K., A.M. Duda. 1999. An ecosystem approach to global assessment and management of coastal waters. *Marine Ecology Progress Series* 190:271–87.

Sydnes, Are K. 2001. New Regional Fisheries Management Regimes: Establishing the South East Atlantic Fisheries Organisation. *Marine Policy* 25:353-364.

Sydnes, A.K. 2001. Regional Fishery Organizations: How and Why Organizational Diversity Matters. *Ocean Development & International Law* 32:349-372 .

United Nations General Assembly. 2001. Report on the work of the United Nations Open-ended Informal Consultative Process established by the General Assembly in its resolution 54/33 in order to facilitate the annual review by the Assembly of developments in ocean affairs at its second meeting. Report A/56/121. 22 June . New York.62p.

UNDP (United Nations Development Program). 1999. Implementation of the Strategic Action Programme of the South Pacific Developing States. Project Document. UNDP/RAS/98/G32/A/1G/99. New York. 93p.

UNEP (United Nations Environment Programme). 2001. Economic reforms, trade liberalization, and the environment: A synthesis of UNEP country projects. Div of Technology, Industry, and Economics, Geneva, Switzerland. 21p.

Large Marine Ecosystems, Vol. 13
T.M. Hennessey and J.G. Sutinen (Editors)

15

The Evolution of LME Management Regimes: The Role of Adaptive Governance

Timothy M. Hennessey

INTRODUCTION

This paper demonstrates how governance and socioeconomics are essential to the establishment and functioning of effective LME management regimes. In the course of our discussion we review the research literature on natural resource regimes in order to identify hypotheses regarding stages of regime development, types of regimes, and regime effectiveness.

We then consider the Global Environmental Facility mandated process and the contribution this makes to the establishment of a management regime for the BCLME, the most advanced program to date. We note that most of the questions and activities associated with this process require an understanding of governance and socioeconomics. And, given the complexity and uncertainty associated with large marine ecosystems, we introduce the concept of *the ecology of governance*. This refers to the complex array of institutions, programs and processes designed to understand and manage the complexity of natural ecosystems. We note in our analysis of the BCLME that the GEF process proved to be an extremely valuable mechanism for guiding the development of the regime, particularly the agenda setting, collective choice making and initial institutional design stages. We next discuss the importance of institutions having a learning capacity provided by adaptive management. We discuss the work of Kai Lee on adaptive management and the opportunities and constraints he identifies for those doing adaptive management in a governance context. Such adaptive governance requires careful attention to issues of institutional analysis and design. Following Elinor Ostrom's work on institutional analysis and design for natural resources management, we relate the conditions she identifies to what is required for effective LME management. What is required is an adaptive governance system which integrates programs at the national, regional and local levels. This would involve, in the case of LMEs, integration at multiple scales with coastal management programs, pollution programs, marine protection area efforts and others. Finally, we argue that governance and socioeconomic concepts and tools should be a fundamental feature of training programs designed to build capacity in each of the LMEs. Moreover, the lessons learned from experience in this regard should be shared among and between the LMEs in order to create and enhance technology transfer via a *diffusion of innovation* throughout the LME community.

THE KEY ROLE OF ADAPTIVE GOVERNANCE IN THE GEF PROCESS: THE EXAMPLE OF BENGUELA CURRENT LARGE MARINE ECOSYSTEM

Currently, there are 64 large marine ecosystems identified around the world. Of these, the following are in various stages of development toward the establishment of management regimes (Table 15-1).

Table 15-1: 121 Countries Participating in GEF/Large Marine Ecosystem Projects

Approved GEF Projects	
LME	Countries
Gulf of Guinea (6)............................	Benin, Cameroon, Côte d'Ivoire, Ghana, Nigeria, Togo
Yellow Sea (2).................................	China, Korea
Patagonia Shelf/Maritime Front (2).............	Argentina, Uruguay
Baltic (9)...	Denmark, Estonia, Finland, Germany, Latvia, Lithuania, Poland, Russia, Sweden
Benguela Current (3).............................	Angola, Namibia, South Africa
South China Sea (7)..............................	Cambodia, China*, Indonesia, Malaysia, Philippines, Thailand, Vietnam
Black Sea (6)......................................	Bulgaria, Georgia, Romania, Russia*, Turkey, Ukraine
Mediterranean (19)...............................	Albania, Algeria, Bosnia-Herzegovina, Croatia, Egypt, France, Greece, Israel, Italy, Lebanon, Libya, Morocco, Slovenia, Spain, Syria, Tunisia, Turkey*, Yugoslavia, Portugal
Red Sea (7)..	Djibouti, Egypt* Jordan, Saudi Arabia, Somalia, Sudan, Yemen
Western Pacific Warm Water Pool-SIDS[a] (13)...	Cook Islands, Micronesia, Fiji, Kiribati, Marshall Islands, Nauru, Niue, Papua New Guinea, Samoa, Solomon Islands, Tonga, Tuvalu, Vanuatu
Total number of countries: 70*	

GEF Projects in the Preparation Stage	
Canary Current (7)..	Cape Verde, Gambia, Guinea, Guinea-Bissau, Mauritania, Morocco*, Senegal
Bay of Bengal (8).................................	Bangladesh, India, Indonesia*, Malaysia*, Maldives, Myanmar, Sri Lanka, Thailand*
Humboldt Current (2).............................	Chile, Peru
Guinea Current (16)...............................	Angola*, Benin*, Cameroon*, Congo, Democratic Republic of the Congo, Côte d'Ivoire*, Gabon, Ghana*, Equatorial Guinea, Guinea*, Guinea-Bissau*, Liberia, Nigeria*, São Tomé and Principe, Sierra Leone, Togo*
Gulf of Mexico (3)................................	Cuba, Mexico, United States
Agulhas/Somali Currents (8).....................	Comoros, Kenya, Madagascar, Mauritius, Mozambique, Seychelles, South Africa*, Tanzania
Caribbean LME (23)..............................	Antigua and Barbuda, The Bahamas, Barbados, Belize, Colombia, Costa Rica, Cuba*, Grenada, Dominica, Dominican Republic, Guatemala, Haiti, Honduras, Jamaica, Mexico*, Nicaragua, Panama, Puerto Rico[b], Saint Kitts and Nevis, Saint Lucia, Saint Vincent and the Grenadines, Trinidad and Tobago, Venezuela
Total number of countries: 51*	

*Adjusted for multiple listings

[a] Provisionally classified as Insular Pacific Provinces in the global hierarchy of LMEs and Pacific Biomes (Watson *et al.* 2003).
[b] A self-governing commonwealth in union with the United States

This circumstance presents researchers with a unique opportunity to conduct comparative analyses of the process of regime formation among these LMEs and others as they come on line over the next decade. A number of these LMEs will emerge as full-fledged management regimes while others may have foundered or failed altogether. Researchers will need to identify the key concepts and criteria in terms of which to evaluate LME regime effectiveness. Following from these observations, the fundamental question in this paper is, "What are the key governance factors which can inform hypotheses about LME regime formation, function and effectiveness?"

GAUGING REGIMES DEVELOPMENT AND EFFECTIVENESS

A comparative research program on LME regimes should take as its point of departure the research literature on the general topic of natural resource regime formation and function. Research into natural resource regimes is a relatively new field of inquiry but important contributions have been made by Oran Young (1989,1993,1999), Edward Miles (2002), Arild Underdal (2002) and Peter Haas (1992).

The first consideration in our analysis is the issue of regime formation. Oran Young, perhaps the leading theorist of international environmental regimes, notes that regimes are devices that interested actors create to ameliorate collective problems (Young 1999). States or groups create regimes with certain expectations about cooperation among interested actors to alter individualistic self-serving behavior, especially when faced with common property problems (Young 1998).

Young (1993) constructs a multivariate model of regime formation. Figure 15-1 below contains the key components and linkages of the model. The independent forcing variables—power, interest, and knowledge—all feed into a process of institutional bargaining which may or may not lead to a constitutional contract. A number of cross-cutting factors such as leadership and context influence the process by directing or channeling the operation of the driving forces. Young reports that this model has demonstrated some explanatory power with regard to Arctic accords and other agreements.

Young (1999) identifies four types of natural resource regimes: regulatory, procedural, programmatic and generative. Regulatory regimes develop a framework of rules that are beneficial to the interests of the parties. These rules and behaviors apply not only to the states that are members of the regime, but also to corporations, NGOs or individuals subject to state jurisdiction. Regulatory regimes may deal with simple coordination issues, such as the specification of sea lanes, or difficult problems such as trade in endangered species. Regulatory regimes require authoritative interpretation of the rules on a regular basis.

Procedural regimes use mechanisms that encourage participants to reach collective decisions on shared problems, such as annual harvest levels of fisheries or forests. Procedural regimes may amend existing rules or establish new rules to achieve expected outcomes. This type of regime differs from the regulatory type because it does not administer rules or regulations, but concentrates on consensus building.

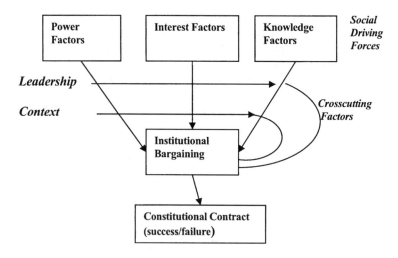

Figure 15-1. A multivariate model of regime formation

Programmatic regimes combine the resources of the participants for projects that could not be accomplished by single participants. This could, for example, mean joint monitoring programs to enhance knowledge of collective problems. Young emphasizes that programmatic regimes involve research and assessments, which require funding on a continuing basis. Such funding can come from members of the regime or from other interested parties.

Generative regimes develop new social practices. They are established around a central theme, such as ecosystem management, and the regime frequently acts as the means to promote the idea. Young notes that this type of regime can result in new constituencies, a realignment of government institutions or a change in worldviews among key actors. Generative regimes may also mobilize unofficial national and transnational groups to take an interest in the regime or theme. Young notes that evidence suggests these groups may ultimately help make the regime successful. As we shall see later in our discussion, the BCLME contains elements of three regime types but cannot be explained by using a single type.

Young (1998) argues that regimes go through several stages of development—agenda setting, negotiation and operationalization. Hypotheses related to these stages are presented in Table 15-2 below. Young observes that, "Efforts to form international regimes may make it through one stage but stall at the next. Some issues fail to get on the agenda. Others have difficulty achieving a high enough priority on this agenda to trigger the initiation of explicit negotiations. Still others stimulate negotiation in which the participants are unable to reach closure on the terms of a regime. Even signed agreements sometimes become dead letters. Only by successfully navigating all three stages can a regime that has real consequences for the nature of collective outcomes come into existence."

In the agenda setting stage, there is an emphasis on openness and the framing of issues. This involves idea-based arguments. Here the role of knowledge based experts is critical as Haas observes in his discussion of *epistemic communities*. He and his associates have done extensive work on identifying the origins of international environmental regimes under conditions of uncertainty. His approach examines the role of networks of knowledge based experts—what he calls epistemic communities—in articulating the cause and effect relationship of complex problems, helping states identify their interests, framing the issues for collective debate, proposing specific policies and identifying salient points for negotiation (Haas 1992). He argues that control over knowledge and information is an important dimension of power and that the diffusion of new ideas and information can lead to new patterns of behavior which may prove to be an important determinant of international policy coordination. These insights underscore the potential importance of emerging LME regimes on the world scene.

Table 15- 2. Hypotheses relating to the stages of regime formation

1. Driving Forces	Ideas are particularly prominent during agenda formation; interests dominate the stage of negotiation; material conditions become increasingly significant in connection with the shift from paper to practice.
2. Players	There are no simple shifts from one stage of regime formation to another in the roles organizations play. With regard to individuals, however, intellectual leadership is particularly prominent during agenda formation, entrepreneurial leadership looms large during the stage of negotiation, and structural leadership is important throughout the process.
3.Collective-action Problems	While stalemate or gridlock is the classic collective-action problem of the negotiation stage, miscommunication is the standard pitfall of agenda formation, and asymmetries in levels of effort are the typical hazard of operationalization.
4. Context	Broad changes in the political environment affect agenda formation; more specific exogenous events influence negotiations, and domestic constraints loom large during operationalization.
5. Tactics	The classic concern with threats and promises is most pronounced during the negotiation stage. Efforts to influence the framing of the problem are characteristic of agenda formation, and the tactics of administrative or bureaucratic politics come to the fore during operationalization.
6. Design Perspectives	Agenda formation is a time for focusing on the big picture; negotiation stimulates a concern for language to be included in agreements, and operationalization leads to a focus on domestic concerns to the detriment of efforts to set up the relevant international machinery.

Perhaps the most important dimension of regimes concerns how effective they are. For Young, this means the extent to which institutions solve the problems which prompted their formation (Young 1998). He suggests that effectiveness be measured by the behavioral change of actors as a result of a regime. Arild Underdal (2002), a well known Norwegian scholar of marine policy, views regime effectiveness as the degree of stringency and inclusiveness of the regime's provisions. When judging regimes he looks to see whether the provisions are comprehensive enough to force change in actors' behaviors, what side effects the rules cause, and the degree of compliance to which members adhere. But Underdal clearly distinguishes changes taking place in the agenda setting stage, from changes in the implementation stage. The former produce rules and regulations while the latter lead to behavioral change and potential environmental impacts.

Edward Miles, in his assessment of regime effectiveness in the management of tuna fisheries, concludes that the regime was effective because it reduced uncertainty in the knowledge base, developed an epistemic community with links at the national and regional political levels, and improved the leadership capacity of the regime (Miles 2002).

According to Levy, Keohane and Haas (1993), regime institutions should create *pathways of effectiveness* through activities which:

1. Increase Government Concern
   - Facilitate direct and indirect linkages of issues
   - Create, collect and disseminate scientific knowledge
   - Create opportunities to magnify domestic public pressure

2. Enhance Contractual Environment
   providing bargaining fora that:
   - Reduce transactions costs
   - Create an integrated decision making process
   - Conduct monitoring of
     - Environmental quality
     - National environmental performance
     - National environmental policies
   - Increase national and international accountability

3. Build National Capacity
   - Create inter-organizational networks with operational organizations to transfer technical and management expertise
   - Transfer financial assistance
   - Transfer policy relevant information and expertise
   - Boost bureaucratic power of domestic allies

We now turn to a real world example of regime development in order to see the extent to which the Benguela Current Large Marine Ecosystem regime meets these conditions of regime development and function.

THE BENGUELA CURRENT LME

In order to address the regime question we have chosen to analyze the Benguela Current Large Marine Ecosystem (BCLME). This system is the most developed of the LME regimes and best illustrates the importance of governance and socioeconomics in regime formation and function. As we shall see, the Benguela case also illustrates some of the issues identified above.

All the LMEs which seek to establish transboundary management regimes must proceed through a process required by the Global Environment Facility in order to obtain funding. The GEF process enhances regime formation by creating an agenda setting process and establishing the basis for choosing an institutional structure to carry out regime objectives. The GEF process provides a forum whereby epistemic communities—groups of knowledge based experts—come together to determine the factors causing marine environmental degradation. Moreover, the process leads to action plans which, when implemented, will deal with the problems identified in the agenda setting process. Additionally, the GEF process requires that institutions be developed which will integrate governance activities at multiple levels so as to treat the problems in an effective and comprehensive manner.

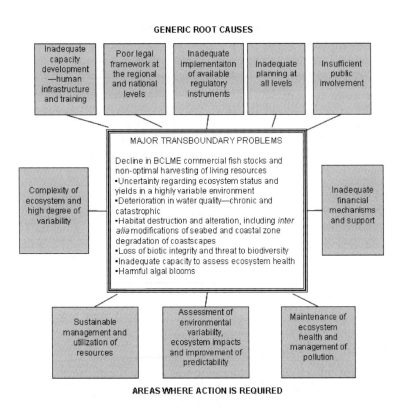

Figure 15-2. Major transboundary problems, generic root causes and areas requiring action

This process is moved forward through two fundamental activities: the Transboundary Diagnostic Analysis (TDA) (Figure 15-2, BCLME example) and the Strategic Action Plan (SAP). The TDA is a scientific and technical assessment which identifies and quantifies the environmental issues and problems of the region, their causes analyzed and their impacts, both environmental and economic, assessed. The analysis includes the identification of causes (Table 15-3, for example), and uncertainties associated with these environmental issues at the national, regional and global scale, and the socioeconomic, political and institutional context within which they occur. The SAP draws on the synthesis matrix and uses the problems identified in the TDA to develop an action plan. A close examination of the Benguela Current Large Marine Ecosystem (BCLME) TDA and SAP, for example, reveals that most of the issues identified in the process are governance and socioeconomic in nature as we see in what follows (BCLME 1999).

Table 15- 3. Main Root Causes

1.	Complexity of ecosystem and high degree of variability (resources and environment)	• Changing state of Benguela • Inadequate information and understanding • Difficulty in monitoring and assessment • Poor variability
2.	Inadequate capacity development (human and infrastructure) and training	• Colonial/political past • Institutional downsizing and brain drain • Limited inter-country exchange
3.	Poor legal framework at the regional and national levels	• Regionally incompatible laws and regulations • Ineffective environmental laws and regulations
4.	Inadequate implementation of available regulatory instruments	• Inadequate compliance and enforcement (overfishing, pollution) • Indifference and poor communication • Posts not filled (some inappropriately)
5.	Inadequate planning at all levels	• Inadequate intersectoral coordination • Poorly planned coastal developments • Limited time horizon of planners • Rapid urbanization
6.	Insufficient public involvement	• Lack of awareness and public apathy • Conflicts about right of access
7.	Inadequate financial mechanisms and support	• Low country GDPs • Ineffective economic instruments • Insufficient funding for infrastructure and management; poor salaries

Source: UNDP, GEF, TDA

The importance of governance and socioeconomic issues is even more pronounced in the three actions areas of the Strategic Action Plan (SAP). For action area A , Sustainable management and utilization of resources, there are five specific actions:
1. Capacity strengthening and training
2. Joint surveys and assessment of shared resources
3. Policy harmonization and integrated management
4. Co-financing with private industry, and
5. Development of new industries.

For action area B, Assessment of ecosystem impacts and improvement of predictability, there are four actions:
1. Capacity strengthening and training for transboundary concerns.
2. Regional networking and international linking
3. Development of regional early warning systems, assessment and predictability capacity and joint response policies, and
4. Cross cutting demonstration projects

For action area C, Improvement of ecosystem health and management of pollution, there are five actions:

1. Capacity strengthening and training
2. Policy harmonization and development
3. Development of a regional framework for assessment
4. Establishment of effective surveillance and enforcement agencies, and
5. Development of stakeholder participation

Again, capacity building, regional collaboration, policy development and harmonization are overarching actions that require knowledge of institutional design collaboration and socioeconomics.

A key role of the SAP is to delineate the institutional structure of BCLME management. Necessary to the objectives of the SAP is the creation of Interim Benguela Current Commission (IBCC), which is to become the Benguela Current Commission with a secretariat within five years of the start of the program. The Commission's purpose is to strengthen regional cooperation through the coordination of regional bodies. Working with the member states the IBCC is to get regional cooperation through the coordination of regional bodies and the development of institutional linkages with the South African Development Commission. In order to obtain sustainable financing, cooperation is encouraged with regional governmental bodies and NGOs, donors, financial institutions, aid agencies and private foundations.

The Commission is also a forum for dispute resolution. It is to be supported by a Program Coordinating Unit (PCU) which will act as a secretariat of the BCLME. The Commission will also be supported by three activity centers which will be located in each state and whose focus will be on marine living resources, environmental variability and predictability, and ecosystem health and pollution. There will also be five advisory groups which are to advise the commission in five areas: fisheries and living resources, marine environmental variability and ecosystem health, marine pollution, legal affairs and maritime law, and information and data exchange.

The Program Coordinating Unit will provide coordination among and between the various advisory groups, activity centers and the Commission. The advisory groups will work with experts and institutions, industry and the NGOs to provide expertise to the commission for decision making. The activity centers will provide technical support to the advisory groups. Each state has a country lead person to provide inputs by each state government and government stakeholders. The figure below illustrates the institutional structure of the BCLME. The IBCC has not been moved from interim to permanent status so the policy stages of implementation and evaluation are not appropriate to analyze at this time. But the agenda setting stage, policy choices and action areas have been successfully established for transboundary marine problems facing Angola, Namibia and South Africa. This is a monumental achievement given the high potential for failure in efforts to create effective environmental management regimes. Many of these pitfalls were identified in the research literature cited above. The success of the BCLME so far can be attributed, in no small measure, to the GEF process which enhanced agenda setting, issue selection, the choice of action areas and the creation of the institutional structure in terms of which actions are to be implemented.

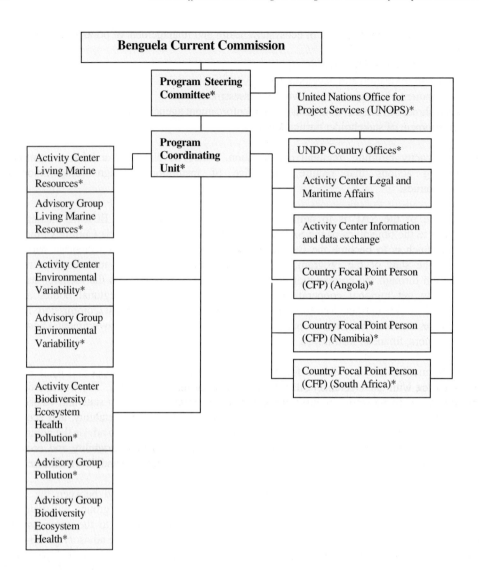

Figure 15-3. BCLME Institutional Structure (***** indicates existing groups)

The IBCC regime has elements of Young's programmatic, procedural and generative regimes. It has procedural elements because it has developed mechanisms that encourage collective decisions on shared problems via consensus building. It has generative elements because it develops new practices around the central theme of ecosystem management and actively promotes the idea around the world. It has programmatic elements because it combines the resources of the participants to support projects that could not be accomplished alone, such as monitoring and research. So far, regulatory elements have not been introduced because rules are not yet in place which could be applied and enforced on a regular basis. Such a regulatory regime may take place in the future but is not in place at this time.

As far as effectiveness is concerned, the BCLME is moving toward satisfying the conditions specified by Levy, Keohane and Haas— increased government concern, an enhanced contractual environment and national capacity. The BCLME also meets the effectiveness conditions mentioned by Miles—reducing uncertainty in the knowledge base, developing an epistemic community with links at the national and regional levels, and improving the leadership capacity of the regime.

But the future of the BCLME depends to a large degree on its institutional design and function. Looking forward it is useful to examine research on institutional analysis and design for natural resource regimes. This research reveals the necessity to have institutions of considerable complexity to match the complexity of the ecosystem being managed. The key element in the design is the capacity of the institution to learn and adapt to changing circumstances. In short, it must demonstrate a capacity for adaptive governance.

THE ECOLOGY OF GOVERNANCE AND DESIGN PERSPECTIVES: ADAPTIVE GOVERNANCE FOR COMPLEX ECOSYSTEMS

The work of Elinor Ostrom and others (Lee 1993, Hennessey 1994, Hennessey and Healey 1994, Juda and Hennessey 2001) has demonstrated the need for governance institutions at a variety of levels to cope with the uncertainty and dynamics of ecosystems, including human uses. In the absence of effective governance institutions at the appropriate scale, natural resources and the environment are in peril. As Elinor Ostrom, perhaps the leading authority on institutional analysis and design, observes in *Science* (2003:12): "The most important contemporary environmental challenges involve systems that are intrinsically global or are tightly linked to global pressures and that require governance at levels from the global all the way down to the local." This observation clearly applies to LME governance needs which will require integration at all levels and sectors from the national to the local and from program to program.

Ostrom suggests the following general principles for the robust governance of environmental resources:
- Devise rules that are congruent with ecological conditions
- Clearly define the boundaries of resources and users
- Devise accountability mechanisms for monitors
- Apply graduated sanctions for violations
- Establish/use low cost mechanisms for conflict resolution
- Involve interested parties in informed discussion of rules (analytic deliberation)
- Allocate authority to allow for adaptive governance at multiple levels from global to local (nesting)
- Employ mixtures of institutional types (institutional variety)

She argues that there are also numbers of activities that enhance effective governance. In order to do so, one should provide the necessary information, deal with conflict, induce compliance with rules, provide physical, technical and institutional infrastructure, and encourage adaptation and change.

Perhaps most important for our purposes here is the notion of "nested institutions." She suggests that institutional arrangements must be complex, redundant, and nested in many layers. She warns that simple strategies for governing resources that rely on one level and centralized command and control have been tried and have failed. Governance should employ a mix of institutional types such as hierarchies, markets, and community self-governance that employs a variety of decision rules to change incentives, increase information, monitor use and induce compliance.

In addition, research on emerging LME regimes should provide governance profiles, which are before and after "snapshots" of governance design and implementation of programs. Juda and Hennessey (2001) are currently undertaking research of this kind.

THE ADAPTIVE MANAGEMENT DIMENSION

Another key feature of adaptive governance is the role of adaptive management. The effective management of LMEs depends on adaptive management whereby the implementation of the program creates opportunities to improve the scientific basis of action. Lee (1993) observes that, "adaptive management is an approach to natural resource management that embodies a simple imperative: policies are experiments; learn from them." He suggests some institutional conditions that facilitate adaptive management and the constraints facing these actions in the real world.

Adaptive governance sees ecosystems as arenas of interdependence for human uses and laboratories of institutional invention. Hence, the management of ecosystems consists of more than passing data from scientists to decision-makers. Management occurs within an institutional setting that attempts to reconcile the differing values of user groups and the general public and then provide the means for implementing chosen objectives. This institutional framework and process we refer to as the governance system.

Table 15-4. Institutional conditions affecting adaptive management (from Kai N. Lee, Compass and Gyroscope: Integrating Science and Politics for the Environment, p.85)

There is a mandate to take action in the face of uncertainty. But experimentation and learning are at most secondary objectives in large marine ecosystems. Experimentation that conflicts with primary objectives will often be pushed aside or not proposed.
Decision makers are aware that they are experimenting anyway. But experimentation is an open admission that there may be no positive return. More generally, specifying hypotheses to be tested raises risk of perceived failure.
Decision makers care about improving outcomes over biological time scales. But the costs of monitoring, controls, and replication are substantial, and they will appear especially high at the outset when compared with the costs of unmonitored trial and error. Individual decision makers rarely stay in office over times of biological significance.
Preservation of pristine environments is no longer an option, and human intervention cannot produce desired outcomes predictably. Remedial action crosses jurisdictional boundaries and requires coordinated implementation over long periods.

Resources are sufficient to measure ecosystem-scale behavior. But data collection is vulnerable to external disruptions, such as budget cutbacks, changes in policy, and controversy. After changes in the leadership, decision makers may not be familiar with the purposes and value of an experimental program.
Theory, models, and field methods are available to estimate and infer ecosystem-scale behavior. But interim results may create panic or a realization that the experimental design was faulty. More generally, experimental findings will suggest changes in policy; controversial changes have the potential to disrupt the experimental program.
Hypotheses can be formulated. And accumulating knowledge may shift perceptions of what is worth examining via large-scale experimentation. For this reason, both policy actors and experimenters must adjust the trade-offs among experimental and other policy objectives during the implementation process.
Organization culture encourages learning from experience. But the advocates of adaptive management are likely to be staff, who have professional incentives to appreciate a complex process and a career situation in which long-term learning can be beneficial. When there is tension between staff and policy leadership, experimentation can become the focus of an internal struggle for control.
There is sufficient stability to measure long-term outcomes; institutional patience is essential. But stability is usually dependent of factors outside the control of experimenters and managers.

CAPACITY BUILDING, TRAINING AND THE DIFFUSION OF INNOVATION

Capacity building is a fundamental imperative for LME management. Such capacity can be built in a number of ways. One important way is through specially designed training programs which focus on socioeconomics and governance and use examples and cases from the real world of LMEs. We at the University of Rhode Island have been designing such a training program for key individuals in the LME community. Our work on "The Human Dimensions of LMEs and the Use of Their Resources" (Sutinen *et al.* 2000) is particularly relevant for training and capacity building in LMEs. We have developed research approaches for the following twelve activities:

Identify:

- Principal uses of LME resources
- LME resource users and their activities
- Governance mechanisms influencing LME resource use
- The public's priorities and willingness to make tradeoffs to protect and restore key natural resources

Assess:

- The level of LME-related activities
- Interactions between LME activities and resources
- Impacts of LME activities on other users
- The interactions between governance mechanisms and resource use
- The socioeconomic importance of LME related activities and the economic and sociocultural value of key uses of LME resources
- Costs of options to protect and restore key resources

<u>Compare:</u>

- The benefits to the costs of protection and restoration options
- Financing alternatives for the preferred options for protecting/restoring key LME resources

Individuals who complete the program will be up to speed on the latest concepts and cases from LMEs around the world. Over time, those who pass through such a program will form a leadership cadre for the future. They will make an important contribution to a process of *diffusion of innovation* between and among LMEs. Moreover, information from adaptive management experiments in the LMEs could form a pool of information for technology transfer between LME programs around the globe.

SUMMARY AND CONCLUSIONS

This paper investigated the degree to which the GEF process helped the formation, function and effectiveness of LME management regimes. We conducted an assessment of the Benguela Current Large Marine Ecosystem's emerging management regime, using concepts and hypotheses from the research literature on natural resource regimes. We found that the GEF process to be an extremely potent mechanism in the development of an epistemic community in terms of which the agenda was established, action items specified and a preliminary institutional structure established. Since the BCLME had not reached the implementation stage, no evaluations could be conducted on this stage of policy formation. The BCLME regime was also effective in the early stages of the policy process although we were unable to determine the degree to which the behavior of the key actors was changed as the result of the emerging regime. This would require a series of field interviews. We also found that the BCLME contained elements of the three types of regimes identified by Young: procedural, programmatic and generative. The fourth type, regulatory, did not apply due to the fact that the regime had not yet promulgated binding rules.

We also found that the GEF process contributed to "paths to effectiveness" (Levy, Keohane and Haas 1993:406) that are increasing government concern, enhancing the contractual environment and building national capacity.

We noted the importance of adaptive governance as a key concept in the design and operation of the institutional structure for the emerging Benguela Current Commission. We also noted the importance of adaptive management in terms of which policies could be examined, lessons learned and policies adjusted accordingly. Finally, we argued for the importance of capacity building and training: lessons learned in one LME could be shared with the LME community. This can be done in a variety of ways, but training programs, such as the one at the University of Rhode Island, would appear to be a very useful venue for investigated critical issues of governance and socioeconomics. Such training could also help create a cadre of LME leaders.

REFERENCES

Benguela Current Large Marine Ecosystem Programme: Transboundary Diagnostic Analysis 2002. Windhoek, Namibia.

Dietz, T., E.Ostrom, and C. Stern. 2003. The struggle to govern the commons. *Science*: 12:1907-1911.

Haas, P. 1992. Epistemic communities and international policy coordination. *International Organization* 46: 1.

Healey, M. and T. Hennessey. 1994. The utilization of scientific information in the management of estuarine ecosystems. *Ocean and Coastal Management* 23:167-191.

Hennessey, T. 1994. Governance and adaptive management for estuarine ecosystems: The case of Chesapeake Bay. *Coastal Management* 22:119-145.

Hennessey, T. and D. Soden. 1999. Ecosystem management: The governance dimension. In Soden, D. and B. Steele, eds. *Handbook of Global Environmental Policy and Administration.* New York, Marcel Decker.

Juda, L. and T. Hennessey. 2001. Governance profiles and the management and use of large marine ecosystems. *Ocean Development and International Law* 32:43-69.

Lee, K. 1993. *Compass and Gyroscope.* Washington, D.C. Island Press.

Levy, M., R. Keohane and P. Haas. 1993. Improving the effectiveness of international environmental institutions. In Haas, P. ed. *Institutions for the Earth.* Cambridge, MIT Press.

Miles, E. *et al.* 2002. *Environmental Regime Effectiveness: Confronting Theory and Evidence.* Cambridge, MIT Press.

Sutinen, J. *et al.* 2000. A Framework for Monitoring and Assessing Socioeconomics and governance of Large Marine Ecosystems. NOAA Technical Memorandum NMFS-NE-158. Washington DC, National Marine Fisheries Service, Department of Commerce.

Underdal, A. One question, two answers. In Miles *et al.*, eds. *Environmental Regime Effectiveness: Confronting Theory and Evidence.* Cambridge, MIT Press.

Young, O. 1989. *International Cooperation: Building Regimes for Natural Resource and the Environment.* Cornell University Press.

Young, O. and G. Osherenko, eds. 1993. *Polar Politics: Creating International Environmental Regimes.* Ithaca, Cornell University Press.

Young, O. ed. 1999. *The Effectiveness of International Environmental Regimes.* Cambridge: MIT Press.

Young, O. 1999. *Governance in World Affairs.* Cornell University Press.

Large Marine Ecosystems, Vol. 13
T.M. Hennessey and J.G. Sutinen (Editors)
© 2005 Published by Elsevier B.V.

16

An Evaluation of the Modular Approach to the Assessment and Management of Large Marine Ecosystems*

Hanling Wang

ABSTRACT

This contribution discusses the modular approach to the assessment and management of large marine ecosystems (LMEs). It addresses the contents and functions of the five modules; the key elements and processes of the Transboundary Diagnostic Analysis (TDA), Strategic Action Program (SAP) and National Action Plan (NAP) in the LME context; the principal common problems facing LMEs and their causes identified in TDAs and action plans formulated in SAPs, as the results of the practical application of the modular approach in LME projects. It also evaluates the significance of the modular approach for international ocean governance. It concludes that this integrated, ecosystem-based approach has rectified some deficiencies of the traditional sectoral approaches and has improved the understanding of LMEs and their management regimes. As a result the integrated, ecosystem-based approach is increasingly being endorsed in international governance of LMEs.

It is the current worldwide practice for large marine ecosystem (LME) projects to adopt an integrated, ecosystem-based approach to the assessment and management of the marine environment and resources.[1] The core of this approach is the application of a five-module assessment and management methodology to the processes of formulation and implementation of the Transboundary Diagnostic Analysis (TDA), Strategic Action Program (SAP), and National Action Plan (NAP). In view of the importance of LMEs to the global environment and socioeconomics, the approach to managing LMEs has significant effects not only on the LMEs themselves, but also on the densely populated human communities that rely on them. Research into this innovative approach to LMEs, which is being used increasingly by LME projects worldwide, has significant practical value. This article discusses and analyzes the latest methodological developments in the assessment and management of the marine environment and resources at the scales of large marine ecosystems (LMEs) and the application of these methodological developments in the LME projects. It examines the key elements and the processes of the modular approach, investigates its practical application, and evaluates its significance for international ocean governance.

---

* First published in *Ocean Development and International Law* 35:267-286(2004). Reproduced by permission of Taylor & Francis, Inc., http://www.routledge-ny.com
[1] For an in-depth discussion, see Hanling Wang, "Ecosystem Management and Its Application to Large Marine Ecosystem Management: Science, Law, and Politics," *Ocean Development and International Law*, 35:41-74(2004).

THE FIVE MODULES

The five linked modules for the assessment and management of LMEs have been defined and illustrated by Dr. Kenneth Sherman, *et al.*[2] and can be summarized as follows.

1) *Productivity module.* This module focuses on an LME's "carrying capacity"[3] for supporting fishery resources. Its measuring parameters are mainly photosynthetic activity, zooplankton biodiversity and oceanographic variability.[4] In this module, systematic measurements are made to monitor and assess the status and changes of these elements. The information obtained not only reflects the natural conditions of an LME, but also indicates the eutrophic impacts on it.

2) *Fish and fisheries module.* The focus of this module is on the changes of biodiversity of fish communities, which not only have impacts on fisheries in an LME, but also affect other components of the ecosystem. In this module, systemic surveys are conducted to obtain information on "changes in biodiversity and abundance levels of the fish community"[5] as well as their causes.

3) *Pollution and ecosystem health module.* This module deals with marine pollution, which is a major cause of the degradation and deterioration of the environment and resources in LMEs.

---

[2] See A.M. Duda and K. Sherman, "A New Imperative for Improving Management of Large Marine Ecosystems, " *Ocean and Coastal Management,* 45 (2002): 797-833; K. Sherman, "Modular Approach to the Monitoring and Assessment of Large Marine Ecosystems," in *The Gulf of Mexico Large Marine Ecosystem: Assessment, Sustainability, and Management,* ed. H. Kumpf, K. Sreidinger and K. Sherman (Malden, MA: Blackwell Science, 1999), 34; K. Sherman and A.M. Duda, "An Ecosystem Approach to Global Assessment and Management of Coastal Waters," *Mar Ecol Prog Ser* 190 (1999): 271-287; K. Sherman and A. Solow, "Modular Strategies for Assessing the Changing States of Large Marine Ecosystems," *ICES C.M.* (1996):2; K. Sherman, "Why regional coastal monitoring for assessment of ecosystem health?" *Ecosystem Health* 6(3), (2000): 205-216; K. Sherman, "Marine Ecosystem Management of the Baltic and Other Regions," *Bulletin of the Sea Fisheries Institute,* Gdynia, Poland, 3 (151), (2000): 89-99; K. Sherman and A.M. Duda, "Large Marine Ecosystems: An Emerging Paradigm for Fishery Sustainability," *Fisheries Management,* 24 (12), (1999):15-26; K. Sherman, "Large Marine Ecosystems: Assessment and Management from Drainage Basin to Ocean," paper presented at the Joint Stockholm Water Symposium / International Center for the Environmental Management of Enclosed Coastal Seas (EMECS) Conference in August 1997; K. Sherman, "Large Marine Ecosystems: Assessment and Management," in *Large Marine Ecosystems of the Pacific Rim: Assessment, Sustainability, and Management,* ed. Q. Tang and K. Sherman, (Malden, MA.: Blackwell Science, 1999), 445-448; K. Sherman, "Food Webs: Large Marine Ecosystems," in *Encyclopedia of Ocean Sciences,* ed. J. Steele, S. Thorpe, and K. Turekian (London: Academic Press, Ltd. 2001), 1462-1469; K. Sherman, *et al.,* "The U.S. Northeast Shelf Large Marine Ecosystem: Zooplankton Trends in Fish Biomass Recovery," in *Large Marine Ecosystems of the North Atlantic: Changing States and Sustainability,* ed. K. Sherman. and H.R. Skjoldal (Amsterdam: Elsevier Science, 2002), 197-211; C. Ibe and K. Sherman, "The Gulf of Guinea Large Marine Ecosystem Project: Turning Challenges into Achievements," in *The Gulf of Guinea Large Marine Ecosystem: Environmental Forcing & Sustainable Development of Marine Resources,* ed. J.M. McGlade, *et al.,* (Amsterdam: Elsevier Science, 2002), 27-39; and K. Sherman, "A Modular Approach to the Monitoring and Assessment of Large Marine Ecosystems," *ICES C.M.* (1997) EE:15, 1-12.
[3] D. Pauly and V. Christensen, " Primary Production Required to Sustain Global Fisheries," *Nature,* 374 (1995): 257.
[4] Sherman, "Large Marine Ecosystems: Assessment and Management," *supra* note 2, at 445.
[5] *Ibid.*

The monitoring and assessment of "the changing status of pollution and health of the entire LME"[6] are conducted in this module. Systemic monitoring of data on water quality and biological indicator species are used to measure pollution effects on the ecosystem and detect emerging disease. The state of the health of LMEs is examined on the basis of indices of ecosystem "biodiversity, stability, yields, productivity, and resilience." [7]

4) *Socioeconomic module.* This module addresses the human dimensions of LMEs. It mainly investigates the state of socioeconomic developments of the human communities connected to LMEs, especially the industries and human activities that are closely related to, or depend on, the LMEs. The output of this module is the social science-based, socioeconomic information of LMEs.

5) *Governance module.* The governance regimes for LMEs are formulated mainly on the basis of the information obtained from the above four modules as well as international rules and systems embraced in relevant global and regional agreements applicable to the areas concerned. Policy, legal, and institutional reforms are made, and other measures are taken at the regional and national levels to improve the governance of the LMEs. The guiding principle of an LME governance regime is the adoption of a holistic, ecosystem-based approach to the management and protection of the marine environment and resources. Attaining this requires the integration of scientific findings with socioeconomic considerations in the management of LMEs. Accordingly, an LME's management plans are developed and evaluated not only on the basis of scientific findings, but also on the socioeconomic elements of the human communities concerned, amongst other things, including their socioeconomic conditions and the socioeconomic impacts of management measures. Mechanisms for integrated management are established with a view to harmonizing the interest of marine environmental resource protection and the long-term socioeconomic benefits of the coastal communities concerned. The governance module, a work-in-progress, is to be based on international experiences gained from LME projects and integrated coastal and ocean management. The trend of the LME governance module is to move away from traditional sectoral, single species approaches to a more holistic, integrated approach to marine management. The general goal is to promote the long-term sustainability of marine ecosystem resources.[8]

Three of the modules (productivity, fish and fisheries, pollution and ecosystem health) focus on the natural state of ecosystems, the outcome of which is science-based information on LMEs. The fourth module (socioeconomics) concentrates on human dimensions, designed to produce social science-based information on LMEs, while the last module (governance) is designed or seeks to adjust human behaviours towards LMEs, and improve the relationship between human society and LMEs. Clearly, all these modules and their components

---

[6] *Ibid.*, at 446.
[7] *Ibid.*
[8] See generally the material in *supra* note 2 and see also L. Juda, "Considerations in Developing a Functional Approach to the Governance of Large Marine Ecosystems," *Ocean Development and International Law*, 30 (1999): 89-125 and L. Juda and T. Hennessey, "Governance Profiles and the Management of the Uses of Large Marine Ecosystems," *Ocean Development and International Law*, 32 (2001): 43-69. For more information on the five modules, see: http://www.edc.uri.edu/lme/intro.htm.

interrelate and interact with each other.[9] The first four models support the Transboundary Diagnostic Analysis (TDA), while the governance module is mainly associated with the Strategic Action Plan (SAP).[10] The adoption of the five modules in the TDA-SAP-NAP processes not only integrates science and socioeconomic elements with management regimes,[11] but also promotes the involvement and collaboration of scientists, managers, stakeholders and the public in LME regime building and implementation.[12] Together the modules facilitate the comprehensive assessment and integrated management of LMEs.

TRANSBOUNDAY DIAGNOSTIC ANALYSIS (TDA)

Definition

Transboundary Diagnostic Analysis (TDA) is a scientific and technical assessment of the environment of an international ocean area. TDA identifies, quantifies, analyzes, and assesses the water-related environmental issues and problems, their causes and impacts in the environmental, socio-economic, legal, political and institutional context at the national, regional and global levels. It normally also identifies and prioritizes solutions to the problems as well as the root causes of the problems.[13]

The identification of the elements that are of transboundary character and the linkages between the problems and their causes is crucial in the TDA process.[14] In the context of TDA,

---

[9] See *A Framework for Monitoring and Assessing Socioeconomics and Governance of Large Marine Ecosystems*, NOAA Technical Memorandum NMFS-NE-158, ed. J. Sutinen, 2000, at 3, online at: http://www.nefsc.noaa.gov/nefsc/publications/tm/tm158/tm158.pdf.
[10] See K. Sherman and A. M. Duda, "An Ecosystems Approach to Global Assessment and Management of Coastal Waters," in *Baltic Sea Regional Project: Project Implementation and Procurement Plan*, Volume 2, Part A, ed. J. Thulin, June 2002, Annex 6, at 27, online at: http://www.ices.dk/projects/balticsea/unzip/vol.2_A_1.doc.
[11] In this connection, Juda and Hennessey, *supra* note 8, at 49 point out:
> Ecosystem-based management needs the integration of contributions and inputs from both the natural and social sciences. Fundamentally, the natural sciences can provide an understanding or the functioning of natural systems, the relationship and dynamics of system components, and the impacts of human use on the operation and changing states of those natural systems. They may also be able to suggest the human use implications of system changes. On the social science side, the focus is on use management and efforts to modify use patterns to advance purposes such as system sustainability.
[12] See K. Sherman, "Large Marine Ecosystem Monitoring, Assessment, and Management across the global North-South divide," a paper presented at The Global Environment Facility Second Biennial International Waters Conference, Dalian, China, September 25-29, 2002, at 24, online at: http://www.iwlearn.net/event/presentations/iwc2002/agenda.php. (hereinafter "2002 Dalian Conference web site").
[13] See *The Benguela Current Large Marine Ecosystem Transboundary Diagnostic Analysis* (BCLME TDA), October 1999, online at: http://www.bclme.org/resources/; *The Yellow Sea Large Marine Ecosystem Preliminary Transboundary Diagnostic Analysis* (YSLME TDA), February 2000, at 7, online at: http://www.gefweb.org/COUNCIL/GEF_C15/WorkProgram.htm (C-21, Part IV); *Transboundary Diagnostic Analysis for the Caspian Sea* (Volume Two), The Caspian Environment Programme, Baku, Azerbaijan, September, 2002, at 1, online at: http://www.grida.no/caspian/additional_info/Caspian_TDA_Volume_Two.pdf; and *Transboundary Diagnostic Analysis for the South China Sea*, 2000, at 2, online at: http://www.unepscs.org/Publication/PDF-B/PDF-B.htm.
[14] *International Waters Managers' Insights Regarding the Global Environment Facility (GEF) International Waters Program Study: Transboundary Analysis, Demonstrations, Sustainability and Lessons Learned*, ed. Al Duda, et al, September, 2002, at 9, online at: http://www.iwlearn.org/ftp/GEF-IW-MGRS-2002-IWPS.pdf.

transboundary means crossing national maritime boundaries. Thus, transboundary environmental issues include, *inter alia*:

Regional/national issues with transboundary causes/sources; transboundary issues with national causes/sources; national issues that are common to at least two of the countries and that require a common strategy and collective action to address; and issues that have transboundary elements or implications (e.g. implications of fishery practices on biodiversity/ecosystem resilience).[15]

TDA identifies the transboundary nature, magnitude, and impacts of the various issues and problems pertaining to water quality, marine resources, biodiversity, habitat degradation, and their root causes, involving socioeconomic problems, policy distortions and institutional deficiencies.

The purpose of conducting a TDA is to assess and scale environmental disturbances and threats to LMEs, and their sources and causes, both immediate, intermediate, and root. It provides information relating to the changing states of LMEs and the causes of their degradation. The goal of a TDA is to provide a sound scientific, technical and factual basis on which to formulate a Strategic Action Plan (SAP) that adopts optimal, cost-effective actions to redress the environmental degradation of LMEs. International experience shows that undertaking TDA prior to the design of a SAP is appropriate and helpful for LME projects.

Process of the TDA

The contents of a TDA process can be classified into two major parts. The first part is a fact-finding process identifying the major perceived problems and their root causes and the second part is an evaluation of the intervention options to deal with the perceived problems.

As regards the fact-finding process and the identification of the major perceived problems, "perceived" problems represent "concerns that may not have yet been identified or proved to be major problems due to data gaps or lack of analyses, and concerns regarding major threats of future degradation conceived in the context of prevailing trends."[16] The major perceived problems of LMEs are normally generic in nature, *e.g.*, pollution of the marine environment and deterioration of marine living resources. However, the extent of such problems in each LME varies. Therefore, the TDA for each LME needs to address the scale of the problems specific to that LME, especially those problems of a transboundary nature in both national jurisdictional areas and international waters.[17]

The next step is tying the problem to its cause. The immediate, intermediate and root causes of the identified issues and problems are investigated. The investigation and analysis should not be limited to the natural environment domain, but move through the chain of cause and effect to the root causes in the management, socioeconomic, legal, and cultural domains.[18] The causal

---

[15] See "Users Guide to the Transboundary Diagnostic Analysis," in the BCLME TDA, *supra* note 13.

[16] J. M. Bewers, *An Evaluation of the Transboundary Diagnostic Analysis (TDA) Approach to the Preparation of Strategic Action Programs (SAPs), International Waters Program Study*, (report to the Global Environment Facility), Annex 8, at 11.

[17] *Ibid.*

[18] John Pernetta, *Working Together: Transboundary Diagnostic Analyses (TDA), Strategic Action Programmes (SAP), and Participatory Processes*, available at 2002 Dalian Conference web site, *supra* note 12.

chain analysis identifies the linkages between human activities and their environmental effects, especially their transboundary consequences, which facilitates the identification of the options for restorative and preventative interventions.

The second part of the TDA process is an evaluation of the options for intervention. Although this can be conducted in the subsequent SAP process, it may also be part of the TDA process. The idea is to assess the net benefits of each of the options for intervention and its possible adverse effects on the natural environment and socioeconomic development since each intervention has costs and benefits. The costs include not only financial costs, but also adverse effects on the environment, resources and socio-economics. Accordingly, various possible interventions should be assessed and compared in order to single out the one that can offer the greatest net benefits in relation to costs and other adverse effects on the communities concerned in the context of the prevailing technical, socioeconomic and political conditions. The results of this assessment allow the incorporation of the optimal interventions in the Strategic Action Plan (SAP).

After the identification of the optimal interventions, the next step in the TDA process is to prioritize the areas for intervention. While each LME has its own priority areas for intervention, the following imminent threats to international waters are considered by the Global Environment Facility (GEF) as priorities for prevention and control: 1) land-based sources of pollution, especially persistent toxic substances, heavy metals and common contaminants such as nutrients, biological contaminants, or sediments; 2) land degradation where transboundary marine environmental concerns result from desertification or deforestation; 3) degradation and modification of critical habitats; 4) unsustainable use of marine resources; 5) ship-based sources of chemicals and alien species. With regard to the methods for addressing the problems, priority will be given to holistic, rather than sectoral, approaches to the management and prevention of these environmental threats.[19]

The final step of the TDA process is to further specify the circumstances pertaining to each action area, including a detailed description of the action areas, relevant data and information, contemporary knowledge and gaps in the understanding of the problems and options for intervention.[20]

In the conduct of a TDA, a number of considerations exist. First, a TDA requires a holistic and multisectoral consideration of the issues and problems of LMEs. The effective participation of the sectors that create stresses on LMEs, stakeholders and the public, is an important component of the TDA and the consequent SAP process. As such, the TDA process is a valuable vehicle for "multilateral consultation and exchanges of perspectives and constraints."[21]

Second, a TDA is to be prepared and agreed upon jointly by the science and management communities as well as other stakeholders in each collaborating country. The TDA process involves work at both the national and regional levels. It is necessary to gather information on

---

[19] *Operational Strategy of the Global Environment Facility*, Chapter 4, International Waters, online at: http://www.gefweb.org/public/opstrat/ch4.htm.
[20] Bewers, *supra* note 16, at 11-13. For an example of this part of the TDA, see the Yellow Sea LME TDA, *supra* note 13, at 32-91.
[21] *International Waters Managers Insights*, *supra* note 14, at 24-25.

relevant issues and problems as well as their causes within national boundaries. An interministerial committee is normally established in each collaborating state to deal with the national elements of the TDA and the consequent SAP, and develop the National Action Plan (NAP). National environmental plans and documents are used to identify environmental priorities. The analysis of the causes of degradation and the subsequent remedial actions and capacity building should take into consideration national economic development plans, sectoral economic policies, and other relevant legislation and policy.

The TDA needs to be conducted on a multilateral basis involving all the states concerned because the issues and problems of LMEs normally have their origins and consequences beyond the boundaries of individual states. Furthermore, the evaluation of options for intervention has to be conducted at the regional level, especially where the effective implementation of interventions requires international cooperation, and where national interventions have potential effects on other countries sharing the same LME.

A last consideration is that the geographical scope of the regional TDA should cover the whole LME and its basins,[22] although some specific problems can be identified in a narrower geographical scope. On the one hand, it is essential to examine linkages among coastal zones, LMEs, and their contributing freshwater basins so that the root causes in upstream basins can be considered in the subsequent integrated management plans.[23] On the other hand, specific problems and their causes in sub-areas and priority hotspots may be geographically identified within the LME so that complex transboundary issues can be divided into "smaller, manageable geographic ones."[24] In this manner, different issues in

---

[22] The justification for a regional and holistic approach to the TDA is well described in some LME TDAs and is worth reproducing.

> Conducting a comprehensive transboundary diagnostic analysis is only possible if an entire water basin or Large Marine Ecosystem and its associated drainage basin is covered under the study. This is required in order that the interactions between the aquatic, terrestrial, and human sub-systems are identified in so far as they are linked through the mechanism of the hydrological cycle. More particularly, the impacts of the land-based activities on water resources and their contribution to water-related environmental stresses can be demonstrated only if all sources, sinks, and shared marine resources are included in the assessment. This successful demonstration requires the commitment of all the countries that are located in the catchment basin or surround the shared marine area to participate in the process (from Yellow Sea Large Marine Ecosystem (YSLME) TDA, supra note 13 at 9).

See, for example, the YSLME TDA, supra note 13, at 9 and Annex 11 of the Caspian TDA, June 1998, at 60, at: www.gefweb.org/wprogram/Oct98/UNDP/caspanex.doc. In the Yellow Sea LME project, although North Korea has not fully participated, it has been actively involved in other related projects and activities concerning marine environmental protection in the region, and its government has indicated that it may participate in the YSLME project at a later stage. See paragraph 44 of *The Project Brief of the YSLME Project*, at 8 and 15, online at: http://www.gefweb.org/COUNCIL/GEF_C15/WorkProgram.htm. (C-21, Part II).

The early phase of the Guinea Current LME (GCLME) Project covered only six countries out of the sixteen countries bordering the GCLME. The GCLME project managed to attract a full participation of all the littoral countries in a later stage. See Edwin P. D. Barnes, *"Large Marine Ecosystems and Coasts: Experiences and Lessons Learned: GEF-UNDP-UNIDO Industrial Water Pollution Control In the Gulf of Guinea Large Marine Ecosystems,"* at 8-9, available at 2002 Dalian Conference web site, *supra* note 12.

[23] Alfred Duda, *"Monitoring and Evaluation Indicators for GEF International Waters Projects (2002),"* Monitoring and Evaluation Working Paper 10, November, 2002, at 4, online at: http://www.gefweb.org/ResultsandImpact/Monitoring_Evaluation/Evaluationstudies/M_E_WP_10.pdf.

[24] *Background Paper – GEF International Waters Focal Area*, para. 3, online at: http://www.wfeo-comtech.org/ConferenceOutcomes/GEFWatersWorkshop/item18.html.

different portions of an LME and its basins can be addressed by geographically specific actions.

Major common problems of LMEs and their causes identified in TDAs

LME case studies show that almost all LMEs suffer from degradation and deterioration of the natural environment and resources. According to the modular-approach-based TDAs, common problems of LMEs are: major declines in commercial fish stocks and non-optimal harvesting of marine living resources; deterioration of water quality both chronic and catastrophic; modification to seabed and coastal zones; habitat destruction and alteration; loss of biotic integrity and decline in biodiversity; loss of endangered species and their genomes; and harmful algal blooms.[25]

The root causes are the same for a large number of these problems: overfishing; climate regime shifts; and pollution and eutrophication.[26] Other causes include alterations of physical habitat and invasions of exotic species.[27] Furthermore, indirect causes are deficiencies and shortages in law, policy, institutions, economy and technology. For example, in the Benguela Current Large Marine Ecosystem (BCLME), causes are identified as: poor legal frameworks; inadequate implementation of existing legislation; inadequate capacity development (human and infrastructure) and training; inadequate planning at all levels; insufficient public involvement; and inadequate financial mechanism and support.[28]

THE STRATEGIC ACTION PLAN (SAP)

Definition and Process

In the context of coastal and ocean management, a Strategic Action Plan (SAP) is a regional framework for actions to manage and protect the coastal and marine environment and resources. In international LME projects, an SAP is an instrument agreed among the collaborating countries, which contains a series of policy, institutional and other socioeconomic actions to be taken at both the national and regional levels aimed at enhancing

---

[25] See Benguela Current LME TDA, *supra* note 13; *Project Brief of the YSLME Project*, Annex A Incremental Cost Analysis, *supra* note 16, at A-1; *A Transboundary Diagnostic Analysis and Strategic Action Programme for the Gulf of Mexico Large Marine Ecosystem*, at 2, online at: http://www.gefonline.org/projectDetails.cfm?projID=1346; M. J. O'Toole, "*Benguela Current Large Marine Ecosystem Programme: Experiences and Some Lessons Learned*," at 2, available at 2002 Dalian Conference website, *supra* note 12, or http://www.bclme.org/resources/; and "*Gulf of Guinea Project*," online at: http://edu.eforie.ro/carmensylva/iwlearn/guineea.html. See also Barnes, *supra* note 22, at 1-2.

[26] *Large Marine Ecosystem (LME) Approach to the Global International Water Assessment (GIWA)*, Working Document of the Technical Workshop for Establishing a Regular Process for the Global Assessment of the Marine Environment, Bremen, Germany, 18-20 March, 2002, online at: http://www.unep.org/DEWA/water/MarineAssessment/reports/germany_report/LME-GIWA.doc.

[27] See Global Environment Facility, *Operational Strategy* (Washington, D. C., February 1996), 47 and *Operational Strategy of the Global Environment Facility*, Chapter 4, International Waters, *supra* note 19. See also M. E. Huber, *et al.*, "Priority problems facing the global marine and coastal environment and recommended approaches to their solution," *Ocean and Coastal Management*, 46 (2003): 479-485 and *Jakarta Mandate: Marine and Coastal Biodiversity – Introduction*, at: http://www.biodiv.org/programmes/areas/marine/.

[28] See the Benguela Current LME TDA, *supra* note 13 and O'Toole, *supra* note 25, at 2.

the management of the environment and resources of the LME concerned.[29] The SAP is to be based on the output of the TDA. In accordance with the priority transboundary concerns and their root causes identified in a TDA, the countries concerned are to collaboratively determine the specific actions that they will take collectively or individually to resolve the identified problems. Normally, these action plans are developed by each collaborating country, often through a national interministerial committee with participation by relevant stakeholders. Then a multinational committee compiles and harmonizes these plans and formulates an SAP for endorsement at the highest levels in governments.[30] The SAP for an LME is a political commitment of the governments concerned to accept agreed management principles and to implement agreed policy actions in order to manage and protect an LME.[31]

An SAP usually sets out the agreed priorities and the array of expected environmental baseline and specific actions needed for resolving priority transboundary environmental concerns. Priority preventive and remedial actions are specifically identified and highlighted. The SAP also provides mechanisms for the long-term preservation, protection and restoration of an LME.[32]

Besides the above-mentioned elements, the following points should be considered in the formulation of an SAP. First, although the problems of LMEs are complex, each LME has its own features. Second, in principle, the SAPs for LMEs generally adopt an integrated management approach, for instance, "integrated freshwater basin—coastal area management measures" are considered to be important for protecting LMEs.[33] Third, although formulation of an SAP relies on the scientific and technical justification provided in the TDA, legal, policy, and socioeconomic elements must also be taken into consideration in the specific combination of activities contained in an SAP. More particularly, activities included in the SAP should be realistically based on the management resources available.[34] Finally, an SAP should embody a philosophy of adaptive management so as to provide for periodic review of the environmental status of the LME, identification of new issues and problems as well as new management measures, and updating of the SAP to better address the ever-changing situation.[35]

Actions Adopted in SAPs

Up to now most LME projects have not finalized their SAPs, thus a comprehensive assembly of actions adopted by SAPs to address the issues and problems of LMEs is not possible. However, a study of some of the existing SAPs show that actions taken to resolve issues and problems of LMEs at both the national and regional levels consist mainly of improvements in

---

[29] See, for example, *The Strategic Action Programme for the Red Sea and Gulf of Aden*, 1998, at: http://www.unep.ch/seas/main/persga/redsap.html.

[30] A. Merla, "*A Commitment to the Global Environment: The Role of GEF and International waters*," online at: http://www.oieau.fr/ciedd/contributions/at2/contribution/gef.htm.

[31] See O'Toole, *supra* note 25.

[32] See *Operational Program Number 8, Waterbody-Based Operational Program*, at 8-4, online at: http://www.gefweb.org/operational_policies/operational_programs/op_8_english.pdf.

[33] *Operational Strategy of the Global Environment Facility*, Chapter 4, International Waters, *supra* note 19.

[34] See, for example, *ibid.*

[35] For more information on the preparation of SAPs, see, for example, A. Hudson, "*Strategic Action Programme Preparation in Complex Contexts: Issues and Best Practices*," online at: http://www.freplata.org/actividades/reuniones/Simposio20020515/Presentaciones/PDF/Andrew%20Hudson.pdf.

the assessment and management methods and measures; increases of management resources and investments; regulatory reform and policy changes; strengthening of institutional mechanisms for implementation; capacity-building; stakeholder involvement and public awareness activities; program monitoring and evaluation; and coordination of priorities with those identified under other focal areas such as climate change, biodiversity.[36] The areas of action can be classified as technical, financial, legal, policy, and institutional.

From a technical aspect, the improvement of LME management relies on advances in the understanding of LMEs and the improvement of management methodologies and techniques, which encompass a number of things. First is improvement of the understanding of LMEs as well as the complex interactions among their components and relevant environmental elements, especially in the transboundary context. Actions taken in this connection include international joint surveys and assessments of shared marine living resources, application of new technologies and equipment in monitoring and assessment, and exchange of information and experience and technical training. Second, methodological improvements of LME management require a shift from traditional and ineffective approaches to innovative and effective approaches, *i.e.* from single-species to multi-species assessment and management; from sectoral management to integrated management; from jurisdictional-boundary-based management to ecosystem-based management; mingling community-based management with regional cooperation;[37] and blending integrated coastal zone management with LME management. Specific measures that might be taken include: protection of fish stocks throughout their migratory range; protection of dependent and associated species; establishment of marine protected areas; and protection of marine environment through the prevention, reduction, and control of land-based pollution, air-borne pollution as well as sea-borne pollution.

Sufficient funding is essential to the successful implementation of an SAP, and addressing financial issues is an important part of the SAP. Based on the cost benefit analysis which is normally done in the TDA process, the SAP normally figures out estimated costs for the agreed actions and lays out funding plans. For example, the South China Sea SAP specifically lists the estimated costs for each of the actions planned in the SAP and indicates possible major sources of funding.[38] With regard to the means of financing the actions adopted in the SAP, the above-noted Benguela Current Large Marine Ecosystem (BCLME) SAP provides that member states are to seek the necessary funding from "national, regional and international sources," through "private and general public funding" or through "the application of specific economic instruments" and "grants and loans."[39] Specific projects for bilateral and multilateral funding are to be prepared and donor conferences held every five years. Specific funding arrangements for the national policies and measures are required to be presented in the National Action Plan (NAP) of every member state.[40]

---

[36] See, for example, *Operational Strategy of the Global Environment Facility*, Chapter 4, International Waters, *supra* note 19.

[37] Global Environment Facility, *International Waters Program Study* (Final Report), at 22, online at: http://www.iwlearn.net/ftp/iwps.pdf.

[38] See, for example, *Strategic Action Programme for the South China Sea* (Draft Version 3, 24 February 1999), at 52-68, online at: http://www.unepscs.org/Publication/PDF-B/PDF-B.htm.

[39] Part V of *the Strategic Action Program for the Benguela Current Large Marine Ecosystem*, revised version of November 2002, at: http://www.bclme.org/resources/.

[40] *Ibid.*

Legal actions to be taken that may be set out in the SAP include formulating a legal framework for the protection and management of the LMEs concerned, for example, concluding a legally binding regional agreement and other arrangements. The SAP may urge the collaborating states to accede to the relevant international agreements and arrangements applicable to the LMEs concerned. Finally, the SAP may call for reform of existing national legislation and the making of new legal regulations. In the SAP for the South China Sea (SCS SAP), for example, member states considered it a priority to endorse a legal framework upon which to facilitate and commit governments to the SAP and relevant regional cooperation.[41] In the SAP for the BCLME, member states agreed to "adopt appropriate legislation."[42]

Collaborating states are to incorporate into national and regional polices the relevant policies and principles that are adopted in the SAPs and other relevant international documents, such as the policy documents emanating from the 1992 United Nations Conference on Environment and Development (UNCED) and the 2002 World Summit on Sustainable Development (WSSD).[43] For example, the SAP for the BCLME requires member states to adopt the following principles: protection of the ecosystem integrity for future generations; application of the precautionary principle where appropriate; taking anticipatory and co-operative actions, such as contingency planning, environmental impact assessment and strategic environmental assessment; use of environmentally friendly technologies and socioeconomic policies; inclusion of environmental, ecosystem and human health considerations into all relevant policies and sectoral plans; and encouragement and fostering of transparency, public participation and cooperation in the process of LME management.[44] While these principles are not new to international environmental law and policy, the SAP develops suites of policy actions based on these principles and approaches to address the identified specific priority issues of the LMEs. Some specific actions formulated in SAPs will be discussed below.

SAPs inevitably recommend the need to strengthen the institutional mechanisms responsible for the management of the LMEs concerned. Nationally, this may involve coordination of all the relevant agencies at both the national and local levels. At the regional level, institutional mechanisms for regional cooperation in LME management may need to be established, coordinated and strengthened. At the global level, the SAP may recommend strengthening the cooperation between national, regional institutions and relevant international organizations such as various UN agencies and NGOs involved in the protection and management of the oceans. Furthermore, it is seen as helpful for different marine regions to communicate, cooperate and exchange information. For example, in the case of the Gulf of Mexico, there is no current institutional arrangement for co-operation among the three riparian states—the United States, Mexico, and Cuba—in the protection of the Gulf of Mexico Large Marine Ecosystem (GOMLME). Thus, one of the main objectives of the GOMLME project is to establish a regional institutional arrangement to provide for co-operation among these three

---

[41] See *Strategic Action Programme for the South China Sea, supra* note 38, at 67.
[42] *The Strategic Action Program for the Benguela Current Large Marine Ecosystem*, Part VI, Principles, *supra* note 39.
[43] These documents are available at: http://www.un.org/esa/sustdev/documents/docs.htm.
[44] *The Strategic Action Program for the Benguela Current Large Marine Ecosystem*, Part II, Principles, *supra* note 39.

states.[45] For the BCLME, the following institutional arrangements have been agreed upon for establishment: 1) a Benguela Current Commission; 2) a program co-ordination unit; 3) three activity centres, which are respectively in charge of environmental variability and predictability, living marine resources, and biodiversity, ecosystem health and pollution; and 4) advisory groups on fisheries and other living marine resources; environmental variability, ecosystem impacts and improved predictability; biodiversity and ecosystem health; marine pollution; legal and maritime affairs; information and data exchange; and training and capacity development. [46]   As noted in the South China Sea SAP, the successful implementation of the actions formulated in SAPs is dependent on coordination of actions by various organizations, agencies, private sectors, and stakeholder groups at both the national and regional levels.[47]

As mentioned above, in an SAP member states determine specific actions to be taken to address the identified issues and problems. Examples of such actions can be seen in the SAPs for the BCLME and the South China Sea.

In the BCLME SAP, the proposed areas for policy actions are as follows: 1) sustainable management and utilization of living marine resources and management of mining and drilling activities; 2) assessment of environmental variability, ecosystem impacts and improvement of predictability; 3) management of pollution; 4) maintenance of ecosystem health and protection of biological diversity; and 5) capacity strengthening.[48]

In the BCLME SAP, policy actions are formulated for each of the above thematic areas. For instance, in order to ensure the sustainable management and utilization of the living marine resources of the BCLME, the member states agree to take the following policy actions:

- establishing a regional structure to assess transboundary fishery resources and ecosystems and to provide advice to governments;
- cooperatively undertaking joint surveys and assessment of shared stocks of key species;
- harmonizing the management of shared stocks through, *inter alia*, addressing technical issues such as fishing gear, mesh size/type, compatible data and assessment methodology;
- cooperatively assessing non-exploited species by gathering and calibration of baseline information on these species, and assessment of the impact of any future harvesting on the ecosystem;
- developing a responsible regional mariculture policy to harmonize national policies in such a manner that actions of one state do not impact negatively on the economic potential of another, nor on the ecosystem as a whole;
- cooperatively analyzing socio-economic consequences of various harvesting methods, the improved use of living marine resources and the economic value of the BCLME as

---

[45] *Project documents of the Gulf of Mexico Large Marine Ecosystem project*, PDF-B Document (Revised), at: http://www.gefonline.org/projectDetails.cfm?projID=1346.
[46] *The Strategic Action Program for the Benguela Current Large Marine Ecosystem*, Part II, Institutional Arrangement, *supra* note 39. See also O'Toole, *supra* note 25, at 3.
[47] See *The Strategic Action Programme for the South China Sea*, *supra* note 38, at 5.
[48] *The Strategic Action Program for the Benguela Current Large Marine Ecosystem*, Part IV, Policy Actions, *supra* note 39.

an ecosystem, with a view to appropriate intervention within the framework of improving sustainable ecosystem use/management and quantifying regional and global benefits;

- harmonizing national policies on protected areas and other conservation measures; and
- complying with the FAO Code of Conduct for Responsible Fisheries.[49]

In the South China Sea SAP, the priority actions focus on: 1) the protection of mangroves, coral reefs, seagrass, estuaries and wetlands;[50] 2) protection of fishery resources from overexploitation;[51] and 3) prevention, control, reduction and elimination of land-based pollution.[52] For each identified issue, specific targets and proposed activities at both the regional and national levels are formulated. For example, to deal with the issue of over exploitation of fisheries, the proposed targets are to achieve the following tasks by 2005:
1) to determine regional catch levels of key economically important species;
2) to have established a regional system of marine protected areas for the conservation and protection of fishery stock and endangered species; and
3) to have developed and implemented a management system for the sustainable development of exploited resources at chosen sites.

In order to meet these targets, the South China Sea SAP proposes suites of activities that are to be carried out at the regional and national level. At the regional level, these activities are:
- developing criteria for selection of marine habitats and areas critical to the maintenance of regionally important fish stocks, particularly those of transboundary importance;
- identifying and prioritizing specific areas for future management and protection;
- formulating regional and national action plans on the development of a regional system of refugia for maintenance of regionally important fish stocks;
- developing and establishing management regimes for the identified areas;
- reviewing destructive fishing activities with the aim of removing and replacing them;
- reviewing fisheries management systems; and
- reviewing compliance to international fisheries conventions.

At the national level, the following activities are proposed:
- establishing marine protected areas in areas identified as critical habitats for fish stock conservation and protection of endangered species;
- implementing programs to provide information on fish stock conservation and sustainable fishery practices among small and artisanal fishing communities;
- conducting resource assessment of fishery resources to determine the level of optimal catch and effort for different fishing grounds in the region;
- developing educational and public awareness materials on sustainable fishery practices for dissemination in member states;
- establishing in selected pilot sites a good management system which can be tested to determine if it is leading to sustainable exploitation of resources; and

---

[49] *Ibid.*, Part IV (A), Policy Actions.
[50] See *The Strategic Action Programme for the South China Sea, supra* note 38, at 13-22.
[51] *Ibid.*, at 22-24.
[52] *Ibid.*, at 24-31.

- promoting the dissemination of and compliance with the Code of Conduct for Responsible Fisheries.[53]

There are some similarities and some differences between the actions formulated in the SAPs for the BCLME and SCSLME. In general, besides some differences in contents, the actions formulated in the SCS SAP appear to be more concrete than those set out in the BCLME SAP, in that the former creates explicit targets for each issue and formulates specific actions to be undertaken at both the regional and national levels, while the policy actions developed in the BCLME SAP are mainly for the regional level. As pointed out in the BCLME SAP, to ensure transboundary cooperation on integrated management of the living resources of the BCLME, the member states have realized "the pressing need to take further concrete actions individually and collectively, at national and regional levels." [54]   In the light of this understanding, the BCLME SAP requires each member state to develop detailed national action plans to further facilitate its implementation.[55]

Despite the inevitable diversity and variety of the actions adopted by different SAPs, there are also similarities. The major common feature of the actions formulated in LME SAPs is that they are focused on the priority issues and problems associated with the modules of productivity, fish and fisheries, pollution and ecosystem health, and socioeconomics, which are identified in TDAs and that their aim is to improve the governance of LMEs. Besides this, since many of the issues and problems different LMEs face are similar, some of the recommended actions to be taken to address them are also similar. This is demonstrated, for example, by previously listed actions which address the same problem of over-exploitation of fishery resources in two different SAPs. The same actions that are proposed in both SAPs are fishery resource assessment, establishment of protected areas, improvement of management systems, and promotion of the compliance with the FAO Code of Conduct for Responsible Fisheries.

THE NATIONAL ACTION PLAN (NAP)

As already noted, a regional SAP is normally a compilation and harmonization of individual (preliminary) national action plans (NAPs). The SAP approach aims at "streamlining the linkage between regional and national priority actions."[56] In order to facilitate member states to act in accordance with the regional SAP, LME projects provide institutional support for relevant national government bodies. At the regional level, there is normally a steering committee and a program coordination unit (PCU) for each LME project to deal with the development and implementation of the regional SAP. At the regional and global levels, the national implementation of the SAP is guided, coordinated, and assisted by the regional and global consulting, coordinating and supporting mechanisms for LME management.

---

[53] *Ibid.*, at 23-24.
[54] *Strategic Action Programme for the South China Sea, supra* note 38, Introduction, paragraph 8.
[55] *Ibid.*, Part IV, National Strategic Action Plans.
[56] UNEP, *Review of Alternative Regional Approaches to Implementation of the GPA*, UNEP/GPA Coordination Office, (2001), at 19.

In order to implement the regional SAP at the national level, each member state has to adjust, modify, and improve its national action plan (NAP) in accordance with the SAP and take necessary enforcement and compliance-enhancing measures at the national, sub-national and local levels. Each member state is to incorporate the contents of the SAP into relevant national legislation, policy, and planning. Specific national measures and investments are required for the implementation of NAPs. Furthermore, each member state should accede to and implement international and regional agreements, arrangements and policy documents that are conducive to the aims and objectives of the SAP.

The requirement for the establishment of NAPs is normally explicitly addressed in a SAP. For example, the BCLME SAP requires each member state to develop a national strategic action plan or other corresponding document, which is to present details of additional national actions, including details of responsibilities and specific projects where possible, to further implement the SAP.[57] In the SCS SAP, the collaborating states are required to integrate the provisions of the SAP into their national plans and policies, and take necessary actions. To encourage and facilitate national agencies to implement the SAP, a set of guidelines for the preparation of national plans are to be developed at the regional level.[58]

To date, most of the LME projects have not produced completed NAPs. Since many NAPs are under preparation and authoritative guidelines for developing NAPs have not been seen,[59] a study of the existing NAPs may afford useful lessons.

As a part of the preparation for the development of the regional SAP, preliminary national action plans are normally formulated. Examples of preliminary national action plans can be seen in the reports of the littoral states of the South China Sea (SCS) on the formulation of a TDA and the preliminary framework of the SAP for the SCS.[60] In these national reports, the NAP was based on the "national TDA" which consisted of: 1) detailed analysis of major water-related concerns and principal issues; 2) national analysis of the social and economic costs of the identified water-related principal environmental issues; and 3) and analysis of the root causes of the identified issues.[61]

The NAPs for the SCSLME member states are uniformly structured as follows.

1) *Constraints to Action*. This part of a NAP analyzes barriers to the member state's national actions for the identified issues. Common major constraints include: lack of information, scientific uncertainties, and lack of public awareness; economic and financial shortages; and

---

[57] *Ibid.*
[58] See *Strategic Action Programme for the South China Sea, supra* note 38, at 74.
[59] However, there are guidelines on land-based pollution, see UNEP/GPA Coordination Office, *UNEP Handbook on the Development and Implementation of a National Programme of Action for the Protection of the Marine Environment from Land-based Activities*, (The Hague, The Netherlands), at: http://www.gpa.unep.org/documents/NPA/NPA%20ENGLISH.pdf.
[60] National reports of Cambodia, China, Indonesia, Malaysia, the Philippines, Thailand, and Vietnam on the formulation of a Transboundary Diagnostic Analysis and preliminary framework for a Strategic Action Program for the South China Sea, online at: http://www.unepscs.org/Publication/PDF-B/PDF-B.htm.
[61] *Ibid.*, the national reports of Cambodia, at 3-69; China, at 10-50; Indonesia, at 11-122; Malaysia, at 10-47; the Philippines, at 4-63; Thailand, at 4-40; and Vietnam, at 7-93.

legal, institutional, and managerial deficiencies.[62] Other constraints are: shortage of capable human resources;[63] lack of political will to apply sustainable development principles in marine resource utilization;[64] and lack of public involvement in decision making on mega projects.[65]

2) *Ongoing and Planned Activities Relevant to the Identified Issues.* This part objectively describes the member state's existing action plans and programs for the nationally identified environmental issues related to the South China Sea.[66]

3) *Specific Action Proposed for Each Identified Issue.* This part is the core of the national report. It formulates specific actions to address each identified issue.[67] The National Report of Thailand is a good example of formulating these action plans. In the Report, there is a description of the major content of the action, the rationale of the action, participating agencies, and costs and sources of funding.[68]

4) *Implications for the Proposed Actions by Sectors.* This part mainly illustrates the effects of the proposed actions on relevant sectors, and the involvement and roles of relevant sectors in the implementation of the proposed actions.[69] A comparison of the SCS national reports shows that the contents of this part in different reports are diverse. Although some of the identified issues in different states are identical or similar, the action plans and programs adopted by each member state are not the same. Discrepancies also exist between the actions proposed in the regional SAP and those in the national reports.[70] As an international study of the SAP approach asserts:

> Due to conflicts of interest influenced by local and national economic and political considerations, priority sites and issues identified on a regional level may be restricted to only those that were politically agreeable to all governments involved. Also national priority setting can differ substantially, because of differences in economic strength and needs between participating countries. This may lead to a gap between national and regional priorities in terms of actions perceived as important. Poverty alleviation and community development are often highly rated as national priorities, but are not listed as priority regional actions.[71]

---

[62] *Ibid.*, the national reports of Cambodia, at 70-76; China, at 51; Indonesia, at 123-126; Malaysia, at 48-49; the Philippines, at 69-73; Thailand, at 41-46; and Vietnam, at 94-95.

[63] *Ibid.*, the national reports of Indonesia, at 125 and Thailand, at 41.

[64] *Ibid.*, the national report of Indonesia, at 125.

[65] *Ibid.*, the national report of Thailand, at 42.

[66] *Ibid.*, the national reports of Cambodia, at 76-80; China, at 52-53; Indonesia, at 126-132; Malaysia, at 50-53; the Philippines, at 63-69; Thailand, at 46-79; and Vietnam, at 95-98.

[67] *Ibid.*, the national reports of Cambodia, at 81-83; China, at 54-59; Indonesia, at 133-136; the Philippines, at 73-81; Thailand, at 80-95; and Vietnam, at 99-100.

[68] *Ibid.*, the national report of Thailand, at 80-95, especially 93-95.

[69] *Ibid.*, the national reports of Cambodia, at 83-87; China, at 59-60; Indonesia, at 136-139; Malaysia, at 53-54; the Philippines, at 73-81; Thailand, at 95-101; and Vietnam, at 101-102.

[70] For example, actions for the protection and conservation of habitats, see the national report of Viet Nam, at 100; the national report of Cambodia, at 83; and *Strategic Action Programme for the South China Sea, supra* note 38, at 22-23.

[71] UNEP, *Review of Alternative Regional Approaches to Implementation of the GPA, supra* note 56, at 20.

In order to promote international cooperation in dealing with common transboundary environmental issues and enhance effectiveness, it is necessary to coordinate and harmonize national actions at the regional level. For member states, this means a need to modify their national action plans and programs in accordance with the regional SAP.

The National Caspian action plans (NCAPs)[72] formulated by the Caspian riparian states can serve as another example of NAPs for the protection and enhancement of the marine environment of a regional sea, although the Caspian Environmental Program (CEP) is not an LME project. Differing from the above-mentioned national reports on the SCS, which are a combination of national TDAs and SAPs, the NCAPs are self-contained. Among them the NCAP of the Azerbaijan Republic[73] is most representative.

The Azerbaijani NCAP consists of eight parts and an annex with a list of priority programs and projects for the conservation and sustainable use of the Caspian Sea.
1) The *Introduction* describes the objectives of the NCAP; the connection of the NCAP with the TDA and the SAP; the connection of the NCAP with the priority investment projects portfolio in the region; the methods for developing the NCAP; the national status of the NCAP (the means of its endorsement and implementation by the government); and the process of revision and amendment of the NCAP.
2) The *National Conditions* describes the current national political, institutional, legislative, and socioeconomic situations and their future development prospects, especially in relation to the Caspian environment; and evaluates the nation's social, institutional and financial capacity for the protection of the Caspian environment.
3) The *Importance of the Caspian Sea for the Country* defines the national geographical and economic areas where human activities and the Caspian environment interact and have significant mutual impacts; identifies potentials for the increasing contribution of the Caspian Sea to national economic development; and shows the environmental, economic, and social significance of the Caspian Sea in a national context in the present and in the future.
4) The *Main Problems and Their Root Causes* is the core of the TDA as reflected in the NCAP. This part identifies, quantifies, and prioritizes existing and emerging national and transboundary issues and problems of the Caspian environment from a national perspective; and provides a causal analysis which links these issues and problems to immediate and root causes of both natural and anthropogenic.
5) The *Strategy and Measures* is the core of the NCAP. This part defines criteria for the ranking of causes and determination of primary strategies and measures; and classifies "long term strategies" and "urgent measures" for the elimination of root causes of the Caspian environmental problems.
6) The *Potential Obstacles and Ways of Overcoming* identifies and examines a range of issues in politics, institutions, socio-economics, human resources, technology, and finance that may hinder successful solution of the problems, and proposes solution to these obstacles.
7) *The Resources Attraction Strategy* identifies the main financial resources for the implementation of the NCAP, including national and external resources.
8) The *Mechanisms of Actions* identifies and establishes the organizational structure for implementing the NCAP and the evaluation system for assessing the state of the

---

[72] The NCAPs are available at: http://www.caspianenvironment.org/report_technical.htm#ncap.
[73] Ministry of Ecology and Natural Resources of Azerbaijan Republic, *National Caspian Action Plan of Azerbaijan Republic*, Baku, 2002, at: http://www.caspianenvironment.org/report_technical.htm#ncap.

implementation. It also establishes a mechanism for making the entire process of the NCAP implementation transparent and accountable to the nation, and for raising public awareness.

As mentioned in the introduction of the NCAP, the previous TDA is used as a technical basis for the NCAP. Thus, the NCAP and the TDA are closely connected. Although the NCAP does not indicate the application of the five modules to its formulation process, the identification of the causes of the environmental issues and problems as well as their solutions has incorporated most of the contents of the five modules. The NCAP deals with environmental issues from a multi-dimensional, multi-sectoral or integrated perspective, involving not only science and technology, but also socioeconomic, political, and legal elements.

SIGNIFICANCE OF THE MODULAR APPROACH FOR INTERNATIONAL OCEAN GOVERNANCE

In the past decades, various mechanisms and programs have been created for international ocean governance. The overall objective of current LME projects, to ensure the sustainable development of marine environmental resources, differs little from other efforts in the protection and management of the marine environment and resources. The most important and unique contribution of LME projects to global ocean governance lies in the developing and consolidating of an integrated and ecosystem-based approach to the monitoring, assessment, and management of the marine environment and resources and their relations with human society. The adoption of the five modules in the formulation of the TDA, the SAP, and the NAP, and the implementation of the SAP in an ecosystem-based context are the core of this approach. Such an approach has led to some advances in international oceans management.

The components of an LME are protected and managed in a holistic, integrated manner. The relations and interactions among components of an ecosystem are comprehensively taken into consideration. This departs from the traditional single-species approach that protects the target species of exploitation without taking into account the dependent and associated species as well as their environment.

The assessment and management of the marine environment and resources are addressed from multiple perspectives, involving natural science, technology, socioeconomics, law, and politics.

The spatial scale of the assessment and management normally extends across different maritime boundaries and jurisdictions to encompass an entire LME. Various political maritime zones and interests are harmonized in the interest of the integrity of marine ecosystems.

In the LME projects, the integrated approach to marine management breaks barriers of sectoral division. Different marine related sectors work in partnership to deal with the same or related issues in the protection and management of the marine environment and resources from a multi-sectoral perspective.

The LME approach deals with marine environmental and resource issues not only across maritime boundaries and sectoral boundaries, but also "across"[74] various global, regional and sub-regional marine-related instruments. Examples of the global instruments include the United Nations Convention on the Law of the Sea,[75] the 1995 Agreement on Straddling Fish Stocks and Highly Migratory Fish Stocks,[76] Chapter 17 of Agenda 21,[77] the 1992 Convention on Biological Diversity (CBD),[78] the Jakarta Mandate of the CBD,[79] the Global Program of Action for the Protection of the Marine Environment from Land-Based Activities (GPA),[80] the 1995 FAO Code of Conduct for Responsible Fisheries,[81] to name a few. The regional instruments include the agreements emanating from the Regional Seas Program,[82] agreements of regional fisheries bodies,[83] and other instruments applicable to the regional and sub-regional seas concerned. Although some of these instruments adopt a multi-species approach or an ecosystem-based approach to the protection and management of the marine environment and resources, others adopt a single-species approach or a sectoral approach to marine issues. In practice, the development and implementation of these instruments are normally undertaken by different organizations and agencies on a sectoral basis. For example, the regional fisheries bodies naturally deal with fisheries issues, while the Regional Seas Program (RSP) focuses on marine environmental protection. Although the RSP "has a comprehensive, integrated, result oriented approach to combating environmental problems through the rational management of marine and coastal areas," and "focuses not only on the mitigation or elimination of the consequences but also on the causes of environmental degradation,"[84] it is executed mainly by UNEP and addresses only one of the marine issues: marine pollution. None of the existing international marine management mechanisms deals with marine issues as a whole in an integrated manner. However, as mentioned in UNCLOS, "the problems of ocean space are closely interrelated and need to be considered as a whole."[85] In order to fill these gaps, LME projects adopt an integrated, ecosystem-based approach to the assessment and management of the LMEs concerned, which is reflected in the five modules as mentioned above. An LME project is expected to be a synergy of the existing regional marine management mechanisms for the LME concerned, rather than a duplication of a part of them or another parallel mechanism. Some LME programs have established close links with other organizations and programs. For example, the BCLME project has established closed links with some existing marine-related organizations and programs in the region.[86] The SCS SAP indicates that all actions are intended to be undertaken in a spirit of collaboration and

---

[74] See Duda and Sherman, *supra* note 2, at 828.

[75] Reproduced in 21 *International Legal Materials* 1261 (1982).

[76] The Agreement for the Implementation of the Provisions of the United Nations Convention on the Law of the Sea of 10 December 1982 Relating to the Conservation and Management of Straddling Fish Stocks and Highly Migratory Fish Stocks, Aug. 4, 1995, reproduced in 34 *International Legal Materials* 1542 (1995).

[77] Available at: http://www.un.org/esa/sustdev/documents/agenda21/.

[78] Reproduced in 31 *International Legal Materials* 818 (1992).

[79] Available at: http://www.biodiv.org/programmes/areas/marine/.

[80] Available at: http://www.gpa.unep.org/documents/about-GPA-docs.htm.

[81] Available at: http://sedac.ciesin.columbia.edu:9080/entri/texts/FAOCode.html.

[82] See http://www.unep.org/water/regseas/regseas.htm.

[83] See http://www.un.org/Depts/los/Links/IGO-links-fish.htm and http://www.fao.org/fi/body/body.asp.

[84] *Regional Seas Programme*, at: http://www.unep.org/water/regseas/regseas.htm.

[85] Para. 3 of the preamble of UNCLOS. This is one of the two fundamental principles of the new law of the sea. Another is the principle of the common heritage of mankind. See E. M. Borgese, *The Oceanic Circle: Governing the Seas as A Global Resource* (Tokyo, New York, Paris: United Nations University Press, 1998), at 133.

[86] O'Toole, *supra* note 25, at 5.

partnership, to enhance synergy between ongoing initiatives at the national and regional levels, and eliminate duplicative and conflicting actions.[87]

Furthermore, the TDAs have led to a more comprehensive, systemic and in-depth understanding of the marine environment and resources, and improved the building of a database of LMEs. This provides a sound scientific basis for marine management and strengthens the integration of science with marine management. The TDA not only enhances the awareness of collaborating states of the common transboundary environmental issues in the LME that they share, but also makes it clear that these issues cannot be effectively resolved without international cooperation. By collaboratively developing and adopting the SAP, member states "commit themselves to cooperate through a set of regional guidelines and plans."[88] On the other hand, the SAP approach also "generates regional and international political pressure on the national governments to act."[89] The successful implementation of the SAP measures that address the priority transboundary environmental concerns should substantially improve the environmental quality of the LMEs concerned. The SAPs have established a new basis for regional cooperation in marine management. They are also expected to promote cooperation in marine management between individuals, stakeholder groups, international organizations and government agencies at the national, sub-regional, regional and global levels.

More countries are involved in integrated, ecosystem-based assessment and management of LMEs. Sixteen LME projects are underway in Asia, Africa, South America and Europe, involving more than 100 states, the majority of which are from the developing world. In these projects, institutional, technical and financial assistance is provided by various UN agencies and some developed countries to the developing countries concerned.[90] The implementation of these projects not only substantively promotes regime building but also capacity building for more effective and efficient marine management in developing countries.[91]

CONCLUSION

The application of the five-module assessment and management methodology to the TDA-SAP-NAP processes constitutes the modular approach adopted in LME projects. This innovative approach moves away from the traditional single-species and sectoral approaches towards an integrated, ecosystem-based approach to the assessment and management of LMEs, and thus is an attempt to rectify some of the perceived deficiencies of traditional approaches. It has led to improvement in the understanding of LMEs and their relations with human society, and has facilitated the building of more scientific and rational LME management regimes. The LME projects are moving toward two important WSSD targets—

---

[87] *Strategic Action Programme for the South China Sea, supra* note 38, at 5.
[88] UNEP, *Review of Alternative Regional Approaches to Implementation of the GPA, supra* note 56, at 19.
[89] *Ibid.*
[90] See Duda and Sherman, *supra* note 2, particularly at 798, 806, and 829 and *Oceans and the World Summit on Sustainable Development: A Large Marine Ecosystems Strategy for the Assessment and Management of International Coastal Waters,* at: http://www.edc.uri.edu/lme/intro.htm. A list of LME projects is available at: http://na.nefsc.noaa.gov/lme/project.htm.
[91] For a discussion on some specific achievements of the Gulf of Guinea LME, for example, see Barnes, *supra* note 22, at 6-8.

to introduce ecosystem-based assessment and management practices by 2010; and to recover depleted fish stocks to maximum sustainable yield levels by 2015.[92]

The author is deeply grateful to Prof. B. Hatcher, Dr. K. Sherman, Dr. F. Bailet, Prof. D. Hodgson, and Prof. K. Hakapää for their helpful comments and assistance, to Prof. L. Juda, Dr. S. Adams, Dr. Y. Jiang, Dr. J. Bewers, and Mr. H. Ghaffarzadeh for valuable information provided, and to the Ocean Science and Research Foundation of Switzerland for generous financial assistance of the author's postdoctoral studies. The views expressed herein are solely those of the author. The internet information cited was available on November 5, 2003.

---

[92] Summit Outcomes A/CONF.199/20: *Report of the World Summit on Sustainable Development*, 4 September 2002 (section 30.d. and section 31.a.), at: http://www.johannesburgsummit.org/html/documents/documents.

Conclusion

The chapters in this volume are continuations of studies employing the Large Marine Ecosystem approach to the assessment and management of marine resources. This approach uses five modules—Productivity, Fish and Fisheries, Pollution and Ecosystem Health and Governance and Socioeconomics—and this volume is focused on two of the five modules: namely governance and socioeconomics.

In the first section, Ken Sherman, the originator of the LME concept, discusses its history, evolution and the importance of the five module approach, particularly socioeconomics and governance. In the next chapter, Ron Baird the Director of National Sea Grant, focuses on the human dimension in ecosystem management as well as the institutional considerations of the LME within the context of the Sea Grant Paradigm. A framework for monitoring and assessing socioeconomics and governance from the LME perspective is given in chapter 3. A team of sixteen researchers from the University of Rhode Island, the Woods Hole Oceanographic Institution and the Northeast Fisheries Science Center in Woods Hole, in a collaborative effort, describe a framework for linking the socioeconomic and governance dimensions to the science driven modules.

Using this framework as a point of departure, the next section deals with governance and socioeconomic issues. In chapter 4, Juda and Hennessey argue for research on governance performance through the use of profiles or "snapshots" of governance institutions over time in each of the emerging LME regimes. This is followed by Dyer and Poggi's paper on the importance of a total capital approach to understanding natural resource crises. The remainder of the first section of the volume brings together important research on economic issues associated with LMEs. Steve Edwards discusses ownership of multi-attribute fishery resources in LMEs. Hoagland, Thunberg, D. Jin and Steinback report the results of their application of the input-output approach in determining the value of economic activity in the Northeast Shelf LME. Edwards, Link and Rountree then report on research into portfolio management of fish communities in LMEs. Upton and Sutinen use a bioeconomic approach to determine the value of fish habitat. Cho, Gates, Logan, Kitts and Soboil, report on the values of Atlantic herring in the Northeast Shelf LME. Grigalunas, Opaluch, Diamantides and Woo link hydrodynamic and economic models to examine the costs of eutrophication in the Northeast Shelf LME. In the concluding chapter in this section Grigalunas, Opaluch and Luo examine the value of fisheries losses associated with sediment disposal in LMEs.

The concluding section of the book focuses on the role of governance and institutions in LMEs. Sherman and Duda discuss the LME approach in relation to World Summit targets. They analyze the progress in 11 LMEs and find that the LME assessment and management approach, combined with Global Environment Facility (GEF) funding process, is moving countries bordering these LMEs forward toward World Summit targets. In the next chapter Hennessey looks at the evolution of LMEs toward effectiveness. His examination of the literature on regime effectiveness suggests that the Benguela Current LME project, one of the more advanced of the GEF-LME projects, has a high potential for developing into an effectively functioning management regime. Moreover, the process of Transboundary Diagnosis Analysis (TDA) and Strategic Action planning (SAP) required by the GEF has enhanced the policy process leading to institutional design and function. In his analysis in the next chapter, Wang reached a similar conclusion about the positive contribution of the five modular approach and the GEF process of TDAs and SAPs. He concludes that these processes have rectified some of the deficiencies in the traditional sectoral approach and have improved the understanding of LMEs and their management regimes.

The results of studies presented in this volume clearly underscore the critical importance of governance and socioeconomics for the establishment and functioning of effective LME management regimes. The study outcomes presented here represent a good beginning but much more needs to be done. We need a large scale study program focused on the establishment, functioning and effectiveness of LME regimes as these evolve through various stages of development, often in very different contexts. This is a fertile field for research into the structure of the institutions chosen by countries pursuing the LME approach, the role of science in the policy process in terms of which problems are identified and addressed and the degree to which adaptive governance can provide the capacity for learning necessary to manage such complex and dynamic systems. Finally, economic valuation of these resources must become a global priority so that governments and citizens will begin to place appropriate values on ocean and coastal resources.

Timothy Hennessey
Jon Sutinen

Index

Adaptive management, 3,7,8,10,11,21,22,31,37,
  66,94,107,108,117,134,183,195,273,278,280,
  288,301,304,312-314,319,330-332,343
  definition, 68
Agenda 21, 5,54,276,279,312-314,353
Agriculture, 33,38,39,42-4,50,52,93,97,98,100,
  103,105,141,203,204,231,232,242,301,310-
  312
Agulhas Current Large Marine Ecosystem, 320
Albania, 320
Algal blooms, see Harmful algal blooms
Algeria, 320
Alien species, see Exotic species
Angling, 201
Angola, 276,299,306,320,327,328
Antarctic Large Marine Ecosystem, see also
  CCAMLR
  6,11
Antarctic Treaty, 96
Anthropogenic impacts, 203,282,309
Antigua and Barbuda, 320
Aquaculture, 38,39,42,44,97,98,100,105,132,
  141,163,164,192,202,203,273,308
Arabian Sea Large Marine Ecosystem, 299
Area closure, 60,61,124,138,,144,145,151,191,229,
  279,286
Argentina, 299,303,320
Atlantic Herring, see Herring
Atlantic Ocean, 183,238,239,276
Australia, 6,194,276
Austria, 300,301

The Bahamas, 320
Bangladesh, 320
Baltic Sea Large Marine Ecosystem, 5,6,14,304,
  312,337
Baltic Sea regional project, 309-310
Barbados, 320
Barcelona Convention, 302
Barents Sea Large Marine Ecosystem, 6,184
Bay of Bengal Large Marine Ecosystem, 6,49,
  299,312,320
Beach use, 32,45,55,69
Belize, 320
Benefit-cost analysis, 51,52,56
  definition, 69
Benguela Current Large Marine Ecosystem, 5,6,
  14,276,303,306-307,320,324,325,327,328,332,
  342,344
Benin, 299,304,320
Bering Sea, see East Bering Sea
Biodiversity, see also Convention on Biological
  Diversity
    biodiversity loss, 5
    definition, 69

Bioeconomic factors, 179,206,208,218,
  220,255,357
Birds, see Marine birds
Black Sea Large Marine Ecosystem, see also
  Danube Basin
  5,6,29,299-301,312,320
Bosnia-Herzegovina, 320
Boundaries, see also Exclusive Economic Zone,
  LME boundaries, Territorial Sea,
  Transboundary concerns
    administrative boundaries, 91
    geographic, 4,67
    jurisdictional boundaries, 27,108,331
    maritime boundaries, 338,352
    political boundaries, 67,74
    transboundary aspects, 7,12,40,52,
      53,57,58,64,66
    transition zones, 202
Bucharest Convention, 301
Buy-back program, 35
Bulgaria, 320
By-catch, 64,96,133,137,138,143,144-
  147,150,177,181,193,276,280,283,284,286-
  290,294

Cadmium, 307
California Current Large Marine Ecosystem,
  5,6,29
Cambodia, 299,304,320
Cameroon, 320
Canada, 42,44,62,105,139,311
Canary Current Large Marine Ecosystem,
  6,299,320
Cape Cod, 120,150
Cape Verde, 320
Capital assets and impacts,
    biophysical, 30,46,49,112-118,126,
      129-134
    cultural, 46,113-115,120,125,126,
      128,134
    economic, 111,113-115,
      119,121,124,126,128-131
    human, 46,113-115.120,124,129,
      130,131
    social, 46,113-115,120,123-125,127,
      128
Caribbean, 278,298
Caribbean Sea LME, 6,29,299,320
Carrying capacity, 13,112,137,186,208,210,
  211,3057,309,310,336
Catch-per-unit-effort, 252,255,257
CBD, see Convention on Biological Diversity
CCAMLR, see Convention on the Conservation
  of Antarctic Marine Living Resources
Chile, 320

China, see also East China Sea LME, South
    China Sea LME
    6,299,300,307-309,320
Climate change, 5,273,297,344
Closed areas, 145,146,149-150,283,287
Coastal erosion, 98,100
Coastal habitats, 3,4,43,103,276,304
Coastal property, 44
Coastal zone, 30,105,176,203,276,307,309,341,
    342
Coastal zone management, 40,42,97,106,273,
    276,309,310,344
Coastal Zone Management Act (1972), 28,39,
    101,164
Cod, 63,123,184,202,205,216,223-225,286,308
    Atlantic cod, 60,61,64,146,147,
        191,193,209,267
    Pacific cod, 284
Colombia, 299,320
Common Fisheries Policy, 281,288,302
Community involvement, see Stakeholders
Comoros, 320
Congo, 320
Conservation groups, 111
Consumer groups, 144
Continental shelf, see Continental shelf Large
    Marine Ecosystems, Outer continental shelf
Continental Shelf Large Marine Ecosystems,
    see Northeast U.S. Continental Shelf LME,
    Southeast U.S. Continental Shelf LME,
    Yellow Sea
Contingent choice, 50,69,74
Contingent valuation, 50,69,74,219,220
Continuous Plankton Recorder (CPR), 8
Convention on Biological Diversity (CBD),
    65,312,353
Convention on the Conservation of Antarctic
    Marine Living Resources (CCAMLR), 11,282
Convention on the Continental Shelf (1958),
    274
Conventions on the Law of the Sea, see
    UNCLOS
Cook Islands, 320
Cooperative mechanisms, 57,58,64,124,137,
    142,147-150,184,207,277,305,306,346
Corals and coral reefs, 45,52,276,301,303,347
Cost effectiveness, 50,51,69,306
Costa Rica, 320
Côte d'Ivoire, 299,304,320
Crabs, 118,284
Croatia, 320
Cuba, 320,345
Culture, 22,48,108,110,113,127,
    129,130,132,133,331
Currents, see Benguela, California,
    Canary, Guinea, Humboldt, Kuroshio,
    Oyashio, Somali Current

Customary International Law, 274,280,281,288

Danube, 300
Danube Basin, 300-301
Database, 159,257,277,354
    IMPLAN, 162,163
DDT, 210
Defense issues, 146,163
Democratic Republic of Congo, 320
Denmark, 320
Developed countries, 354
Developing countries, 5,17,19,20,23,51,
    297,298,301,312,354
Direct use, 32,45,69,71
Discount rate, 177,208,210,211,258,260,
    261,263
Djibouti, 320
Dogfish, 12,144,146,183,184,192,216,
    223,225,267,284
Dolphins, 144,216-218,267
Dominica, 320
Dominican Republic, 320
Drainage basins, 29,112,301
Dredging, 64,98,203,249-251,255,262
Drift net, see Fishing gear

Earth Summit (1992), 5,279
East Bering Sea Large Marine Ecosystem,
    6,29,184
Eastern Europe, 298
Economic factors, see Capital assets and
    impacts, Forcing, Market considerations,
    Socioeconomic trends
Economic valuation, 10,45,69,358
Economies of scale, 37,124
Ecosystem change, 98-100,286
Ecosystem health, see also Health indices,
    Pollution
    8,9,14,29,38,39,42,84,104,109,131,283,304-
    306,310,326-328,336,337,345,346,348,357
Ecosystem-based management, 3,5,8,12,14,
    17,18,20-23,43,90,93,101,109,110,194,
    283,286,289,301,307,309,344
Ecosystem reserve, 215,219,222-224,226,227
Educational aspects, 17,20-22,37,46,50,87,89,
    104,125,130,239,304,347
EEC, see European Economic Community
Eelgrass, 56
EEZs, see Exclusive Economic Zones
Effluent, 137
Egypt, 320
El Niño-Southern Oscillation (ENSO), 189
Energy resources, 60,66,150,274,306
Enforcement, 19,35,43,53,60-63,65,67,68,
    72,74,88,103,140,141,143,145,147-151,184,
    186,188,191-194,283,286,297,302,326,327,
    329,348

England, 122
ENSO, see El Niño-Southern Oscillation
Environmental impacts, 18,43,87,95,103,129,
    323
Environmental impact assessment, 103,287,345
Environmental impact statement, 43,103,206,
    255
Equatorial Guinea, 320
Erosion, see also Coastal erosion
    33,66
Estonia, 320
Estuaries, see also National Estuary Program,
    Peconic Estuary
3,9,28,31,38,43,55,83,103,201,203,241,257,
    275,285,300,311
        definition, 70
        South China Sea estuaries, 347
European Economic Community (EEC), 281,
    288
European Union (EU), 42,106,281
        Common Fisheries Policy, 281,288,
        302
Eutrophication, see also Northeast U.S.
    Continental Shelf LME
5,8,31,72,98,100,300,310-312,342,357
        definition, 70
Exclusive Economic Zone (EEZ), 123,138,217
        US EEZ Proclamation (1983), 145,274
Exotic species, 342
Externalities, 30,35,40-43,62,63,71,74,95,102-
    104,115,116,134,139,140,158,203
        negative, 40,95,98
Extractive industry, 31,46,195
Exxon Valdez, 146

Factory vessels, 115
Failure, 70
        institutional, 57,70
FAO, see Food and Agriculture Organization
FAO Code of Conduct for responsible fisheries,
    137,286,298,306,3468-348,353
Faroe Plateau Large Marine Ecosystem, 6
FCMA, see Magnuson Act
Fecal coliform, 230
Federal agencies, 21,27,28,91,128,147
Federal government, 17,21,27,83,145,147,
    165,257
Ferries, 33
Fertilizer, 50,93,229,232,242,311,312
Finland, 320
Fish, see also Groundfish, IFQs, ITQs, TAC
        age structure, 137,145,146,186
        diet, 37,137,143,144,193,216
        distribution, 48,137
        gene pool, 137,143,183,193
        growth rate, 137,143,181,182,185,186,
        193,210

        habitat, 39,43,103,138,146,177,
        201-213,357
        juvenile, 183,188
        kills, 141,150
        larvae, 62,184
        nursery areas, 67,279,310
        pelagic species, 8,14,59,144,183,215,
        217,277,280,284,286,294,300
        quotas, 130,133,1357,136,138,
        144, 145,147,148,192,286,287,309
        recovery, 14,111,112,115,121,129,
        133,210,211,250,252-254,258,
        262-265,282,288,300,315
        recruitment, 206
Fish meal, 188
Fisheries. see also By-catch, Fish, Foreign
    fleets, Illegal fishing, Overfishing, Poaching
        artisanal, 38,95,134,347
        carrying capacity, 13,112,137,186,208,
        210,211,305,309,310,336
        closed area, 146,147,148,283,287
        closed season, 287
        collapse, 122-126
        commercial, 12,33,44,66,116,125,
        130-132,141,145,157,158,162-164,
        168-171,176,203,220,223-225,250-
        252,256,258,261,280,284,342
        fishing effort, 12-14,67,126,177,186-
        188,191,206,209,212,213,218,
        252,290,300,308,315
        international, 347
        products, 67,201,208
        recreational, 32,38,39,55,66,69,95,
        111,121, 125,131,143,146,148,162,
        163,201,229,249-256,258,259,265,273
Fisheries management, see also Ecosystem-
    based management
        multi-species approach, 64,181-
        185,192,194,344,453
        single-species approach, 3,58,59,64,
        147,181,182,187,191,193,195,287,
        289,344,352-354
        and social science, 73,93,204,263,273,
        337
Fishermen, 30,38,39,41,46-48,60,61,63,
    70,74,95,99,102,106,117,119,120,122-
    125,128,132,143-145,147,148-150,183,
    189,191-193,203,206,222,223,261,288
Fishing, see also By-catch, Ghost fishing,
    Overfishing
        distant-water, 217
        foreign fishing, 12,13,123
        poaching, 220
Fishing fleets, 121-123,135,161,191,
    217,224,283, 303
Fishing gear, 38,46,48,60,61,63,64,70,
    95,98,120,121,123,124,126,129,133,137,140,

143,145-147,150,185,186,191,193,203,206,
208,212,283,284,287,306,346
    drift net, 280,286
    gill nets, 48,123
    purse seine, 123
    trawl gear, 144,286
Fishing industry, 46,112,117,119,124,125,145,
148,150,161,162,171,176,177,193,203,231,310
Fishing license, 35
Fishing technology, see also Fishing gear
149-151,181,188,209
    line fishing, 63
Flounder, 12-14,145,146,183,191-193,201,209,
255,256,258,259,266,267,284,300
Food and Agriculture Organization (FAO), see
   also FAO Code of Conduct for responsible
   fisheries
138,279,280,285,304
Food chain, 8,9,37,59,74,143,186
Food web, 9,12,13,37,56,66,144,145,147,177,
183,186,1891-191,249,250,254,255,259-
263,265,284,297
Forcing, 42
    economic pressures, 104
    environmental and physical, 242
    human,73,321
Foreign fleets, 123
France, 276,2991,320
Fuji, 320

Gabon, 320
Gambia, 320
Gas, see Oil and gas
Gear, see Fishing gear
GEF, see Global Environment Facility
Geneva Convention on the Continental Shelf
   (1958), see Convention on the Continental
   Shelf
Georges Bank, 12,13,29,60-63,124,144,146,
147,182,183,193,206,212,217,275,284
Georgia, 320
Geopolitical trends, see also Boundaries,
   Governance
63
Germany, 276,299-301,320
Ghana, 299,304,320
Ghost fishing, 284,286
Gill nets, see Fishing gear
Global Environment Facility (GEF), see also
   SAP, TDA
11,12,29,276,277,280,297-309,311-315,319-
329,332,340,357,358
    approved GEF projects, 299
    GEF projects in preparation, 299
    role, 5-7
Global Program of Action for the Protection of
   the Marine Environment from Land-Based

   Activities (GPA), 300-303,322,324,353
Global warming, 134
Goods and services, 3,7,10,30,33,35,36,46,
51,59,61,63,65,67,71, 72,74,89,93,114,128,
139,157,158,201,202,207,219,297
    definition, 70
Governance, see also LME boundaries,
   Management, Marketplace, NGOs, Political
   regimes, Stakeholders
17,18,20,21,23,27-75,83-110,115,127,130,
143,147,150-152,188,191,194,195,249,273,
274,276,277,280-283,189,319-332,335,
348,357,358
    definition, 70
    international, 21,41,42,106,335,352
    institutional, 18,19,59,89,189,329,357
    module, 8,10-12,14,29,30,70,84,
    131,134,181,285,304,305,337,357
    national, 22,41,42,44,106
    governance regime, 7,8,11,104,
    304,337
    in Northeast U.S., 263
    regional institutions, 41,42,106
Government
    local, 21,47,88,305,313
    state, 60,151,161,327
    tribal, 91
GPA, see Global Program of Action for the
Protection of the Marine Environment from Land-
Based Activities
Greece, 320
Grenada, 320
Groundfish, 12,49,63,64,111,112,122,
   133,134,144-148,201,206,208-212,
   283,284,286,289
    collapse, 122-128,320
Guatemala, 320
Guinea, 320
Guinea-Bissau, 299,320
Guinea Current LME, 6,10,14,277,299,303,304
Gulf of Guinea, see also Guinea Current LME
6,276,277,299,304-306,320
Gulf of Maine, 12,29,41-43,49,60,63,103,105,
106,115,122,124,144,146,170,206,212,215,
217,218,275
Gulf of Mexico, 6,29,299,311,312,320,345
Gulf of Thailand Large Marine Ecosystem,
   6,300

Habitats,
    fish habitat, 39,43,103,138,146,
    177,201-213,357
    habitat alteration, 4
    habitat loss, 4,7,28,203,307
    habitat protection, 206,276,286
    habitat recovery,
HABs, see Harmful algal blooms

Haddock, 12-14,64,123,146,182,184,192-
193,209,2168,225,267,300
Haiti, 320
Hake, 12,145,192,216,223-225,266,267
Halibut, 123,184,268,284
Harmful algal blooms (HABs), 9,229,276,
310,312,342
Hazardous substances, 53
Hazardous waste, 175
Health indices, see also Ecosystem health
disease, 9,143,220,337
Heavy metals, 340
Helsinki Commission, 309,311
Herring, 12-14,62,115,116,123,
184,192,259,284
Atlantic herring, 215-228,256,266,357
and Northeast U.S. Continental Shelf
Large Marine Ecosystem, 215-228
Pacific herring, 308
Honduras, 320
Hotspots, 301,303,341
Human behavior, 30,34,58,70,73,84,85,87,
90,93,99,230,244
Human impacts, see also Anthropogenic
impacts,
18,119
Humboldt Current LME, 5,6,299,320
Hydrocarbons, see also Oil and gas, Petroleum
8
Hydromodification, 43,103
definition, 70

Iceland Shelf LME, 6
IMO, see International Maritime Organization
India, 6,132,299,320
Indians, 139
Indicators, 10,11,14,18,101,102,109,283,
285,313
biological and physical, 283,284,336
coastal condition, 9
economic, 45,49,221
scientific, 289
social, 48,125
Indian Ocean, 6
Indigenous peoples, 22,48
Indirect use, 32
definition, 71
Individual Fishing Quotas (IFQs), 137,138,145,
148,192
Individual transferable quotas (ITQs), 135
Indonesia, 6,24,299,320
Indonesian Sea LME, 6,299
Industrial siting, 42,44,97,98,100
Industry, see also Extractive industry, Fishing
industry, Manufacturing
53,54,105,121,151,152,159-164, 167-172,175-
177,184,326,327

International Convention for the Prevention of
Pollution by Ships (MARPOL), 39,71,78
Intergovernmental Oceanographic Commission
(IOC), 297,304
International Maritime Organization (IMO),
297,304
International Union for the Conservation of
Nature and Natural Resources (IUCN),
297,304,306,307
Intergovernmental approach, 28,42,105,304
IOC, see Intergovernmental Oceanographic
Commission
Israel, 320
Istanbul Convention, 301
Italy, 123,125,280,320
ITQs, see Individual transferable quotas
IUCN, see International Union for the
Conservation of Nature and Natural
Resources
Ivory Coast, see Côte d'Ivoire

Jamaica, 320
Japan, see also Sea of Japan
62,64,227
Jeddah Convention, 301
Jordan, 320
Jurisdictional issues, see also Boundaries,
Governance
19,21,276,281,303,308,339

Kenya, 320
Keystone organisms, 184
Kinship, 46
Kiribati, 320
Korea, 24,229,299,308,309,320
South Korea, 307
Kuroshio Current LME, 6
Large Marine Ecosystems (LMEs), see
Continental shelf, Fisheries management,
Governance, Health indices, Pollution,
Predator-driven, Scale factors,
Socioeconomic trends
Latin America, 298
Latvia, 320
Lebanon, 320
Legal frameworks, 17,19,326,342,344,345
Liberia, 320
Libya, 320
Lithuania, 320
Litter, see Marine litter
LMEs, see Large Marine Ecosystems
LME boundaries, see also Boundaries
34,275
LME partnerships, 313
Lobster, 60,63,64,113,146,218,222,255-259,266
die-off, 116-129
Lobster fishery, 74,111,114,130,133

Long Island Sound, 111,113,114,126,128,129,
    133,238,239
        lobster fishery, 116-123,126,133

Mackerel, 13,14,123,186,267,268,284,300,308
        Atlantic mackerel, 12,215,267
Madagascar, 320
Magnuson Fishery Conservation and
    Management Act (FCMA), 43,103,123,134
Magnuson-Stevens Fishery Conservation and
    Management Act (1996), 28,43,123,138,
    145,146,181,183,195,344,353
Malaysia, 320
Maldives, 320
Mammals, see Marine mammals
Management, see also Adaptive management,
    Coastal zone management, Ecosystem-based
    management, Fisheries management,
    Management Councils, Monitoring,
    Watershed management
        management choices, 90,91,92
        plans and planning, 11,43,101,103,
        140,145,146,182,193,205,218,
        227,240,282,288,310,337,341
        public management, 158
        strategies, 84,113,117,133,287,289
        units, 83
Management areas, 124,273
Management Councils, 13,43,103,146,151,289
        Mid-Atlantic, 14
        New England, 14,112,124,133,205
        North Pacific, 283-285
Mangroves, 44,45,96,276,302,303,305,347
Manila Bay Declaration, 303
Manufacturing, 31,39,306
Mapping, 41,104
Maps, see also LME boundaries
    234,281
Mariculture, 54,229,306-309,346
Marinas, 33,43,103,131,247
Marine birds,27,32,69,216,218
Marine debris, 286
Marine mammals, 9,27,32,52,61,62,66,72,
    144,146,147,175,183,215,216,218-227,
    229,283,284,290,307
Marine parks, 44,52,303
Marine pollution, see also Pollution
    27,40,52-54,58,177,302,327,336,346,353
Marine protected areas, 141,279,286,297,
    301,315,3446,347
Marine sanctuaries, 28,39,101
Marine transportation, 33,38,55,150,157,273
Market failure, 30,49,57,203
Marketplace, 44,85-88,103,200
Market value, 55,56,73,74,202,215,221, 227,
    251,252,255
MARPOL, see International Convention for the

Prevention of Pollution from Ships
Marshall Islands, 320
Mauritania, 320
Mauritius, 320
Maximum Sustainable Yield (MSY), 6,145,146,
    149,181,182,191,193,224,286,288,297,298,315
        definition, 71
Mediterranean Sea Large Marine Ecosystem,
    299, 302
Mekong Basin Project, 303
Menhaden, 215,267
Mercury, 307
Metals, see Cadmium, Heavy metals
Mexico, 6,9,29,311,320,345
Micronesia, 320
Military uses, 42,44,97,98,100,105
Mining, 32,41,42,64,105,106,147,150,273,
    306,307,346
Models and modeling, see also TETRA TECH
        bioeconomic model, 177,206,208,220,
        250,252,254,255
        economic models, 229-231,243,244,
        357
        IMPLAN model, 130,159,162-164,
        168,170-172,175-177
        input-output, 36,37,71,157-163,176,
        177,287,357
        Pella & Tomlinson, 209
        total capital response model, 117,126,
        127
Morocco, 320
Mozambique, 320
MSY, see Maximum Sustainable Yield
Multi-species approach, see Fisheries
    management
Mussels, 308
Mussel Watch, 9
Myanmar, 320

Namibia, 276,306,320,327,328
Narragansett Bay, 249,251,253,256-259,261,
    262,265
National Action Plan (NAP), 335,340,344,347-
    349
National Environmental Policy Act (NEPA,
    1969), 43,206
National Estuary Program (1987), 28,43,55,103,
    231, 232
National Marine Fisheries Service (NMFS),
    124,137,144-146,205,206,216
National Oceanic and Atmospheric
    Administration (NOAA), 9,14,20,28,54,105,
    129,164,178,205,274,276,277,283,297,304,
    306,307
Natural resource communities, 45-48,112,113,
    126,133,137
Nauru, 320

Navigation, 57,98,163,164,168-171,203
New England, see also Georges Bank
  13,14,29,63,64,113,115,128,144,150,174,
  175,177,191-193,205,219,224,275,284,311
        groundfish fisheries, 49,111,112,122-
        126,133,134,201,206,208,210,212
New Zealand, 148,276
Nicaragua, 320
Nigeria, 299,304,320
Nitrates, 8
Nitrite, 8
Nitrogen, 7,62,228-232,235,237,241-244,297,
  300,301
Niue, 320
NMFS, see National Marine Fisheries Service
NOAA, see National Oceanic and Atmospheric
  Administration
Non-governmental organizations (NGOs),
  28,42,44,72,86-88,105,301,304,321,327,345
Non-native species, see Exotic species
Non-renewable resources, 46,66,70,137,195
Northeast Australian Shelf Large Marine
  Ecosystem, 6
Northeast Brazil Shelf LME, 299
Northeast U.S. Continental Shelf Large Marine,
  Ecosystem, see also Eutrophication,
  Governance, New England
        eutrophication, 229-247,357
        fish and fisheries, 12-14,61,143,144,
        175,201,208,250,273,282,284,357
        herring, 215-228, 357
        overfishing, 201
        pollution, 229
        socioeconomic trends, 159-177,208,
        229-247,249,357
North Sea Large Marine Ecosystem, 6,184,188,
  231,311
Norway, 184,276
Norwegian Shelf Large Marine Ecosystem, 6
Nutrient loading, 8,242,311

Oceans Act (2000), 273,274
Oil and gas, see also Hydrocarbons, Petroleum
  102,143,145,146,195,273,307
Oil spills, see also Exxon Valdez
  40,52,53,58,102,126,137,147,176
Oil spill response, 57
Open access, 40,72,184,186,207-212
Opportunity cost, 69,73,145,147,150,157,207,
  208,213,222,223,227,232,239,252
Outer continental shelf, 60,150
Outputs, see Goods and Services
Outreach, 17,87,130
Overfishing, see also Northeast U.S.
  Continental Shelf LME
  12,28,64,88,124,138,146,149,182,201,208,
  211,284,287-290,307,309,326,342

Oyashio Current LME, 6
Oysters, 204,308
Ozone depletion, 5

Pacific Ocean, 286,302
Panama, 320
Papua New Guinea, 320
Paralytic Shellfish Poisoning, 8
Passive use, 32,45,50,54,72
Patagonian Shelf Large Marine Ecosystem, 6,
  299,303,312,320
Pathogens, 8,98,100,229,230
Peconic Estuary, 231-244
Peru, 320
Pesticides, 33,93
Petroleum, see also Hydrocarbons, Oil and gas
  3,8,60,143,150,249
Philippines, 320
Phosphorus, 62,229,244
Plankton, see Continuous Plankton Recorder
Poaching, 220
Poland, 320
Political ecology, 49
Pollock, 184,215,216,225,267
Pollutants, 29,67,302,307,309
        persistent organic pollutants, 5
Pollution, see also Eutrophication, Marine
  pollution, Pollutants, Heavy metals
        land-based, 33,38,276,297,298,300,
        302,303,310,315,340,344,347,353
        non-point source, 43,103,230,232,309
        point source, 204,230,241,309
        transboundary, 53,57,58
Ports, 33,42,48,97,100,122,123,125,220,
  249,307,308
Portugal, 125,299,320
Property rights, 30,35,40,48,53,58-68,72,137-
  143,145,146-147,148-151,181,187,189,191-
  195,203,208
Protected areas, see Marine protected areas
PSP, see Paralytic Shellfish Poisoning
Public good, 35,50,57,62,71,150,203,230
        definition, 73
Public participation, 102,104,110,345
Puerto Rico, 320
Purse seine, see Fishing gear

Quotas, see Fish

Real estate, 33,41,106,163,164,168-174,176
Recreation, see also Fisheries, Whale-watching
  18,28,30,33,39,41,42,44,45,55,66,97,98,100,
  105,106,137,145,147,148,157,163,164,
  168-171,219,230-234
Recreational boating, 43,95,103,176
Red Sea Large Marine Ecosystem, 299,
  301,302,320

Red tides, 8,308
Regime, see Governance
Religion, 48,88,125
Renewable Resource, 7,46,62,66,70,112,185,
    207,212,297
Rent seeking, 73,109,227
Resource Management, see also Ecoystem-
    based management, Property rights
    17,18,21-23,28,45,48,50,51,58,
    67,68,111,138,148,149,195,290,330
Resource use, see also Stakeholders
        direct use, 32,45,69,71
        indirect use, 32,71
        non users, 52,72
        passive use, 32,45,50,54,72
        resource users, 31,33,48,67,113,130,
        333
Resource valuation, see also Valuation
    44,45,51,73
Resources
        living, 8,11,43,102,103,112,276,278,
        279,284,290,300,304,306,307,329,
        341,344,346,350
        non-renewable, 70,137,197
        renewable, 70,209,214
Restaurants, 33,118,119,131
Rhine River, 311
Rio Declaration, 276
Romania, 320
Russia, 62,299,320
Russian Federation, 320

Saint Kitts and Nevis, 320
Saint Lucia, 320
Saint Vincent and the Grenadines, 320
Salmon, 62,151,202,205,284
Salt marsh, 56,201
Samoa, 320
Sampling data, 8,9,231,234,235,244,256-258,
    262
Sand and gravel, 64,273
Sandlance, 217,218,284
São Tomé and Principe, 320
SAP, see Strategic Action Program
Sardine, 308
Saudi Arabia, 320
Scale factors
        decadal, 8,12,253,286,298,305
        geographic, 18,20,23,62
        LME scale, 273,284,319,331,335
        spatial, 3,9,28,63,68,145,290,352
        temporal, 9,28,67,68,145,241,283,
        290,330
Scallops, 60,63,64,134,144,146,148,191,
    193,203,206,212,231,240
Sea of Japan, 6
Seafood, 61,66,116,119,137,142,164,168,

170,171,175,185,188,201,207-209,261
Sea Grant, 17,20-23,357
Seagrass, 229,347
Seals, 62,144,216,284
Seaweed, 70
Semi-enclosed LMEs, 29
Semi-enclosed seas, 4,29
Senegal, 320
Septic systems, 50,229,230,242
Sewage treatment, 204,229,232,305
Sewerage, 57,163,164,168-171
Seychelles, 320
Sharks, 183,267,268
Shelf, see Continental shelf
Shellfish, see also PSP
    31,32,44,56,69,116,143,185,204,229,
    231,253,257
Ship building, 164,168,170,171
Shipping, 30,32,42,47,57,58,60,97,100,105,
    141,147,163,164,168,170-176,203,307
Shrimp, 61,96,144,194
Sierra Leone, 320
Single-species analysis, see Management
Skates, 12,116,183,266
Slovenia, 320
Social benefits, 74,110,207,208,210
Social conflict, 40,73
Social costs, 40,41,52,73,93
Social networks, 35,41,73,115,122,125,131
Sociocultural factors, 31,34,44,45,48,49,62,
    68,332
Socioeconomic trends, see Anthropogenic
    impacts, Economic factors, Ecosystem-based
    management, Fishermen, Management,
    Market, Recreation
        socioeconomic benefits, 8,30,74,
        283,305,309,337
Solomon Islands, 320
Somali Coastal Current LME, 6,299,320
Somalia, 320
South Africa, 276,299,306,320,327,328
South China Sea Large Marine Ecosystem, see
    also Estuaries
    6,299,303,312,320,344-347,349,350
Spain, 194,299,311,320
Spillover effects, see also Externalities
    62,74,102,138-145,147,148,150,181,185,193-
    195
Sri Lanka, 320
Stakeholders, 7,8,14,20,21,28,39,41,48,51,
    89,90,99,102,104,106,114,124,126,131,134,
    193,194,281,283,285,289,306,313,314,327,
    338,340,343,346,354
Stockholm Declaration on the Human
    Environment, 274
Straddling stocks, 277,279,2868,287,352
Strategic Action Program (SAP), 7,8,11,112,

276,277,300-304,306,313,325-327,335,337-
340,342-354,358
Stratton Commission, 41,90,104,274
Sudan, 320
Sulu-Celebes Sea LME, 299
Sustainable Fisheries Act (1996), 39,43,
103,205,287
Sweden, 320
Swimming, 234,235,237-240,243-245,308
Syria, 320

TAC, see Total allowable catch
Tanzania, 320
TDA, see Transboundary Diagnostic Analysis
Territorial Sea, 274
TETRA TECH, 230-232,241,244
Thailand, 299,320,350
Third World, 17
Togo, 299,304,320
Tonga, 320
Total Allowable Catch (TAC), 145,215,298,300
Tourism, see also Recreation
33,44,54,55,163,164,168,170-174,176,
203,231,273,302,306,308,310
Toxins, 8
Training, 20,57,58,119,120,125,127,129-
132,305,3191,326,331,332,342,344,346
Transboundary concerns, see also Boundaries
pollution, 53,57,58
resources, 306
Transboundary Diagnostic Analysis (TDA),
7,8,11,300-304,306,313,325,326, 335,337-344,
348,349,351-354,358
Transportation, see Marine transportation
Trawlers, see also Fishing gear
8,123,124
Trawling, 48,191,283,305
surveys, 8,12,216,305
Trinidad and Tobago, 320
Tuna, 62,67,141,143,144,184,267,268,302,324
Tunisia, 320
"Turf", 42,104
Turkey, 320
Tuvalu, 320

Ukraine, 320
UNCED, see United Nations Conference on the
Environment and Development (1992)
UNCLOS, see United Nations Convention on
the Law of the Sea (1982)
UNDP, see United Nations Development
Programme
UNEP, see United Nations Environment
Program
UNIDO, see United Nations Industrial
Development Organization
United Nations Conference on the Environment

and Development (UNCED, 1992), 3,21,
276,297,345
United Nations Convention on the Law of the
Sea (UNCLOS, 1982), 274,276,279,282,352
United Nations Development Programme
(UNDP), 5,276,277,301-304,307,326,328
United Nations Environment Program (UNEP),
5,65,67,301-304,353
United Nations Industrial Development
Organization (UNIDO), 277,298,304,306
United States of America (USA), see also
Exclusive Economic Zone, Northeast U.S.
Continental Shelf LME
12,13,19,23,24,27,53,60,290,105,111,134,
145-147,150,152,157,162,163,191,2157,216,
254,255,274,279-281,283,284,286,289,
300,311
Universities, 10,17,20-23,304,331,332,357
Urban development, 43,103
Uruguay, 320
User groups, 40,48,64,74,92,127,330
Usufruct rights, 137,148-152

Valuation,
bequest value of resources, 32,72,227
existence value of resources, 32,72,227
Vanuatu, 320
Venezuela, 320
Vietnam, 320
Viruses, 117

Waste disposal, 31,35,41,42,66,97,98,
100,105,106,137,141,147
Wastewater, 50,307
Watershed management, 43,103
Western Pacific, 299,302,320
Western Pacific Warm Water Pool-SIDS, 302
West Greenland Shelf Large Marine Ecosystem,
6
Wetlands, 32,43,55,71,103,203,
301,303,311,312,347
wetland loss, 38,99
wetland productivity, 55,56
Whales, 32,52,69,144,146,147,202,216-
220,224,225,227
Whale-watching, 46,141,219-223,227
Wildlife viewing, 32,69
World Bank, 5,29,280,301-303,305
World Conservation Union, see International
Union for the Conservation of Nature and
Natural Resources (IUCN)
World Summit on Sustainable Development
(WSSD), 3,277,297,298,315,345,354,357,
358

Yellow Sea Large Marine Ecosystem, 5,6,10,14,
299,300,303,307-309,312,320

Yellowtail, 12-14,146,191,193,209,267,284,300
Yemen, 320
Yugoslavia, 320

Zoning, 40,87,98,141,150,229